COSMONAUT

COSMONAUT

A Cultural History

Cathleen S. Lewis

University of Florida Press

Gainesville

28 27 26 25 24 23   6 5 4 3 2 1

Library of Congress Cataloging-in-Publication Data
Names: Lewis, Cathleen S., 1958– author.
Title: Cosmonaut : a cultural history / Cathleen S. Lewis.
Description: First edition. | Gainesville : University of Florida Press,
    2023. | Includes bibliographical references and index.
Identifiers: LCCN 2022058142 (print) | LCCN 2022058143 (ebook) | ISBN
    9781683403708 (hardback) | ISBN 9781683403814 (pdf) | ISBN 9781683403944
    (ebook)
Subjects: LCSH: Astronautics—Russia (Federation)—History. |
    Astronautics—Soviet Union—History. | Astronautics—Social
    aspects—Russia (Federation) | Astronautics—Social aspects—Soviet
    Union. | Astronauts—Russia (Federation)—History. | Astronauts—Soviet
    Union—History. | BISAC: SOCIAL SCIENCE / Popular Culture | SOCIAL
    SCIENCE / Media Studies
Classification: LCC TL789.8.R8 L49 2023 (print) | LCC TL789.8.R8 (ebook)
    | DDC 629.40947—dc23/eng/20230109
LC record available at https://lccn.loc.gov/2022058142
LC ebook record available at https://lccn.loc.gov/2022058143

UF PRESS

UNIVERSITY
OF FLORIDA

University of Florida Press
2046 NE Waldo Road
Suite 2100
Gainesville, FL 32609
http://upress.ufl.edu

In loving memory of my mother, Jacqueline Lucille Adams Lewis,
1926–2022

Contents

Figures

Acknowledgments

In 1997, the Institute for European, Russian, and Eurasian Studies awarded me a George Hoffman grant that allowed me travel and resources that contributed to the original research that shaped this book. I would first like to thank Muriel Atkin, Hope Harrison, and Ronald Spector at the George Washington University, who were instrumental in helping me shape this jumble of ideas into a presentable and coherent work. Dr. Atkin in particular has demonstrated enduring kindness and interest over the years.

My colleagues at the Smithsonian Institution National Air and Space Museum have provided intellectual comfort and support during this process; I would like to thank them for that. Within my department a succession of chairs and supervisors have pushed me to continue with this project over the years and have provided release time, tuition assistance, research support, and ardent encouragement. For those efforts, I would like to thank Robert Smith, Ted Maxwell, Michael Neufeld, Paul Ceruzzi, and Margaret Weitekamp. Among them, I owe an irredeemable debt to Michael Neufeld, who has read too many drafts of this book to count and patiently edited and helped me shape my ideas into a book. Jo Ann Morgan, Toni Thomas, and Amanda Young have provided technical, administrative, and moral support over the years. My former colleague Frank Winter continues to be gracious and generous with his time and vast accumulation of material and knowledge on early Russian rocketry history and his acquaintances in the field throughout the world. Special and heartfelt gratitude is due to Joanne Gernstein London and David H. DeVorkin, whose thorough readings and astute comments have helped me focus my thoughts. My colleague Lisa Young has supported my work and has been willing to carry some of my load at work while I write. I hope to return the favor.

Smithsonian Journeys, the educational tourism branch of Smithsonian Enterprises, provided me with the opportunity to visit spaceflight-related sites in Russia and the Kazakh Republic with new eyes. Traveling with non-historians and non-specialists has provided me with new perspectives on the cultural impact of human spaceflight on the USSR.

Outside of the Smithsonian and George Washington academic community, I would also like to thank the following for their insights, suggestions, and guidance on earlier versions of parts and ideas for this book: James Andrews, Harley Balzer, William Barry, Jonathan Coopersmith, Alexander Geppert, Slava Gerovitch, Steven Harris, Andrew Jenks, Betty Ann Kevles, Alex Kojevnikov, Amy Nelson, Asif Siddqi, Roshanna Sylvester, and the late Richard Stites.

My research would not have been possible without the access to libraries and archives. I have made use of three libraries in the Washington, DC, area and called on the assistance of their staffs extensively: the staff of the Smithsonian Libraries, especially Carole Heard, Bill Burr, Chris Cottrill, and Leah Smith. Grant Harris and the staff of the European Reading Room of the Library of Congress have been fully attentive. The staff at the Global Reading Room and the former Slavic Reading Room at Gelman Library of George Washington University have done much to apply twenty-first-century technology to historical research on the Soviet Union and their efforts eased the path to obtaining important materials for this study.

The Academy of Sciences' Institute for the History of Science and Technology has acted as a home base for me when I have been in Moscow. Lena Eberle welcomed me into her family and shared her insights into the Russian culture of flight. Over the years, the staff and faculty at the institute have freely offered their insights and exhibited kindness, generosity, and patience. Among them are Igor Drovennikov, Larisa Belozerova, Valentina Ponomareva, and Dmitri Sobolev. Viktor Sokolsky, the late director of the aviation and spaceflight section of the institute, was one of the kindest people that I have known. I miss him. Archives and research facilities in Russia also granted me access for research. I thank the following individuals who assisted me at their respective organizations over the years: Tat'iana Golovkina at RGANTD in Moscow; Sergei Gromov at RKK Energiia in Korolev; Marina Stepanova in Gagarin City; Andrei Maiboroda and Natalya Talanova in Star City; Anatoli Grigor'ev, Oleg Gazenko, Evgenii Shepelev, and Abraham Genin at the Institute for Biomedical Problems in Khimki; Tat'iana Zhelinina at the Tsiolkovsky museums in Kaluga; Nikolai Afansenko, Nikolai Dergunov, and Boris Mikhailov at Zvezda in Tomilino; Roman Artemenko at the Polytechnic Museum in Moscow; the staff of the Baikonur Museum in Kazakhstan; and the staff of the Sergiy Korolyov Astronautics Museum in Zytomyr, Ukraine. I am also indebted to Elena

Pereplenkina and Art Dula for their administrative assistance and hospitality while in Moscow. I look forward to a time when we can renew our collaborative ties.

I have a special appreciation for what a book editor does: navigating that fine line between cheerleader and taskmaster. Sian Hunter at the University of Florida Press has mastered that fine balance of encouraging this very slow-moving writer to make progress and having a clear vision of the potential of my proposal.

On a more personal note, I am particularly blessed with a circle of family and close friends who have supported me over the years with cheers, nourishment, humor, spiritual guidance, and the occasional gentle nudge of a polite inquiry into the status of my project. All of them have demonstrated that the proper application of prayers and thoughts can produce results. My brother, James, has always been the ideal elder brother. Barbara Roesmann's sharp eyes, dry wit, and intelligence have been with me every step of the way, and for that I am grateful. I would like to thank my son Alex, who is my lifecycle monitor and bedtime alarm clock. Every day I see him grow into the man I have always hoped him to be, all the while keeping his promise to his grandmother to "take care of his mother." And finally, I thank Mingus and Rollins, and Dio, who have channeled the spirits of all the German Shepherds who have preceded them by patiently waiting under and around my desk, exuberantly greeting any indications of physical activity, and faithfully keeping all of my secrets.

Of course, even with all this support, all errors of omission or commission in this book are indeed my own.

Abbreviations

Buran	Snowstorm, the name of the USSR's space shuttle program
Cheka	Chrezvychainaia komissiia po bor'be s kontrrevoliutsei i sabatazhem [All-Russian Extraordinary Commission for Combating Counter-Revolution, Speculation, and Sabotage]
Cosmodrome	The site of a launch facility and its support facilities
CPSU	Communist Party of the Soviet Union
DOSAAF	Dobrovol'noe obshchestvo sodeistviia armii, aviatsii i floty [Voluntary Society of Assistance to the Army, the Air Force and the Navy]
GAO	Gosudarstvennoe aktsionernoe obshchestvo [state stock company]
GMT	Greenwich Mean Time
GUK	Gosudarstvennoe upravlenie kinografii i fotografii [State Directorate for Cinematography and Photography]
FAI	Fédération Aéronautique Internationale [International World Air Sports Federation], also known as the International Aeronautics Federation
ICBM	Intercontinental Ballistic Missile
IGY	International Geophysical Year
IMBP	Institut Mediko-biologicheskikh Problem [Institute for Biomedical Problems]
imeni	the name of, named after
INKhUK	Moscow Institute of Artistic Culture

KB	Kostruktorskoe Biuro [design bureau]
KGB	Komitet gosudarstvennoi bezopastnosti [Committee for State Security]
Komsomol	Kommunisticheskii soyuz molodezhi [Communist Youth Union]
Lennauchfilm	Leningrad Popular Scientific Films
Mezhrabpom	Mezhdunarodnaia rabochaia pomoshch' [International Workers' Aid]
Mir	Soviet and later Russian modular space station, 1986–2001
NASA	National Aeronautics and Space Administration
NEP	New Economic Policy
NII	Nauchno-issledovatel'skii institut [scientific research institute]
NKVD	Narodnyi komissariat vnutrennikh del' [People's Commissariat for Internal Affairs] (1922–1946)
OKB-1	Osoboe konstrutorskoe biuro No. 1 [Special Design Bureau, No. 1]
OMON	Otriad militsii osobogo nazhacheniia [Police Special Weapons and Tactical Unit]
Osoaviakhim	Soiuz obshchestv sodeistviia oborne i avaiatsionno-khimicheskomu stroitel'stvu SSSR [Union of Societies of Assistance to Defense and Aviation-Chemical Construction of the USSR]
OVIR	Otdel viz i registratsiia inostrannykh grazhdan [Department of Visas and Registrations for Foreign Citizens]
Pioner	Vsesoiuznaia pionerskaia organizatsiia imeni V. I. Lenina [Lenin All-Union Pioneer Organization]
Proletkult	Proletarskaia kultura [Proletarian Culture]
RGANTD	Rossiiskii gosudarstvennyi arkhiv nauchno-tekhnicheskoi dokumentatsii [Russian State Archive for Scientific and Technical Documentation]
RGNIITsPK	Rossiskii gosudarstvennyi nauchno-issledovatel'skii ispitael'nyi tsentr podgotovki kosmonavtov [Russian State Scientific-Research Test Spaceflight Training Center]

RKK Energiia	Raketno-kosmicheskaia korporatsiia imeni S. P. Koroleva [S. P. Korolev Energiia Rocket and Space Corporation], the legacy corporation of Korolev's design bureau
RSDWP	Russian Social Democratic Workers' Party
Salyut	Salute, name of the first generations of Soviet space stations, 1971–1986
Soyuz	Union, name of Soviet and Russian maneuverable ferry spacecraft
Sovkino	Sovetskoe Kino, Soviet film production house
TASS	Telegrafnoe agenstvo sovetskogo soiuza [Telegraph Agency of the Soviet Union]
TsDSA	Tsentral'nyi dom sovetskoi armii [Central House of the Soviet Army]
TsPK	Tsentr podgotovki kosmonavtov [Spaceflight Training Center]
UTI	Uchebnyi trenirovochnyi istrebitel' [Pursuit Training Craft]
VDNKh	Vystavka dostizhenii narodnogo khoziaistva [Exhibition for Economic Achievements]
VSKhV	Vsesoiuznaia selsko-khoziastvenaia vystavka [All-Union Agricultural Exhibition]
VPV	Vsesoiuznaia promyshlennaia vystavka [All-Union Industrial Exhibition]
VVS	Voenno-vosdushnye sily [Soviet Air Force]
VVTs	Vserossisskii vystavochnyi tsentr [All-Russian Exhibition Center]
Voskhod	Sunrise, renamed multi-passenger version of the *Vostok* spacecraft
Vostok	East, first-generation Soviet spacecraft
Zvezda	Otkrytoe aktsionernoe obshchestvo nauchno-proizvodstvennoe predpriiatie "Zvezda," OAO NPP "Zvezda" [Open Stock Society (LLP), Research and Development Production Enterprise "Zvezda"], enterprise that has manufactured all Soviet and Russian spacesuits

Note on Transliterations

All Russian words and names have been transliterated with the Library of Congress (LC) System Russian Transliteration Table without the diacritical marks. Exceptions to this transliteration scheme occur in the cases of those Russian words and names that have alternative spellings that are more familiar to English speakers. For example, Yuri is more familiar than the strictly transliterated Iurii, and Tsiolkovsky is the commonly accepted form of the scientist's name. There are some resulting inconsistencies. Proper nouns might appear in the text in the familiar English transliteration and then be footnoted with the LC transliteration. This eases retrieval of sources from library databases that use the LC Russian transliteration.

1

Introduction

After the collapse of the USSR in 1991, people throughout Russia seemed to participate in the reevaluation of the seventy-four-year experiment. The failure spared nothing from reassessment, including the most traumatic experience of the nation—World War II. For an all-too-brief period of the 1990s, town halls, television, cafés, and dining rooms overflowed with arguments over the true meaning of the USSR. These post-Soviet discussions emerged out of the human need to evaluate and edit the experience and to determine how to celebrate and remember the past. Former Soviet citizens sought to balance the memory of communist rule between the nostalgia of being a great power and the regret of a failed and corrupt state. The image of the Cold War hero, the cosmonaut, could not evade this reevaluation despite its seemingly uncontroversial past. The Russian population experienced the stress of choosing between nostalgia for a romantic and hopeful past and a need to blame someone for the failure to fulfill the promises of postwar life. As a result, two images of the cosmonaut exist today—that of the hopefulness of youth and that of the forlorn victim of a cynical state.

This book presents new avenues and perspectives from which to understand how the image of the Russian cosmonaut has changed over time. It relies heavily on sources outside of the traditional archives, incorporating visual and material culture, films, and literature, and introduces a range of actors in the space program that traditional histories of the space program rarely address. Doing so realizes two goals. The first is to discern the limitations of state-sanctioned images to perpetuate an idealized concept. The USSR could only hope for control and manage a vision for a short period, after which individual interpretations inevitably emerged. The second goal is to understand better the process through which the Russians have reexamined their past in the aftermath of the collapse of the USSR. This book does not describe a decline and fall of the cosmonaut, nor even a withering of the image. It does explain a change of public and artistic opinion once freed from close state direction. From this perspective, one can understand the transition and link between the images of the smiling, happy face of a young cosmonaut of the 1960s through the twenty-first-century portrayal of a perplexed hero who must surmount the state and bureaucratic forces that constrain him.

The public unveiling of the cosmonaut as a state instrument occurred with the first announcement of a human being orbiting the Earth. The official press agency of the USSR, TASS, publicized the flight of Yuri Gagarin on 12 April 1961, just a few minutes after his rocket launched him into space. Upon landing, Gagarin's prepared formal thank-yous to the Communist Party of the Soviet Union (CPSU), the Soviet government, and Nikita Khrushchev contained tried-and-true phrases from Soviet heroic rhetoric. He paid tribute to the integrated party and state structure that built his equipment. The one surprising note in this relatively low-key proclamation was a final line, stating that the "accomplishment of a piloted space mission opens up new vistas for humanity's conquest of space." Gagarin's message made no mention of the Cold War competition with the United States and situated the accomplishment of his flight beyond the borders of the USSR. It implied that he was a peaceful pioneer who was leading the world into a new era. That was the birth of what I term "the Red Stuff"—the result of generations of gestation of the idea of sending a Russian into space. Almost as soon as it was born, however, the elements of the Red Stuff began to unravel, as though the carefully managed accomplishment that took generations to create succumbed to an inevitable entropy. Within fifty years of his flight, the role and the shapers of the public understanding of Gagarin were quite different from those of the 1960s.

The origins of the Russian cosmonaut image resulted from extensive contributions of past generations and immediate needs of contemporary politicians. Immediately after Gagarin's landing, the state, party, and Nikita Khrushchev's leadership began to expand his image beyond the noble cause of leading humanity to conquer space. Gagarin and subsequent cosmonauts quickly became Soviet Cold Warriors, exemplars of a new Soviet generation and the visible representatives of thousands of anonymous space workers. To stretch the Red Stuff to cover so many purposes within the party and state, Khrushchev relied on the creative energies of many sectors of society to augment the material, visual, and literary culture that supported the cause. After the mid-1960s, the propaganda utility of the Red Stuff dwindled at home and abroad, leaving many who had played a supporting role to their own devices to interpret the meaning of Russians in space. The party state and leadership changed their attitude and personnel after 1961. Russians in space remained a reality, even as the meaning of the Red Stuff had changed with the changing world around it.

"The Red Stuff"

The phrase "Red Stuff" has dual origins. The word "Red" in Russian is more than the name of a color. Its connotations go beyond the best-known and relatively recent political distinction between the Reds (Bolshevik) and Whites (anti-Bolshevik) of the Russian Revolution. In Russian, "red" has an ancient meaning associated with beauty, goodness, and honor. Historical examples of this usage include the Red Square in Moscow, which originated as a public market space situated between St. Basil's Cathedral and the Spassky Tower (Christ, the Savior) of the Moscow Kremlin. The word referred to the beautiful square over which hung the country's most cherished icon and the church-monument to Ivan IV's (commonly known as "the Terrible") defeat of Kazan. The word conveyed both physical and spiritual beauty. For the last thousand years, Russian Orthodox households have had a "red corner" that featured icons and talismans of Russian Orthodox prayer and scripture. During a brief period in the twentieth century, secular icons of Soviet party leaders (also "Red") temporarily replaced religious artifacts. In the last quarter century, the overtly religious character of the red corner has returned, even on board the secular International Space Station. During the seventy-four-year period of Soviet rule, Bolshevik leaders took advantage of the positive connotations of the Russian and revolutionary associations with the color red to buttress their power. "Red" in this book refers not only to that which is communist, but also to that which is intrinsically Russian, and which reforms itself within the contexts of political, economic, and social change.

The second part of the phrase gestures to Tom Wolfe's book about the *Mercury* astronauts, *The Right Stuff*. Wolfe used the term to encapsulate the mental and physical characteristics of test pilots and their willingness to take risks, as exemplified by Chuck Yeager. In contrast, the Red Stuff describes the process of adding layers of meaning to the idea of Russians in space from the end of the nineteenth century through the first decade of the twenty-first. It includes the act of selecting a different and broader collection of personality, family, and cultural characteristics that best reflected the authors' vision of human spaceflight and deleting others that do not match contemporary expectations. These traits do not solely define techni-cal prowess, but the possessor's ability to act as a guide to society into the future or to serve as a mirror from which to reflect on the past. The Red Stuff is a Russian version of human spaceflight, rendered in the image of the Russian cosmonaut.

Placing Soviet Human Spaceflight into Historical Context

This book attempts to reconcile the history of the late USSR and Russia in the twenty-first century with the historiography of the Soviet human spaceflight program. The two fields frequently referenced each other for the last sixty years but have never offered a resolution to the intricate inter-action between the two histories—a function that the public image of the cosmonaut fills.

Knowledge of the Soviet space program has changed dramatically over the last fifty years, reflecting changes in the availability of archival resources. Initially, Soviet secrecy encouraged extreme methodologies. Early space historians, lacking traditional archival and documentary evi-dence, relied on remote observations and close comparisons to the more open American space program to understand what had gone on within the USSR. They began by presuming similarities between the early Soviet human spaceflight program and the American *Mercury* program, almost as though the latter program induced the former's actions, adopting the binary superpower competition theme that dominated US-Soviet politics of the era. More sophisticated historians relied on theoretical political-historical analysis to place Soviet space history within a comprehensible context. In all cases, historians such as James Oberg and Walter McDougall studied cosmonauts within the context of the Cold War and made direct comparisons to their American counterparts.[1] Their arguments are under-standable. Oberg, among others, judged Soviet activities based on some incomplete official statements, observations from specialists, incomplete published records, and his technical knowledge of the American program. McDougall was assessing the technocratic origins of the space program and its implications from the perspective of the politics of the United States during the Space Race. Although this comparison served a purpose from a political perspective, it fell short of recognizing the indigenous Russian cultural origins of the idea of spaceflight and local motivations and strate-gies for exploring space.

Even as the earliest historians began their work without access to traditional sources, cracks in the image of the Soviet cosmonaut were apparent. The effort to create an idealized vision of a Soviet citizen in space relied on very high expectations of the Soviet population as eager to accept the immediate creation of a communist state. Moreover, some of the evident openings for public criticism resulted from recruiting actors to create their version of the Red Stuff. The state shared responsibility for producing unanticipated consequences of the public exploitation of human spaceflight. The significance of these unintended results increased over time. Pieces from Gagarin's actual life that did not match an idealized image, although not necessarily disastrous to the theme of Soviet conquest of space, did not make it into the picture. These were his boisterous and vulnerable extremes—the human side of the man. The act of excising them because they were not ideally suited to an archetype set in motion a decade of habits that flattened and made suspect the image of all Soviet cosmonauts. The forced, two-dimensional image did not adapt well to historical changes in the USSR and the world in the ensuing years. It aged poorly and demanded individual and public reinterpretation. Many came forward to answer that demand. Early historians of the Soviet space program seized on these flaws as evidence of a crumbling system.

Two things have changed in the historiography of Soviet human spaceflight in the last forty years. The first was the significant and unanticipated result of Mikhail Gorbachev's late 1980s policies of Glasnost and Perestroika—the broad opening of previously firmly held secrets of the Soviet Union, including the space program. Individuals and then institutions opened their memories, diaries, and archives as economic conditions worsened and the political repercussions for doing so diminished. Sergei Korolev's deputy, Vasilli Mishin, was one of the earliest to break the silence, publishing a selective and somewhat self-serving account of his role in the failed Soviet effort to send humans to the Moon.[2] His recollections were followed by a cascade of memoirs and diaries, including the four volumes of Boris Chertok, an engineering deputy of Korolev.[3] The offspring of Nikolai Kamanin, one of Stalin's Falcons (aviation heroes from the 1930s) and later the first commandant of the Soviet cosmonaut corps, chose to publish their father's diaries from his time in the space program as another four-volume set.[4] Public and corporate archives followed suit, publishing collections of founding documents and photographs from its history.[5] The game-changing consequence of these publications and open archives was the publication of Asif Siddiqi's major revision of the historiography in his NASA-published *Challenge to Apollo*.[6] In ways that upended the field, Siddiqi established a documented history of the establishment and operations of the Soviet space program that had been missing during much of the Cold War.

Just as there has been a rewriting of the Soviet space experience considering archival and personal revelations, so have Russian social analysts propelled a reassessment of Soviet political leadership. Historians once relied on secondhand reports, domestic amateur historians, and the Kremlinology to interpret motivations and operations inside the Kremlin. Over the decades, trained Soviet historians, national archives, and aggressive journalism have revealed some of the mysteries of Moscow politics. Archives that opened to Western scholars after the fall of the Berlin Wall have contributed to a global insight into the motivations of the period. Politicians' thoughts and aims remain ambiguous, even with archival documentation, but open archives have allowed historians to construct a model of the once-veiled institutional organizations.

Cosmonaut: A Cultural History is not an institutional history of the space program, but a broader, cultural history of Russians in space that supplements the internal accounts. Those who contributed to the initial Gagarin-idealized cosmonaut image did not have a background in prior iterations of Soviet technological heroes. They were not the visionaries, engineers, or politicians who had invested much in defining what a Russian man in space might mean. They were outsiders—small-scale manufacturers, filmmakers, storytellers, artists, and architects. They approached the cosmonaut as a new commodity for the public. These outsiders, who had worked their crafts to augment the state-sponsored image, held divergent interests in the perpetuation of the official portrait. Their interests had little to do with the promotion of Soviet science and technology or the Cold War. Soviet policies in the 1960s state had pulled back from relying on tightly regulated and state-sanctioned purveyors of the images to buttress its profile at home and abroad. They also unleashed public demand to participate in state activities in a personal way, creating a need for a material culture of spaceflight that had not existed in a previous form. The state recruited small armies of artisans, artists and illustrators, architects, and film directors, among others, to augment the vision of the Russian man in space, all bringing in their perspectives. As the providers and the market for personal copies of the images expanded, the state control of the content dwindled in effectiveness even more. These independent makers and thinkers began to make their contributions to the image, presaging an overall reassessment of the Soviet experience that was to occur later. The situation set the stage to encourage others to appropriate the Red Stuff as raw material for their products. The smallest cracks and peels of the image trace back to the initial well-ordered idea of the cosmonaut almost from the beginning.

2

The Birth of the Cosmonaut

"Real Cosmonaut Lives" as a Guide to Youth, the Nation, and the World

By 6:07 UTC (9:07 a.m. Moscow time) on 12 April 1961, after a sleepless night in a small cottage on the premises of the launch facility near the Kazakh city of Leninsk, a young pilot of the Soviet Air Force put on a heavily reinforced, high-altitude flight suit alongside his colleague, German Titov.[1] The pair rode in a bus that would carry them a few kilometers to the launch site where *Sputnik* had launched three and a half years earlier. Two hours before his morning liftoff, Gagarin parted ways with his backup pilot and climbed the service gantry to the elevator that would take him to the top of the 30.84-meter rocket with only the company of engineers and technicians. They were there to buckle, plug, and seal him inside his spacecraft and to offer final words of support before his journey began. As the rocket rose from the launch pad, the young pilot had only one word to say, "Poekhali!"—"Let's go!" After about nine-and-a-half minutes of the first stage of the rocket firing, his craft achieved the velocity and direction to orbit the Earth. Yuri Gagarin became the first human being to enter space.

A few minutes later, the official press agency of the USSR, TASS, announced the flight to the world, stating that the culmination of centuries of humans dreaming of leaving the Earth's surface was an "accomplishment of a manned space mission [that] opens up new vistas for humanity's conquest of space." One hour and eighteen minutes after his launch, the *Vostok* spacecraft engines began a retrofire to send Gagarin back to Earth. Thirty minutes later, Yuri Gagarin was on the ground in Russia, near Saratov, and on his way to becoming the most famous man on Earth. That simple, neat story was the one recounted throughout the world for many years. Of course, this uncomplicated story did not reveal the whole truth.

Figure 1. Yuri Gagarin begins his journey to become the first human to orbit the Earth and shouts "Poekhali!" ("Let's go!") to his comrades listening from Earth.

While reentering the Earth's atmosphere, the *Vostok* reentry module, the sphere that was to carry him back to Earth, failed to separate from the service module that carried the oxygen, fuel, and reentry rockets. The two sections of the spacecraft remained attached via a cable, causing them to gyrate wildly around their center of gravity as they entered the atmosphere. Twenty minutes after the components of the spacecraft separated and ten minutes before landing, the pilot ejected from the spacecraft to parachute to Earth. The ejection was not a result of equipment error. The *Vostok* spacecraft could not slow sufficiently for a safe landing with a human inside. To survive landing, Gagarin had to eject and parachute about 7,000 meters from the Earth's surface. These two facts were not included in official Soviet proclamations. If they had been, they would have indicated equipment failures and inadequacies that were not in keeping with the idealized picture of the Soviet space program.

The Soviet population warmly welcomed the news of Gagarin's flight. Villagers near Saratov claimed to have witnessed his landing. Spontaneous celebrations took place in Moscow when the public first read of his flight in the newspapers on 13 April 1961 and of the welcoming celebration in Red Square the next day.[2] Valentin Gagarin, brother of Yuri, described the Moscow scene: "Muscovites hurried to Red Square, playing accordions and

Figure 2. Gagarin and Khrushchev stand overlooking Red Square with raised arms celebrating his flight.

dancing on the bridges. 'This is how it was in 1945 on Victory Day,' a driver told us."[3] Khrushchev ordered MiG fighters to escort the plane carrying Gagarin back to Moscow.[4] The newly promoted pilot, who was not yet at ease with his fame, walked down a red carpet to greet his family members at Moscow's Vnukovo Airport with an untied shoelace.

In contrast to the pilots of the 1930s, the cosmonaut did not thank a single political leader like Stalin for his flight but attributed his success to Soviet society. In fact, the cosmonaut was not given complete credit for the success of the flight. During the initial period following Gagarin's mission, accounts did not stress the individual achievements of the cosmonaut but heralded the scientific potential of human-crewed spaceflight. It had been a collective achievement, not an individual one, as fully articulated in his essay that appeared in the popular science journal, *Priroda* [Nature]:

And thus, a great heroic deed has been accomplished, turning another shining page in the history of the civilization of man. This achievement is one of the Soviet people, led by the Communist Party and the Soviet leadership. This is a heroic deed of a collective of scientists, designers, engineers, technicians, and workers, this is an achievement of all those researchers who took part in the never-ending preparation of the spaceship for launch, an achievement of all of those who insured the normal flight and landing of the spacecraft, this is the achievement of the great son of the Soviet fatherland—Yuri Alekseevich Gagarin. His name has already become legendary.[5]

Shared kudos were not unique to the popular scientific press; the readership would likely have been scientists, designers, engineers, technicians, and workers. The role of Soviet scientists and engineers received significant attention in the Communist Party's newspaper, *Pravda,* in celebration of the "glory to Soviet scientists, engineers, technicians, and workers—the masters of space."[6] This professional support for the space program remained anonymous if not entirely secret. The secret legacy of space scientists and engineers grew out of secret projects that went officially unacknowledged even after the launch of Gagarin. The engineers who worked on the public space programs were the same ones who worked on programs, such as spy satellites, that remained secret.[7] For example, work continued with Intercontinental Ballistic Missile (ICBM) development, reconnaissance and communications satellites, and other national security programs. Using a distant historical legacy of the sciences and engineering, attention shifted from attempts to identify current participants to celebrations of the past. Gagarin's flight joined the official history that placed the origins of the spaceflight in the nineteenth century and the early works of Russian rocket scientists, starting with nineteenth-century revolutionary Nikolai Kibal'chich and early twentieth-century mathematician Konstantin Tsiolkovsky.[8]

For all this celebration and proclamations, these early, golden years of the Red Stuff were very brief. Little more than two years separated Gagarin's flight and the final *Vostok* flight of Valentina Tereshkova in June 1963. Consolidation of the launches placed limits on the press impacts of the flights. German Titov became the first human to spend a day in space in August 1961. The flights of *Vostoks* 3 and 4 took place a year later. Pilots

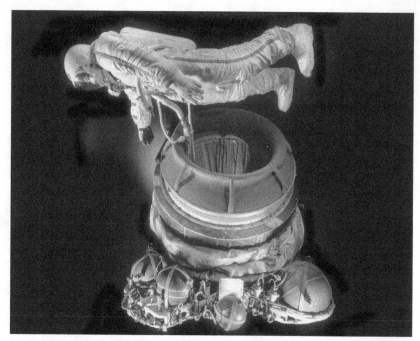

Figure 3. This is the training model of the spacesuit and airlock that Aleksei Leonov used to prepare for the world's first spacewalk that occurred on 18 March 1965. Courtesy the Perot Foundation.

Andriian Nikolayev and Pavel Popovich entered Earth orbit one day apart to give the illusion that their craft were capable of maneuvering near one another. Ten months later the illusion repeated when Valeri Bykovskii entered orbit inside *Vostok 5,* followed two days later by the first woman in space, Valentina Tereshkova. Following a sixteen-month gap, the first multi-crew ship launched from Baikonur, the *Voskhod.* This was not a new spacecraft but a stripped-down version of the *Vostok* to accommodate three passengers. A repeat *Voskhod* launch placed Pavel Belayev and Aleksei Leonov into space, where the latter performed the world's first spacewalk. From start to finish, fewer than four years separated the beginning and end of the period. A politician had defined this one.

Khrushchev at the Controls

No single individual from the USSR is more identified with the Space Race than Nikita Khrushchev. He had a leading role in defining the Soviet space program and its public mission. Moreover, he incorporated it into his balancing act that maintained his domestic political position and his country's role in the Cold War. There is a famous but apocryphal story about the immediate response to Nikita Khrushchev's secret speech at the Twentieth Party Congress in 1956 during which he denounced the crimes of Stalin.[9] During a question-and-answer session after the speech, one delegate anonymously asked of Khrushchev, "Where were you when these crimes were taking place?" Khrushchev immediately demanded that the questioner step forward, and when he did not, Khrushchev responded, "That was precisely where I was at the time." The message of that story is that Khrushchev had been anxious to dismantle the personal apparatus of terror that Stalin had erected around himself, but he was not entertaining any more profound changes in the Soviet system, nor was he prepared to implicate any of those who had supported Stalin unless it served a purpose. Khrushchev was prepared to choose the good and the bad from the Stalin legacy to accomplish his two jobs. The first was to dismantle the cult of Stalin. The second was to maintain the brinksmanship against the United States in the Cold War.[10] Both jobs shared an imaginary component. Khrushchev chose to carry out his job by conforming to Stalinist forms and simultaneously creating an elaborate set of complex rituals through which he seemingly denounced Stalinism while maintaining pressure against the United States. If the facts did not conform, then insist on the ritual even more forcefully. The emerging arena of spaceflight offered a new sphere in which to do this. The new Soviet cosmonauts were ideal replacements for the personality cult of Stalin, as they represented the Red Stuff.

In January 1962, Nikita Khrushchev made a second and lesser-known secret speech. His 1956 speech condemned the crimes of Stalin. The latter speech outlined his strategy for maintaining an illusion of domestic progress for political gain. Using an analogy of the liquid meniscus of a full wine glass, Khrushchev likened the bulge of the liquid over the edge of the glass to the pressure that he sought to maintain against his enemies. The force of the surface tension of the liquid held the wine in an overfilled glass. In real life, the surface tension was the ritual and repetition that reinforced faith in the political illusion. This trick was not Khrushchev's invention. Stalin had established a similar use of illusion a generation before Gagarin in the field of aeronautics.[11] Stalin's Falcons' flights established a model of a distracting ritual for domestic Soviet consumption.[12] Stalin exerted his influence on the creation of heroes after an extraordinary series of events and a remarkable number of aviation images had emerged out of Russian revolutionary culture. The ship *Cheliuskin* carried an expedition of 104 people, including women and children, and became stuck in the Arctic ice during its expedition through the Northeast Passage.[13] Aerial rescue attempts began in February 1934 and continued through April, when the final group of explorers came home.[14] The rescue of the crew, followed by repeated transpolar missions, provided a backdrop to repeated ceremonies culminating in awarding pilots the "Hero of the Soviet Union" medal. The theater of technological progress was reassuring to the public.

Khrushchev first adopted this method when he saw the startling international response to the *Sputnik* launch in 1957. *Sputnik* had provided no additional evidence of the Soviet Union's rocketry prowess beyond the first successful launch of an Intercontinental Ballistic Missile (ICBM), which occurred two months earlier, in August 1957. The second event overshadowed the first in public. The *Sputnik* launch became a propaganda event, even though Soviet planners had not anticipated the public response to the launch throughout the world.[15] *Sputnik*, like Stalinist Falcons, had in it the potential to distract the domestic population from its current problems. As an added benefit, spaceflight could distract both domestic and international attention from the weakened infrastructure of the postwar Soviet economy, drawing attention to an illusion of a modern future and the potential for a new symbol of the technological program. All that Khrushchev had to do was to find a way to ritualize spaceflight in the way that Stalin had ritualized aviation. New space rituals had the potential to supplement Khrushchev's foreign policy strategy for maintaining the perception of strategic parity with the United States.[16]

Khrushchev's use of the symbolic power of *Sputnik* was consistent with his political style for the postwar geopolitical and domestic situation. Khrushchev had to create a new strategy for confronting his Cold War opponent that would mollify the nation. Having denounced Stalin's cult of personality, he had to substitute new imagery that was sufficiently novel as to in no way resemble a paean to Stalin to which a previous generation had had grown accustomed. The new party leader had to find a way to build on Stalin's successes while avoiding any mention of the man. Khrushchev's understanding of stunning Soviet successes in the development of nuclear weapons in the 1940s encouraged him to carry his own new policies to the extreme.[17] The goal of his preemptive strategy was to make an attack on the Soviet Union style of government popularly unacceptable throughout the world, thus strategically compensating for the reality of the military and fiscal inferiority of Soviet forces. Along with this, Khrushchev had to convince his domestic audience that an open conflict was distant and unlikely.

Figure 4. Monument indicating the location of the launch of the world's first artificial satellite, Sputnik. This place is adjacent to the launch pad from which Yuri Gagarin was launched in 1961 and from which Russia continued to launch cosmonauts in the twenty-first century at the Baikonur Cosmodrome in the Kazakh Republic (2015).

To be successful, this strategy had to sustain credible deterrence, considering the enemy's full knowledge of its strategic superiority—a highly risky game of "chicken." Even after the USSR lost that illusion to the American reconnaissance aircraft, listening stations, and satellites, Khrushchev still could maintain the illusion for much of his public.

In no way is Khrushchev's use of Stalinist forms in space more apparent than in the presentation of Soviet cosmonauts to the public in the 1960s. He demonstrated this skill in his application of the new technology of ballistic missiles to the global and domestic politics of the USSR in the form of the space program. For eight years, Khrushchev applied his style to existing Stalinist forms, apparatuses, and rituals. From it, he created the Soviet space program and its brief and convincing challenge to the United States. Khrushchev did this by adopting Stalin's technologically driven ideology to create a direct challenge against the US industrial and technological dominance after World War II, leaving Stalinist regnant myths and taboo topics in place despite Stalin's personal removal from favor.

After *Sputnik*, Sergei Korolev's position depended on his ability to deliver a supply of symbolic launches. Khrushchev understood the risks of awakening the giant but wanted to keep poking him at a safe distance. Cogent technical arguments notwithstanding, Khrushchev was implacable. He was the first to recognize the political value of space events. Khrushchev understood that a near-hollow silver ball in orbit around the Earth had a more significant influence on world politics than an arsenal of ICBMs.[18] To this end, he chose more demonstrations of Soviet space shots to make a serious effort at advancing Soviet ICBM technology. Soviet rocket engineer Boris Chertok later wrote of Khrushchev, "The political successes that we brought, and an additional sensational space launch was of more importance to Khrushchev than the benefits of intercontinental nuclear rockets."[19] Korolev immediately approached the Army Institute of Aviation Medicine, which had conducted previous high-altitude experiments sending dogs inside rockets to test their response to the upper atmosphere in the late 1940s and early 1950s.[20] He pressed these scientists to adapt their high-altitude hardware quickly for a spaceflight. The resulting flight of *Sputnik 2* with the ill-fated dog Laika occurred 3 November 1957, in time for the fortieth anniversary of the Bolshevik Revolution. Laika did not survive for the celebrations. She died from heat and stress within hours of first orbiting the Earth.[21] They hid that fact from the public through a series of false stories and doctored reports.[22]

With much of the public culture of Khrushchev's triumphalism in place before the launch of the first man in space, the Soviet population could anticipate the message that cosmonauts would send. Those cosmonauts received the same honors and celebratory rhetoric that aviation heroes had a generation before. In addition, the honors, celebrations, and rhetoric repeated time after time. What differed for the cosmonauts was the political culture of the time. Instead of Stalin, cosmonauts thanked the party for the opportunity to demonstrate Soviet superiority. Evidence of Stalinist practices and Khrushchevian modifications exists in the vast literature of cosmonaut biographies, their numerous public statements and appearances, press coverage of their missions, and the diaries and memoirs of those who had been privy to the decisions about the cosmonauts' lives at the time. These ritualized cosmonaut stories served a purpose. Their primary focus had nothing to do with the technical aspects of spaceflight. Cosmonaut stories and their images were about the Soviet state. Spaceflight was merely an attention-grabbing method with which they could gain worldwide notice.

The profession of a first-generation spaceman, too, lent itself to cultural manipulation. Piloting a spacecraft in the early 1960s did not require great aeronautical skill. The Soviet and American first-generation spacecraft, *Vostok* and *Mercury*, were not very complex machines compared to the state-of-the-art supersonic aircraft of the time. Each craft had attitude control but no ability to change orbits. During the most dangerous parts of flight—takeoff and landing—the pilot did not have much control of the craft, relying on automated systems. Because of the limited number of missions, there have been far more individuals capable of completing pilot training in the USSR and the US than there have been opportunities to do so. In both countries, the skills acquired during the ordinary course of flight training exceeded those required to maintain attitude control in a *Mercury* or *Vostok* capsule. As a result, the primary purpose of the selection process for astronauts and cosmonauts was not to weed out the vast number of technically unqualified individuals but to make selections based on other, non-technical traits.[23] Although these non-technical traits are less evident within the context of the Space Race, they reflect specific national social and cultural values. Moreover, the personification of these values resonated among the domestic population far longer than proficiency in technical accomplishments. The Americans drew from the test-pilot culture of the US Air Force and Navy. This culture is what became commonly known as "the Right Stuff." In the case of the Soviet cosmonauts, the values are an articulation of previously established cultural expectations of heroism and spaceflight. The pilots who found themselves among the twenty selected to fly represented a vision of Russian nationalism and idealized Soviet male behavior for a real purpose. In contrast to the Americans' Right Stuff, the Red Stuff represented only the outer, public layer of Russian spacefaring. Their duty had been to symbolize the forward-looking Soviet state and distract from the deeper costs of challenging their Cold War foes.

In 1959, a commission of Soviet aviation physicians, aviators, and military officers undertook a two-stage process to select cosmonauts. First, they determined the technical selection criteria. Second, they had to judge the cultural measure of the ideal candidate. Complete details of the selection process and the cultural criteria used to pick cosmonauts during this period remain unavailable; there are no standard official records. On occasion, Russian historians reported data of an individual cosmonaut's medical tests.[24] However, they did not provide a complete set of results that would provide an opportunity for detailed analysis of the screening process. Firsthand accounts of the process reflect the subjective opinions of the writers and blur the lines between measuring technical expertise and perceptions of social and cultural strengths. In the end, the complete picture of the attributes sought in a cosmonaut is the public biography and cultural portrayal of him or her, which conflates both technical and nontechnical qualifications to create an enduring model of heroism.

The Soviet news agency TASS announced spaceflights and the identities of cosmonauts only after the agency was confident that the flights were successful. Photographic introductions to the cosmonauts did not appear in the popular press until well after their missions.[25] In comparison to the American program, the selection process remained a mystery throughout the 1960s. Even after recent publications of first-person accounts of the recruitment process and the release of training photographs and film collections at the Rossiiskii gosudarstvennyi arkhiv nauchno-tekhnicheskoi dokumentatsii (Russian State Archive for Scientific and Technical Documentation, RGANTD), reconstructing the precise parameters those Soviet scientists and physicians used to measure the qualifications of pilots as cosmonaut candidates is next to impossible. The archival evidence remains elusive. One can assume that the process allowed some subjective judgment that enabled the commission to factor in a *je ne sais quoi* element. As a result, there remains a single, unified if undetailed story of how the first cosmonauts made their way from aviators to cosmonauts through testing. This widely publicized official narrative provided the framework for all cosmonaut autobiographies and biographies. And, as a result, a self-fulfilling prophesy remains. Those salient characteristics of the first cosmonauts have become the specific characteristics that the commission sought. Yuri Gagarin, the cosmonaut who could not stop smiling, became the archetype of this new Soviet man in space.[26]

As is the case with so much of the Soviet space program, its greatest contrast to the United States in the selection of its first cosmonauts was the level of secrecy. There had been no public announcement about the recruitment. Absent was a public relations campaign to endear the (yet-to-fly-in) spacemen to the public. Once they had flown in space, Soviet cosmonauts were the subjects of an elaborate publicity campaign that provided few technical details of their mission but were to appeal to a set of cultural and social values that the public shared. The Soviet public accepted these values as readily as Americans later accepted the public image of the *Mercury* astronauts and joined in public celebrations of their accomplishments. Whereas Wolfe used the well-documented bureaucracy of NASA as the backbone of his story, it was the nature of the Soviet bureaucracy to be secret. It was through the portrayal of the cosmonauts that the public could discern the intentions and nature of the bureaucracy. The Soviet "right stuff" was as unique to Soviet culture as the American version was to its own.

Cosmonaut Emerges from Years of Preparation

The launch of Yuri Gagarin was the beginning of the public manned spaceflight program in the Soviet Union. It also marked the remaking of the image of the new post-Stalinist Soviet man. For this reason, it is essential to look at the men of space, who lived out the science-fiction fantasies of previous generations, fulfilled the myths of Soviet legends, and reflected the greatest optimism for Soviet life in the post-Stalin era. The creation of the image of the cosmonaut also represented the creation of an updated version of the heroic myth in the Soviet Union, which drew on classical heroic traditions. The cosmonauts carried messages that did not deviate from official doctrine, proving the control that managers had over the cosmonauts. Under this tight control, the cosmonauts were able to take their message beyond the Soviet sphere to the West and the non-aligned world. There, they translated a domestic appeal into an international one. In the case of Yuri Gagarin, the Soviet press announced that he was in orbit.[27] Immediately upon landing, he partook in the ritualized celebration of his accomplishment. The second stage of the construction of the biography was the publication of life stories that appeared during the prime of the cosmonaut's celebrity.[28]

We now know the details of how Gagarin's flight deviated from the press accounts thanks to the diligent work of historians. From this, we can understand what the Soviet public did and did not know and why. The details of the failed upper-stage disconnection emerged only from archival records and firsthand accounts many decades later. It was a hardware failure that did not cause a fatality. The issue of parachuting from the spacecraft before landing erupted as a little international public relations brushfire months after Gagarin's flight but went unnoticed in the Soviet press. Gagarin's ejection from the spacecraft before landing was an insignificant detail that could not diminish the flight's accomplishments. As far as the Soviet public was concerned in 1961, the only flaw in Yuri Gagarin's mission was his untied shoelace on the red carpet.

Given the US-Soviet rivalry, one might assume that the Soviet cosmonauts followed the examples of their American counterparts concerning qualifications and personalities. However, if one compares a photograph of the Soviet *Vostok* cosmonauts to one of the *Mercury 7*, the apparent difference is that the two groups did not pose or dress alike. The Americans lined up in their high-tech-looking flight suits that had been painted silver. Even on the rare occasion that the seven wore military uniforms for photographers, they did not wear the same uniform, as they represented three separate branches of the armed forces, the Navy, Marine Corps, and Air Force.[29]

In their pictures, the Soviet cosmonauts, except for Valentina Tereshkova, wore uniforms of the Soviet Air Force, proudly displaying a military affiliation that the Americans had brushed aside.[30] As a group, the cosmonauts were younger, almost all in their twenties, and none was a military veteran of World War II.[31] With only two years in age separating the oldest cosmonaut from the youngest astronaut, it might seem that the differences between the two groups were minor. However, the split in the emphasis was over experience. The Americans emphasized piloting experience. The Soviets emphasized childhood experiences during the war. Each side chose its approach in managing the public personae of these frontline soldiers in the Space Race. The Soviet cosmonauts are not imitations of the American astronauts but part of their own heroic traditions.

Even though there was no formal public announcement of recruitment, Soviet memoirs published after the 1980s have stated that in 1959, a special commission approached young Soviet pilots and asked them if they might be interested in flying a different type of craft.[32] These memoirs indicate that at the time, the candidates who went through the screening process endured a rigorous selection process with a higher than 90 percent washout rate.[33] In many ways, the selection of the cosmonauts mimicked the previous development of heroes in the Soviet Union. They not only had to fulfill professional qualifications but also had to possess additional ideological and social credentials and meet subjective expectations of the selecting officials that were like those required for heroes of previous generations in the Soviet Union.

All the recruits were recent flight school graduates, taken from the most substantial portion of the recruitment pool. At least a portion of the selection process sought to identify candidates who would best represent the public ideals and cultural influences of the nation. Three major cultural influences of the late 1950s and early 1960s cut across all aspects of Soviet life. They included the legacy of the triumph of USSR in World War II against Germany; the continuing importance of science and technology in Soviet life; and the shift away from the cult of personality of Stalin toward a cult of the Communist Party.[34]

Despite the vigorous and thorough presentation of cosmonauts in this manner, the question remains whether this promotion of space travelers as national heroes was successful in the end. The global acknowledgment of Gagarin's flight, the crowds at Red Square, the international magazine covers, and their general reception attested to this. From 1961 to 1965, the world accepted the role of the Soviet cosmonaut. However, how enduring was the image? Until very recently there was almost no information that supported an argument one way or another. Recently, a Russian demographer published a fascinating chart online. The chart recorded the popularity of first names in Moscow during the twentieth century.[35] The chart challenged common assumptions about how the popularity of cosmonauts infused Moscow families. The first name Yuri was nineteenth overall during the twentieth century. Even though there is an anticipated peak of the use of that name that brought its popularity up to number nine in 1961, the first name was the sixth most popular name in 1937. Yuri is the only cosmonaut-associated name that experienced an upsurge during the space program. The most common first names for boys in the twentieth century in Moscow were Alexander, Sergei, and Vladimir, and those names witnessed surges in the late 1950s, prior to the launch of *Sputnik*. Similarly, there was no great interest among Moscow parents to name their boys Pavel, Valeri, Aleksei, Andrian, or German, the names of the first male *Vostok* cosmonauts.[36] The name Valentina, which was the most common female name in 1939, saw a minor bump in popularity in 1963, but then only to number twenty-two even as it remained the eleventh most common name in the twentieth century. The Soviet management of public identities in the early 1960s had been very effective at the time, but its effectiveness seems to have been greatest in its contemporary era (late 1960s and 1970s).

Given the paucity of archival documentation of details on the screening of pilots for the cosmonaut program, cosmonauts' public accounts of this phase provide the most readily available documentation. Unfortunately, most of the cosmonauts and their biographers remained cryptic on this matter. These omissions might not be part of an effort to conceal the screening protocols but are likely due to assumptions about the natural animosity that a pilot might have felt toward the flight doctors. The doctors were able to determine who could fly and who could not. When German Titov recounted this period to journalists Pavel Barashev and Yuri Douchayev, he was frank about his anxieties: "Actually, I feared doctors, as many another flier did. Doctors, looking after your health, are always looking for something that they can use to keep a man from flying."[37] Later biographies provided more details on the process, from the cosmonauts' perspective, without shedding a great deal of light on the physicians and the psychologists' thought processes. For example, even Viktor Mitroshenkov's chronicle of Yuri Gagarin's life devoted far more pages to his party membership interview and his test on the history of astronomy than on medical screening.[38] In light of the lack of results and a vague mention of the process, one can only speculate whether cosmonaut screening always remained an opaque process. Another, equally biased source of information are the training films from that period. Documentation of cosmonauts' pre-flight exercising is fascinating. Camera operators of the time focused on outdoor physical activity that included bicycling, gymnastics, and swimming; in contrast, the Americans focused on the astronaut as an engineer, pouring over hardware and engineering diagrams in preparation for their flight. Their clear motivation was to provide B-roll of the aspirational new Soviet man.

There were two audiences for the Red Stuff messages—one received the outward message that projected the USSR's position in the world and one received a lower-level message that addressed Soviet society as it existed after Stalin. These messages provided the cosmonauts with reasons to address the public. In general, there were two types of messages from cosmonauts: some reinforced the domestic policy of the USSR; others reinforced its foreign policy. This added role differed from the presentation of past heroes, who addressed their fellow citizens exclusively.

Tsiolkovsky's image and words were present at every opportunity when mention of Gagarin's flight appeared in print. Mention of past scientists, including Tsiolkovsky, included overt references to the scientists and ones that appeared to be almost subliminal. One example of the subliminal appearance of Tsiolkovsky is in artist V. S. Vasilenko's illustration "Yuri Gagarin," which placed the face of Tsiolkovsky in the background at the top left-hand corner among the clouds and rocket as Gagarin stands in the foreground, prepared for flight. The face is almost imperceptible among the clouds and launch vehicle. The invocation of pre–World War II and even prerevolutionary rocket enthusiasts satisfied the need for a tradition while keeping the identities of current Soviet rocket scientists unspecific. Tsiolkovsky represented the strong Russian tradition in spaceflight and thus lessened the need to identify the current Soviet experts in the field. After Korolev's identity became public, he, too, achieved this revered position of the image behind the cosmonaut. The lacquer painting *Portrait of Yuri Gagarin and Sergei Korolev* placed Korolev at the center of space activities. Gagarin, *Sputnik,* and his R-7 launch vehicle surround him. At the time of the painting, Korolev, like Tsiolkovsky, was part of the Soviet past, again diverting attention from living individuals.

The invocation of the nameless, faceless, and long-dead scientific and technical heroes of the space program also served to reinforce the political status quo of the Soviet Union and served the conservative message that the cosmonauts carried. Even though the cosmonauts did not identify with a single political leader, in contrast to the Stalinist Falcons of the 1930s, their message linked them carefully to the collective of Communist Party rule. Both words and symbols described this link and emphasized the necessity of the party to spaceflight success. *Pravda* explained the relationship between Gagarin, the space community, and the party as follows:

Human-crewed spaceflight is the result of the successful execution of the active programs of large-scale communist construction, the constant work of the Communist Party and its Central Committee and under the leadership of N. S. Khrushchev for the continuous development of science, technology, and culture for the good of the Soviet people.[39]

As though the point required further emphasis, an illustration surrounding coverage of Gagarin's flight in the newspaper *Izvestiia* depicted the flight as the culmination of Soviet power, with him sitting atop the achievement of Soviet power and the victory over Germany in World War II.[40] Other cosmonauts were equally explicit, invoking their gratitude to Khrushchev and the party for their role in the space program. Gagarin's message to the Soviet public carried broader promises than reinforcing the war memory and gratitude to the party. He also offered a promise to his fellow citizens of a bright future. In a political cartoon in *Pravda* soon after his flight, an artist depicted Gagarin leading the way for a new Soviet spaceflight-based reality. People lined up for routine transportation, shopping, and tourism in space.

Propaganda gain for the party notwithstanding, Khrushchev saw Gagarin's flight as an opportunity for national celebration that could cover up other shortcomings.[41] Yuri Gagarin's face, name, and the details of his flight dominated the Soviet press for weeks. Not only did his face or name appear on the front page of *Pravda* for much of the rest of April, but the inside pages of the newspaper also contained articles from leading scientists and technicians on the technical background of his flight.[42] In contrast to the shock over the international reaction to *Sputnik,* the Soviet press was well prepared for the public reaction to Gagarin's flight. Page 4 of *Pravda,* the traditional location for foreign news, carried stories of international reaction to his flight for weeks as well. *Pravda* proudly reprinted a Picasso drawing of the cosmonaut that had first appeared in the French communist newspaper *Humanité.*[43] The only event that displaced Gagarin from the front pages in April 1961 was the ninety-first anniversary of Lenin's birth on 22 April.[44] Immediately, after the *de rigeur* celebrations of Lenin's birth, Gagarin returned to the front page, alternating with May Day and Victory Day (May 1 and 9) celebrations and perhaps even providing cover against the knowledge of the defection of Kirov Ballet dancer Rudolf Nureyev in June 1961.

As well as Gagarin's domestic message, he almost immediately took his message to the world outside the USSR. His flight also directly supported the foreign policy of the USSR of the time. The immediate public exploitation of his flight occurred outside of the Soviet Union. The pilot-cosmonaut barely had time to meet with his family before he went on an international goodwill tour. During the months from the end of April through August 1961, Gagarin traveled to Czechoslovakia, Bulgaria, Britain, Poland, Cuba, Brazil, Canada, India, Japan, and countries throughout Africa.[45] His post-flight travel schedule was so vigorous that one journalist asked which was more difficult, the spaceflight or his foreign travels.[46] The eternal diplomat, Gagarin merely smiled in response.

The mission of his trip was twofold. First, because of the earlier Stalinist habit of claiming scientific firsts based on little more than ideology, observers in the West regarded Gagarin's flight with some skepticism. Visits in person went a long way to support Soviet claims. At the end of April, Yuri Gagarin boarded an airplane and flew to Prague, Czechoslovakia, where he began the first leg of a world tour.[47] Upon his return and after visits to Russian cities such as Kaluga, Gagarin again went abroad to Bulgaria.[48] In June, he traveled to Finland,[49] followed by another trip in the middle of July to Britain.[50] Later that month Gagarin went on a trip to Warsaw,[51] and then to Cuba,[52] Brazil,[53] and finally to Canada, where he nearly missed a meeting with American astronaut Alan Shepard. At the end of this Canada trip, Gagarin asked to return to the USSR to continue training with German Titov.[54] Subsequent cosmonauts followed Gagarin's itinerary. Valentina Tereshkova traveled first to Prague and Sofia, following the path that Gagarin had taken, and thus she reemphasized Czechoslovakia's and Bulgaria's importance to the Soviet Union; this also associated Tereshkova with Gagarin in the minds of foreign allies.[55]

Gagarin's appearances throughout the world reassured many doubters that the flight did indeed occur. The second purpose of the trip was for reassurance in another sense. Until Gagarin's flight, space exploration manifested itself as a competition between the American and Soviet military hardware. The flight of *Vostok* marked the beginning of the humanization of the Space Race, adding a civilian patina to the existing public space programs. Despite his affiliation with the Soviet Air Force, Yuri Gagarin was a peacetime pilot.[56] His uniform signified professionalism.[57] His postspaceflight rank of major signified the quality of his training, not the military intentions of his mission. Along these lines, Gagarin's speeches emphasized the peaceful intentions of his mission and Soviet exploration of the universe, further separating cosmonauts from the military intents of missile development. In the film *Yuri Gagarin,* he makes a typical speech to an unspecified young audience: "I bring you the sincere wishes from Soviet youth, and all the Soviet people, that this festival will strengthen friendship among the youth of different governments, and that this festival will eliminate the necessity for the word 'war.'"[58]

There is no doubt that Gagarin's trips throughout the world supplemented Khrushchev's "Peace Movement." The movement offered peaceful language to contrast with the American Cold War rhetoric. What is significant is that Gagarin's rhetoric was the same within the Soviet Union as outside. The slogans of peace applied to foreign audiences as well as Soviet ones. Following his return from Czechoslovakia, Gagarin visited Kaluga, where Tsiolkovsky had lived, and spoke similar words.[59] He sought to convince his compatriots that he advocated peace. As far as a peace offensive is concerned, Gagarin's words were moderate. The popular press had harsher words of peace, which were far more antagonistic. One political cartoon portrayed Gagarin lecturing missile-nosed "Atomshchiki" (atomic weapons holders) from space: "I advise you to take up peaceful work and watch your noses"—an onerous reference to the anti-Semitic caricatures in the cartoon.[60] The message of shame to the US for their militarist activities was equally apparent in domestic and international activities.

Russian/Soviet Identity

In addition to the political message, the Red Stuff carried with it the elements of standardized Russian and Soviet nationalism. These men passed the screening process through their ability to represent an ideal that distinguished them from the Americans in not only rhetoric and discipline but also culture and identity. They were all Soviet by citizenship, but their selection also followed the hierarchy of nationality that underlay the Soviet identity. Of the first twenty recruits, sixteen were Russian. Two, Bondarenko and Popovich, were Ukrainians. Nikolayev was the sole Chuvash cosmonaut. Rafikov, the lone Tatar in the group, left the program in 1962 without a spaceflight. It is worth noting that there were no Jewish cosmonauts among the first twenty recruits.[61]

As if to underscore the separation between the Red Stuff and the Right Stuff, one element of nationalism that was essential to Khrushchev's contribution was his establishment of the distinction between Soviet and American space travelers. In Vasilli Zhuravlev's 1936 science-fiction movie, *Kosmicheskii reis* [Spaceflight], the academician Sedykh and his young friend Andrushka, upon return from their flight to the Moon, refer to themselves as "astronauts."[62] Twenty-five years later, the Soviet press announced that Yuri Gagarin, the first man in space, was a cosmonaut.[63] Although Soviet journalists did not explain this linguistic shift from *astros* to *kosmos*, this was a definite step to reinforce the delineation between the Soviet and American programs. Both roots were of Greek origin—*astros* meaning stars and *cosmos* meaning universe. This distinction without a difference emerged as an artifact of the Cold War where there had previously been no reason to distinguish nationalities among space travelers. "Astronaut" had become the preferred usage among space enthusiasts during the 1920s and 1930s.[64] However, the meaning not only included space travelers but also encompassed all those who were interested in the new realm of exploring beyond the Earth. Soviet writers did not use the term "cosmonaut" until 1959.[65] Linguist Morton Benson notes that the term *kosmonavt* (cosmonaut) won quick acceptance in the American lexicon to distinguish Soviet space travelers.[66]

Beyond mere national and linguistic identity, the Soviets relied on the familiar mythology tropes to describe the transformational stages of the Red Stuff life. The cosmonauts told a story with their lives that was a necessary part of their example to the public. They followed the pattern that folklorist and anthropologist Joseph Campbell collected in his writings. In his famous studies of mythology and heroism, Campbell has defined stages of the heroic life that are common to many mythic traditions, both from the East and the West. Briefly outlined, Campbell's three stages begin with childhood, continue through the transformation of the hero, and conclude with the message that the hero brings back after his separation from society. The childhood of a hero is usually a period of obscurity,[67] during which the parents act as guardians.[68] Mythology portrays this childhood as a period of nourishment, during which the hero acquires the character traits essential for the transition to the next stage. The transformation stage begins with the separation from previous surroundings.[69] The separation is both physical and intellectual, often taking the form of a physical separation for enlightenment.[70] In many ways, the physical separation is a screening process, by which those worthy of further enlightenment proceed to the next station.[71] A new father figure carries out enlightenment or training. Often a severe, godlike figure, this new father provides the knowledge necessary for the hero to continue his/her journey through a trial of fire to bring the message back to conventional society.[72]

The apotheosis of the hero takes place when the hero ultimately overcomes all fear and takes the final step to enlightenment. This last step is often dramatic, literally a trial of fire, which emphasizes the significance of the step.[73] It is the attainment of this knowledge that is the hero's goal. The hero's experience of obtaining the truth justifies the message to men. This highly schematic summary of Campbell's thought on the life of the hero closely approximates the life sequence of Soviet cosmonauts as portrayed in official Soviet biographies. Although the form remained the same over three decades, the emphasis on the stages declined as time passed and the number of cosmonauts increased. The idealized lives of early cosmonauts, most notably those of Yuri Gagarin, German Titov, and Aleksei Leonov, followed these patterns. In later years, the official narratives shifted from hero myths to an emphasis on the cosmonaut's professionalism.

The new heroes contrasted sharply with the superhumans of Russian folklore and modern Soviet mythology. These new stories emphasize that the cosmonauts are not superhuman at all. For that reason, biographers pay a great deal of attention to the early family life of the cosmonaut. While the cosmonaut's role in society has changed over time, certain assumptions about his family life have remained constant. Three ideals of family life, those of Yuri Gagarin, German Titov, and Aleksei Leonov, were examples for the Soviet population. The most common method of instruction about these lives was through popular biographies, especially biographies for young children.[74] Publishers and editors frequently reprinted and recompiled these books for subsequent generations. As a genre, biographies consisted of idealized accounts of the cosmonauts' youth. Invented conversations with cosmonauts in their youth conveyed their model upbringing. Nevertheless, no matter to what extent the published lives were fiction, two traits remained constant in the biographies. The first was the family structure. Close-knit and nurturing, the family contributed to the upbringing of the hero-to-be. In the case of all three, Gagarin, Titov, and Leonov, it was an older brother who played an essential role in the development of the younger. The second trait was the central role that the experience of World War II played in cosmonauts' lives. This emphasis reinforced attention paid to the war in other areas of Soviet public life. The role of World War II is especially significant when one considers that the cosmonauts, even the first group in the *Vostok* era, were young boys during the war and had no World War II combat experience. Nonetheless, biographies of the earliest cosmonauts took great pains to point out the legacy of the war, an event that pervaded the consciousness of the Soviet population.

The boyhood family situations of Gagarin and Leonov were similar. Both met the standard heroic requirement of obscurity. Gagarin's family was *kolkhozniki* (collective farm workers)—the father a carpenter, the mother a milkmaid.[75] Titov's father was a schoolteacher, but his grandparents, too, were peasants.[76] Leonov's were miners.[77] In each case, both parents survived the war. In all biographies, the parents were obscure but virtuous. The mother was always patient, silent, and a constant throughout the cosmonaut's life. The father, too, was silent and played an undemonstrative protective role. The older brothers played the role of an active older male. The exaggerated influence of Gagarin's brother resulted from the fact that he was a biographer and the chief spokesperson for the family. Aleksei Leonov, too, had an older brother, Petr, whom he claimed had a more significant influence on him than his father did. The characteristics of independence and pride passed from older brother Petr to Aleksei to youngest brother Boris.[78]

Following the pattern of classical heroism, at a specific point in his life, the man became a cosmonaut by stepping outside of the mainstream of society. This transformation brings expectations of behavior that shares characteristics common to many Russian and Soviet heroes. Although cosmonauts were not reproductions of previous heroes, their heroism had to be apparent to the population at large to be successful. The first step of the transformation was separation from previous familial forms. At this point, the hero left the nurturing world and joined what Campbell refers to as "the world of specialized adult action."[79] Separation represents a step from the world of childhood to the world of the father. In the example of Soviet cosmonauts, the transition took the son from his biological or surrogate father to a spiritual father. In the biographies of early cosmonauts, the new spiritual father was Sergei Pavlovich Korolev, the chief designer of the Soviet space program.

The final transition from family life and life experiences, which has become standard in the mythical life of the cosmonauts, was the bonding in a cosmonaut "family," thus leaving the biological family behind. In a 1966 biography of Aleksei Leonov, his father took a visible secondary parental position to Korolev. The father's diminished role appears with references to him as Starik Arkhip (the venerable Arkhip), rather than Aleksei's father, once Leonov joined the cosmonaut corps.[80] The point of shared paternity is driven home in a scene following Leonov's spacewalk. The two "fathers" meet at the Kremlin, in the biography's semi-fictional account, and outline their respective positions in Leonov's life:

> Korolev: Thank you for your son, Arkhip Alekseevich.
> Arkhip Alekseevich: This is all thanks to you, Sergei Pavlovich.
> They grasped each other's hand, two experienced men, two fathers of cosmonaut Leonov.[81]

Either as a surrogate father or teacher, Korolev's role as gatekeeper to the status of cosmonaut was absolute. It was Korolev in either of his roles that marked the pilot's recognition as able to pass to the next stage of his journey. The first stage of recognition occurred during the first phase of selection for cosmonaut training. The introduction to the spiritual father or master instructor, Korolev, marked the presentation of his role as a teacher. He introduced the cosmonauts to the way to a new world. In his reminiscences, the third cosmonaut, Andriian Nikolayev, described a tour of a magical new world: "Sergei Pavlovich himself led the tour of the workshops and large halls where the spacecraft and launch vehicles stood. He took us to the completed *Vostok* craft, stood, and described the construction and the guidance principles of the craft in detail."[82]

Of course, the transformation did not come about by passive observations on the part of the cosmonauts. Only through hard work and study did they accomplish a higher level of enlightenment. Recognition of this accomplishment was evident after the spaceflight. However, more subtle recognition of the cosmonaut's destiny occurred before the flight. Others wrote about the early recognition of Gagarin's destiny more than for any other. His knowledge and proficiency in tasks happened before he distinguished himself from all others. Gagarin's superiority was evident to his colleagues and more distant observers. Once again, Nikolayev described the magical scene during which Gagarin revealed his destiny, through graceful and immediate mastery of the course of study of the *Vostok* spacecraft:

> As it happened, the first one to take the exam was Yuri Gagarin. He took his place in the spacecraft and clearly and distinctly answered numerous questions. I simply admired him, was proud of him, and admired his resources, keenness of wit and imperturbability. Everyone liked Yuri's answers and his knowledge was rewarded with the grade of outstanding.[83]

Pavel Popovich gives another example, in which he describes a scene of recognition in his *Beskonechnye dorogi vselennoi* (Unending Road to the Universe), in which he described a scene immediately before the flight of Gagarin when technicians asking for his autograph surrounded him.[84] This scene represents a celebration of Gagarin's skills that led to his selection. There were no immediate physical indications of the transformation. The spiritual clues are subtle. Extraordinary and apparent bravery and perseverance made the cosmonaut-in-training a true cosmonaut, and those close to him could immediately recognize those traits.

Related to the traditional mythology, the Red Stuff created a framework that substituted the Stalinist paternal framework to an anonymous father that was nested closely inside the party and state framework. The displacement of the father figure fit into 1960s Soviet reality in other ways. First, it was a Soviet reality that many fathers had died in the war, leaving millions orphaned. The biographies could have been written as an acknowledgment of the absence of biological fathers and the routine substitution for fathers for that generation. Second, the diminution of the father role left room for his replacement with a higher authority. In the 1930s, Stalin had played the father role outright in similar biographies. The cosmonauts had other father figures in their biographies: the anonymous collective of the party and the anonymous "chief designer" of the space program.

Another consistency in the recounting of the family structure was the incorporation of the national scientific tradition into the narrative of the cosmonaut's life. This consistency passed off from one paternal figure to another. In her brief biography of her son, Anna Timofeevna Gagarina recounted the importance that a schoolteacher had in her son's life. As if to mark the significance of the teacher and his role as a link to that tradition, she reported, "In school, Lev Mikhailovich Bespalov read stories about the life of Tsiolkovsky to Gagarin."[85] Anna Gagarin's life was one of many deliberate links made in the biographies between the early Soviet past and postwar USSR. As the cosmonauts were carrying on the legacy of the Soviet scientific tradition, they had to maintain bonds to that past.

Bonds to the Soviet social structure were also a necessary element of the biographies. Gagarin's biographies and the press coverage of his life repeatedly alluded to his peasant origins. This aspect of his background is predominant in his national publicity throughout the Soviet Union. His brother stated that although his profession was pilot-cosmonaut, his true allegiance was to his parents' profession: "Yuri always considered himself to be connected to the land, and throughout his life valued working people . . . Kolkhozniki were proud to be able to count him among their number."[86]

In the early 1960s, World War II was the common bond among all Soviet families. With one exception, all cosmonauts during the 1960s were children during the war.[87] However, the sacrifice common to war knew no age distinction. The personality characteristics of the adult Gagarin are said to have been shaped by his experiences during the war. As his brother writes:

It seems to me that some of the character traits of Yuri, in fact the character of the future pilot and cosmonaut Yuri Gagarin, which showed persistence in the achievement of goals, compassion for the suffering of others, readiness to come to the assistance of others, ability and readiness in the unexpected, no matter what the risk; took shape at that time, during the war.[88]

The war was also the useful rationalization of the peaceful rhetoric of the first group of cosmonauts. The cosmonauts provided a link between the Soviet wartime experience and Khrushchev's peace offensive to appeal to the non-aligned world during the Cold War.[89] Gagarin frequently referred to the war in his statements about peace. At a press conference immediately after his flight, Gagarin invited the world's population to join him in his experience of spaceflight and see the beauty of the Earth from above the atmosphere. After experiencing flight then, he said, "I believe . . . all people would look each other in the eye, hug one another, and would live in eternal peace and friendship."[90] It was from the position of one who had witnessed the privations of war that Gagarin could offer a credible proposal for peace, even though he represented one of the two military superpowers.

The cosmonaut biographies published before 1966 presented Sergei Korolev in an unusual light. Until his death in January 1966, Korolev's identity as the organizer behind the Soviet space program was an official secret. The reasons behind the secrecy are not clear. Another military chief designer did not share Korolev's fate of anonymity. Atomic bomb designer Igor Kurchatov came out in support of Khrushchev at the 20th Party Congress in 1956.[91] The 1958 *Bol'shaia sovietskaia entsiklopiia* included a biography of Kurchatov that indicated his role in the atomic bomb program.[92] Given the symbiotic relationship between the nuclear warheads and ballistic rockets, it does not make sense that Korolev's anonymity was a national security concern that applied to him alone. Weapons and aircraft designers received accolades from the time of their initial successes. For example, Il'iushin figured prominently in the 1930s biographies of Soviet test pilot Vladimir Kokkinaki.[93] Moreover, in repeated Soviet media coverage of the program, Korolev went unnamed but mentioned in obscure terms. Among cosmonaut biographies, German Titov, the second man in space, took pains to describe the "chief designer" in detail,[94] going as far as to allude to Korolev's time in prison in typically veiled Soviet language:

The fact is that the Chief Designer had a hard time of it at first. Many people considered him a wild dreamer, had no confidence in him, and often he was left alone with his projects, plans, and blueprints.

He never talked to us about those past difficulties, but we learned of them and felt all the more respect and affection for this man's mighty spirit. It is easy and very hard to do his portrait. He is not very tall, but strong and broad-shouldered. He holds his head in such a way that he seems to look at you sullenly, but when he looks you in the eyes, you see not only the iron will and the lucid mind of the designer and mathematician, but also the attentive, heartfelt goodness of a strong man with a generous spirit.[95]

Titov provided enough details about the chief designer to convince the reader that he was a single individual and not an amalgam of faceless bureaucrats. He came close to taunting his readers into speculating about the engineer's identity. Titov's memoirs did not stop with an intellectual biography and physical description of Korolev but continued with a complaint about the unfairness that the engineer did not receive due acknowledgment for this work:

If I have any regrets in respect to my training program, and later my operational flight as a cosmonaut, it is the lack of public attention given to the single most important man in the Soviet Union's cosmic flight program. His name must still remain hidden from the world until our government decides that the time is right for him to be identified, but I for one regret personally (as do all the others who have worked for and with this outstanding person) his long anonymity. He is a man who walks with the giants who have affected and even made history, and he can be known as yet only by his title—the Chief Constructor.[96]

Titov's mention of regret over Korolev's anonymity made it clear that even during the early 1960s his anonymity inspired some controversy among the spaceflight establishment. After his death, Korolev gained a face and name in subsequent cosmonaut biographies. His official unveiling took place at the time of his death in the form of his obituary.[97] Issues of national security did not seem to be enough to explain why Korolev remained unnamed until his death and the real motivations have remained uncertain. His anonymity served the purpose of shifting attention from an individual to a broad collective of unnamed scientists and engineers. This shift paralleled Khrushchev's shift from Stalin's personality cult to the more secret cult of the party.

Gagarin's biographers do not make as much of Korolev's role as a father.

Valentin Gagarin's 1972 biography of his brother did not mention or even hint at Korolev at all. Recent biographer Viktor Stepanov placed Korolev in a central role in Gagarin's training, but not as a father figure.[98] In fact, Stepanov's book styled Korolev as a live replacement for the Tsiolkovskian ideal of the scientist. In contrast to Leonov's father figure, Stepanov portrayed Gagarin's Korolev as a taskmaster who inspired him to study the principles of spaceflight.[99] The different assessments of Korolev's role might well be due to the timing of the biographies. Avdeev wrote his biography of Leonov at the beginning of the dry spell of Soviet manned spaceflight activities in 1966, just after Korolev's death, when his identity became public. Valentin Gagarin's biography was first published in 1972. Written by the brother of a cosmonaut, Korolev appropriately appeared to the family to be no more than a shadowy figure in the cosmonaut's training. Stepanov's biography is the most recent of the three and represented the current trend to place increased emphasis on cosmonauts' professional training. The image of Korolev was ambiguous.

New Soviet Man as an Aspirational Ideal

Changes in attitude toward maleness in Soviet society reflected the new relationship between the Soviet state and society after the initiation of the first Five Year Plan under Stalin (1928–1932). As a distraction from the purges and the absurd hardships of the economic plan, this new attitude placed new emphasis on the accomplishments of the Soviet Union. The record-setting aviation flights provided technological legitimization of the Soviet state. Historian Kendall Bailes placed the flights in the setting of the struggle for technological legitimacy during the prewar years of Stalin. Literary historian Katerina Clark discussed them in the context of the emerging literary style of socialist realism.[100] Historian John McCannon examined the specific example of the rise of the Arctic in Soviet popular culture during the 1930s.[101] Each treatment of this brief period in Soviet history is rigorous on its merit, but together, provided a way to see the connection between the cultural and technological elements of the Stalin flights. They also point out that the primary audience for these accomplishments had been the domestic Soviet population.

According to Clark, during the first Five Year Plan era, Soviet ideological novelists began to move away from discussion of the traditional conflicts between machine and nature. They perceived danger in the obsession with machines, as it obscured the human character of the social transformation that was supposed to be taking place at that time. The worship of machines would interfere with the Stalin-era emphasis of the heroic man, and man could become indistinguishable from a machine. It was at this time that metaphors for nature in movies began to replace the machine metaphors of recent years.[102] By juxtaposing man to nature instead of a machine, the man regained his traditional life cycle, going through the typical phases of childhood and passage to adulthood. Now, the machines of industrialization aided him in his struggle instead of being a surrogate. At this point, culture (Soviet power) gained an advantage over nature. That advantage supported the Soviet ideology of industrialization. Man could use machines in his ideological struggle. The machines were tools now that they were no longer the weapons of capitalism.

The symbolism of Stalin's Falcons fit nicely into this new metaphor. The pilots overcame nature (the harsh environment of the North Pole) with the aid of Soviet technology (airplanes), and thus proved that man could vanquish nature with initiative, the right tools, and, most importantly, the correct culture. Moreover, the myth of the great family supported the image of a man overcoming nature. These pilots possessed courage and daring, but their ultimate source of motivation lay in the "Father" (Stalin), who moderated the high-spiritedness of the pilots, taught them discipline, and thus provided them with the skills to overcome nature. As Clark recounted the roles of the 1930s pilots, Stalin played the role of father or teacher.[103] Stalin's role as the father is not solely a literary technique. Historian John McCannon has recognized the role of Stalin as family head in his study of the cult of the Arctic during the 1930s that included films and other forms of popular culture:

> In keeping with the socialist realist vision of Soviet society as a pyramidally patriarchal "Great Family," Stalin was also portrayed repeatedly in the Arctic myth as the symbolic father of polar explorers and Arctic pilots, just as he was depicted as the father of all Soviet heroes and, ultimately, the "Father of Nations." During the greater part of the 1930s, Arctic heroes were among the favorite sons in Stalin's "Great Family." There were his fledglings (*pitomtsy*), reared with infinite care and love. As they grew and matured, Stalin bestowed fatherly wisdom upon them, tempering their heroic energies with discipline and concern. In all ways, he supported them and made their heroic exploits possible.[104]

Equating these national heroes with model sons created a new image of the Soviet man. The necessary characteristics of undisciplined energy find a focus on the discipline of Stalin. This need for focus is an example of the Stalinist doctrine that states only through moderation of spontaneity (undisciplined energy) by consciousness (Stalinist discipline) could the Soviet Union enter the heroic age, leaving behind the shortcomings of the past.

The archetype of the Stalinist hero biography is Nikolai Kamanin's autobiography, *Moia biografiia tol'ko nachinaetsia* [My Biography Has Only Begun].[105] As one of the pilots who helped rescue the civilian exploration crew of the *Chelyuskin* from the Arctic ice, Kamanin was one of the first recipients of the Hero of the Soviet Union award in 1934, an award that Stalin designated for civilian heroics.[106] Throughout the thin volume, Kamanin repeatedly thanked the Party and Stalin for the opportunity to fly, to the point of squeezing out all but the barest details of his flight for which he earned the honor. Kamanin's autobiography reads like an acceptance speech for the Hero of the Soviet Union medal. This ironic episode had one set of pilot-heroes emerging to rescue another group of famous explorer-heroes. The rescue was a demonstration of the Soviet claim to be able to act autonomously, without dependence on outside help. Kamanin's recounting of his experience set the stage for stories about cosmonauts and human spaceflight, as it made clear that the Soviet Union was prepared to demonstrate technological parity with the rest of the world just as it had tried to do before World War II. Thirty-five years later, Nikolai Kamanin became the commander of the newly formed cosmonauts' corps. The image of the Red Stuff continued this metaphor of man using machine to conquer nature. Technology was a tool to combat an enemy that every peasant understood.

While the image of cosmonauts has varied over the last forty years, one cannot assume that the appeal was artificial or readily manipulated. Heroes are essential to any society and share characteristics across borders. Nonetheless, the Soviet space experience remained unique. The heroism of cosmonauts was distinguishable from other heroism in two ways. First, Stalin, the Party, or an anonymous collective, and not the popular imagination, created the ideals of heroism. Therefore, the public life of the cosmonaut followed distinct forms of heroism. The meticulousness of its creation did not necessarily mandate acceptance, but the pervasiveness of the message and the degree to which the Soviet government immersed the population in propaganda did not allow the public to ignore it. The second distinguishing feature of Soviet hero cosmonauts is the conservative message that their heroism conveyed. Cosmonaut heroes did not exist to present new and innovative solutions to current problems or to initiate changes in society, but to reinforce previously articulated messages and mandates and demonstrate the viability of the current political course.

As a member of Soviet society, the cosmonaut had a dual role as a leader, representing the ideal for society, and yet a conformist member of that society. To be a compelling hero and inspire a positive response from the public, he had to draw on accepted values of the society, while at the same time challenging that society to a new level of effort using the established values. Initially, it was simple to accomplish these two ends with the selection of a handful of individuals. Once the corps of the cosmonauts grew, however, it was difficult to serve these ends with numbers exceeding one hundred. For that reason, the cosmonaut's heroic role in Soviet society wore thin after Gagarin's flight. The next generations of cosmonauts did not receive the same degree of attention or have monuments erected in their honor. Gagarin epitomized the ideal favorite image of the cosmonaut during that first period (1961–1965). Nonetheless, certain character traits were standard among cosmonauts in both this period and later. The change that took place is the emphasis on the stages of the heroic lives. Cosmonauts' importance to Soviet society was not as apparent in 1961 as it became later in that decade. Gagarin exemplified the ideal favorite image of the cosmonaut. The idealization of his image increased after his death when there was no possibility of incidents to tarnish it. In death, Gagarin became a more celebrated hero than he had been in life.

In Yuri Gagarin and German Titov, the USSR had two versions of the new Soviet man. The former did well at ingratiating himself with the Soviet establishment and public while carrying on with his real life. The latter, far more loquacious, adhered firmly to a party line, but his willingness to talk threatened to lift the veil on Soviet secrets. In his book, *The Cosmonaut Who Couldn't Stop Smiling*, Andrew Jenks has peeled away the layers that composed the legend of Gagarin.[107] German Titov's story has remained in the hands of the press and space enthusiasts.

Gagarin was the first to articulate the cosmonaut's message, but as was so often the case with propaganda, others repeated and refined it. The refinement of the message became more specific and refined with each subsequent cosmonaut. German Titov, whose flight coincided with preparations for the Twenty-Second Party Congress in 1961,[108] waxed eloquent about the role Khrushchev played for him and the pride that he showed in being elected to full party membership as the reward for his twenty-five hours orbiting the Earth. The party, according to Titov, was responsible for the mission:

> I replied that all the glory for this new victory belonged to the Party, the people, and, of course, the builders of the spaceship. If there hadn't been a spaceship I shouldn't have been able to fly in space. If there hadn't been Titov, there would have been Ivanov, Petrov, Nikolayev, or Sidorov to make the flight. We have thousands of people capable of doing what the two first cosmonauts have done.[109]

Khrushchev bestowed a great honor upon him by shortening his probationary period for party membership, just shy of his twenty-sixth birthday, the minimum age for membership.[110]

Despite his strident support for the party, or perhaps because of it, German Titov emerged into the spotlight of spreading the word of the Red Stuff unexpectedly. Gagarin had injured his face badly while trying to escape from his wife finding him with another woman in early October 1961.[111] This situation left Titov as the sole Red Stuff spokesperson for both the 22nd Party Congress and an obligatory visit to Berlin later that month. The previous summer, the German Communist Party had followed a determined campaign to permanently resolve the issue of the permeable border with the West despite Khrushchev's determined efforts to keep the Western powers off balance with his continuing threats against the German and Berlin borders. Ulbricht and the GDR wanted a permanent economic solution for their problems of population flight. Khrushchev saw global political value in an unresolved situation. The conflict between Khrushchev and the GDR over sealing the border heated between 1960 and 1961. The GDR informed the USSR of its plan to close the borders in July 1961. To save face, the USSR sought to assure the world that this had not been a unilateral decision.[112] It was no coincidence that German Titov's first post-flight trip abroad was to East Germany.[113] It was there that he appeared on Berlin television at the newly constructed wall.[114] His performance left a lifelong impression of his personality as well as official Soviet support for the construction of the wall. Archival research has relieved Khrushchev of that association, but that taint remained with Titov for the rest of his life.

One of the three international messages that supported Khrushchev's retreat from wartime concessions toward the Orthodox Church was also ambiguous. At one time, Gagarin's most overt political message was his position on the question of the existence of God. Not surprisingly, according to all contemporary accounts, Gagarin was an atheist.[115] He believed that his flight dispelled religious superstitions among the Soviet population. According to his brother, he received many letters questioning what he saw in space. The most reassuring letters to him were those that accepted that he had proven the nonexistence of heaven: "Much later Yuri said that in the enormous numbers of letters that he received after his flight, there were hundreds of letters from former believers, who said that after his flight had been announced, they renounced their old beliefs."[116] This hero's message reinforced the established atheism of Stalin's rule, which had only moderately relented during World War II. Titov reiterated this position a year later during a visit to the United States when he stated that he saw no angels in space.[117] By that time, Titov's reputation for outspokenness and his statement about not seeing angels or God in space became a further provocation against the United States. Few non-Russians knew at the time, but Titov was not parroting the party line on religion, but instead was quoting early twentieth-century Russian and Soviet poet Mayakovski. "Sky inspected inside and outside. No gods, no angels are found" was a line from Mayakovski's 1925 poem "The Flying Proletariat," set in the futuristic year 2125.[118] Titov had repeated a call for secularism, but it had not been Khrushchev's secularism. It had been the revolutionary call from the earliest days of the USSR. This call had very little effectiveness and even lacked a definite audience, as its simple message of inspecting the skies for God carried a metaphorical meaning but did not disprove the existence of God.[119] While not an adequate measure against religion in the USSR, these messages reiterated Soviet efforts to secularize space that had begun in the 1920s and 1930s. As early Soviet science-fiction literature and film had employed space travel to find revolution or science, Gagarin's flight was a final demonstration that the Red Stuff was a secular effort.[120]

A year after his trip to Berlin Titov traveled to the United States.[121] En route to the 1962 Seattle World's Fair in Seattle, Titov visited Washington, DC, where he toured the US capital with astronaut John Glenn and met President John Kennedy. In Seattle, Titov had the opportunity to see John Glenn's *Friendship 7* spacecraft that had just arrived at the fair after its world tour. While Titov had been an excellent representative of the Soviet Union in Berlin, his appearances in the United States were not as smooth. At the time of his visit to the United States, German Titov already had a reputation in the West for being prickly; this was much different from his Soviet persona as a good-natured communist professional. One potential source of his perceived terseness might have been a controversy that the cosmonaut had sparked himself. During *Pravda's* coverage of his flight, he candidly described how he had ejected from the spacecraft at twenty-thousand feet. He had volunteered that the first time *Vostok* had two landing options and that while Gagarin had chosen to land inside his spacecraft (he had not), Titov had chosen to try the ejection option.[122] The report to the Fédération Aéronautique Internationale (FAI) reflected his statements. By the beginning of March 1962, cosmonaut commander Kamanin had to brief a delegation to consult with the FAI in Paris to reconsider their decision to reject the petition for the record of having spent the first day in space. This event occurred at the same time as the FAI had approved Glenn's mission. As Kamanin noted in his diary:

> Prepared directives for Kokkinaki and Skuridin, who were departing on 9 March to fly to Paris to meet with the FAI to gain confirmation for the flight record of Titov. The FAI sent comments that it was not possible to affirm the record of Titov because he did not land inside his spacecraft.[123]

Of course, none of the calls to retract Titov's records ever made it to the Soviet press. Still, there was never an effort to refine or retract Titov's original statements conceding that he had ejected from the craft in contrast to the continuing policy that concealed the fact that Gagarin had done the same.

The efforts to keep the cosmonauts on message while traveling abroad required more than the security retinue. Among the vast records of KGB foreign activities records that Vasilli Mitrokhin brought to the West, historian Christopher Andrew found a rather bizarre account of the forces brought to bear against the possibility of a photograph of Tereshkova or Gagarin during a trip to Mexico in October 1963 with a bottle of Coca-Cola in the frame.[124] This was a response to a photograph of Khrushchev drinking Coke the previous year:

> All went well until a banquet in Mexico in 1963 when an alert KGB officer noticed a news photographer about to take a picture of Tereshkova with a waiter holding a bottle of Coca-Cola in the back-

ground. A member of the Mexico City residency wrote later: "The provocation prepared with regard to the cosmonauts did not slip past our vigilant eyes. The first female cosmonaut, a Soviet woman, featuring in an advertisement for bourgeois Coca-Cola! No, we could not permit this. We immediately turned to our Mexican colleagues for help."[125]

In retrospect, the thought that a bottle of Coke could be the source of a diplomatic incident seems almost ridiculous, but this flurry of activity and accusation of provocation indicate the importance of the cosmonaut message to the public image of the Soviet space program.

Conclusion

Nikita Khrushchev spared little to firmly portray the 1960s cosmonauts as the new, post-Stalin Soviet man. Although he modeled them on Stalinist tropes, Khrushchev's human spaceflight program was Stalin's Falcons with rockets and without Stalin. Their national identities followed a specific brand to fit a uniform mold. His messages, tone, presentation, and even the extent to which women were singled out as a demonstration of Soviet prowess were each modern adaptations of what Stalin had done in aviation two decades before. Their presentations to the public were no less well-tended.

The *Vostok* era cosmonauts represented a far more modest change in Soviet society than had Stalin's Falcons. There was no state-sponsored inducement to adopt spaceflight as a national pastime in the name of civil defense. It was only presented as a cultural fantasy. Neither did the cosmonauts represent a shift in Soviet industrial might. Although the press did present them as representing the legions of unnamed scientists, engineers, and technicians who worked on the program, they did not call for new forces to join them. The one way in which the *Vostok* cosmonauts resembled the Stalinist aviators the most was in their public and avowed allegiance to the state. This time, however, there was no Stalin, and cosmonauts declared their allegiance to the party instead of an individual.

The one medium in which it is most evident to see the extent to which Soviet space planners patterned the image of cosmonauts on the Stalinist Falcons is the biography. Cosmonaut biographies proliferated from the time of Gagarin's successful mission in the form of newspaper stories, magazine articles, and later, books. These biographical and autobiographical stories told highly ritualized tales that resembled the heroic biographies of the 1930s. In the absence of Stalin and rapid industrialization, the cosmonauts thanked the party and frequently referred to their childhood experiences during World War II. The biographies carried over in tightly managed identities that space managers, especially Commandant of the Cosmonauts Kamanin, orchestrated to convey a message of hope for the near future. In a circular argument, the party, in turn, pointed to the existence of the cosmonauts as an indication that the immediate future would be bright. With this, the government tied the Khrushchev-promised 1980 utopia to the cosmonauts.

The cosmonaut's role as a diplomat in the early 1960s was very tentative. These space-traveling cosmonauts were playing multifaceted roles. First, they acted on behalf of the hardware that was absent for reasons of secrecy. During the 1960s, the Soviet Union shifted from demonstration and exhibition of hardware and relied on cosmonauts to convey their messages of technological prowess and peace to the world. They used the men and the woman of spaceflight as evidence that they had, in fact, flown into space. Second, they reinforced the message of the Communist Party of the time. The messages supported Soviet domestic and foreign policies of atheism and Khrushchev's peace offensive. These were modest messages, not exhortations to revolution, and had the dual duty of reassuring the Soviet population of stability at home and projecting a professional image abroad. The trips that they took primarily cultivated relationships with the aligned and friendly nations but recognized the need to contact the West. The third aspect of the cosmonauts' diplomatic role was the extent to which it was well rehearsed and controlled. They were visits that had exact choreography, avoiding all perceptions of impropriety. Soviet cosmonauts in the early 1960s were the ultimate ambassadors of reassurance.

Even with the successful introduction of the Red Stuff into Soviet public life, there were early signs of flaws in the presentation. These cracks grew over the next decades that led to decoding the Red Stuff and the destruction of its pedestal. There were four cracks. Not only their existence but also their treatment was vital to how they led future generations to reinterpret them. Two emerged from the culture of secrecy and deception that defined Khrushchev's post-Stalinist illusion. Moreover, two emerged from the inherent contradictions of attempting to create idealized and aspirational life models from real human lives. None of these flaws would prove inherently fatal to the Red Stuff, but they each assured that the official presentation of the cosmonauts would not be accepted without skepticism.

The culture of deception was not new in the 1960s. Censorship, embellishment, and denial of the truth were a feature of the Russian state that dates back at least a thousand years. Imperial Russia and the USSR both lacked a free press and open criticism to challenge these bad habits of a state. The hardware failure of Yuri Gagarin's spacecraft that led to the wild gyrations that he experienced at the end of his flight was embarrassing and potentially catastrophic. It remained a deeply held secret for thirty years. Similarly, the cover-up over the inability to land Gagarin inside his spacecraft was embarrassing and threatened the public appreciation of Gagarin's flight. In each case, however, it was not the sin of the cover-up that caused the most significant damage to the legacy, but it was the scope of the cover-up. Lying and disinformation were no longer the purview of government officials. The space program enlisted thousands of lower-level bureaucrats into these deceptions, creating a culture of lying.

Aspirational characters as real-life human beings are impossible to control and perfect. Take the two conflicting personalities of Gagarin and Titov, neither of whom was suited as the ideal Soviet man. To the public, Gagarin was the easygoing, sweet Russian boy who was always smiling. His loose shoelace on the red carpet at Vnukovo Airport defeated the staged event and endeared him to the public. Yet, also, there was another side of this man's carelessness that defied the conservative, ideal family life that officials sought to portray. His womanizing forced the hands of his handlers and thrust his backup and *Vostok 2* cosmonaut into the limelight in 1961 and 1962. Although Titov adhered to the well-promoted family devotion model, his outspoken nature betrayed the lies about Soviet hardware and diverged from the message of peace that Khrushchev had so carefully constructed in his post-Stalinist illusion. These were not the only structural flaws in the presentation of the Red Stuff, but their early presence added to the development of others that were to surface over time.

3

The Women

The Collison of Expectations in Domestic and International Politics

The story of the first woman in space merits a separate discussion. The iconography of male cosmonauts drew from the traditions of Stalin's Falcons of the 1930s. Soviet mastery of aviation technology was a political, economic, and military mission. Women cosmonauts had a more nuanced tradition from which to draw their parallels. From the time of the Bolshevik Revolution women, too, had demonstrated their prowess in technology and participated in Soviet preparations for war. The role of feminism in revolutionary politics in Russia and the USSR had echoes throughout the twentieth century. Sometimes the echoes were harmonious, and other times dissonant.

On 16 June 1963, Valentina Vladimirovna Tereshkova became the first woman to orbit the Earth. She flew on the *Vostok 6* spacecraft, one virtually identical to the spacecraft that carried the five previous Soviet cosmonauts, beginning with Yuri Gagarin. Tereshkova followed the rigid protocols for launch that her predecessors had followed. She spent the previous night in isolation, save for her backup cosmonaut. She spent the hours before her launch waiting out the routine equipment checks. Like her male predecessors, she had an assigned call sign. Hers was *Chaika,* the Russian word for seagull. Unlike the previous cosmonauts, Tereshkova had not one, but two backup pilots. Irina Baianovna Solov'eva was her primary backup, and Valentina Leonidovna Ponomareva was the second. As German Titov had done before Gagarin's flight, Solov'eva had suited up in case a last-minute substitution was necessary. Ponomareva wore a simple dress to the launch pad. The pattern of her dress made it inside the tightly cropped official photos, but her face did not, which left her identity unknown for decades.[1] Tereshkova climbed into her capsule and followed the equipment checks that proceeded to launch. Tereshkova's spacecraft left the launch pad at 9:30 UTC; she entered orbit almost nine minutes later. Her flight lasted just under three days.

Valentina Tereshkova's mission was different from the first two flights of *Vostok*, but very similar to the *Vostok 3* and *4* flights of the previous year. Two days prior to Tereshkova's flight, Valeri Fedorovich Bykovskii entered the Earth's orbit in the *Vostok 5* spacecraft on 14 June 1963. *Vostok 6* entered an orbit that would bring the two craft close to one another through their trajectories rather than through orbital maneuvers. As had been true with *Vostok 3* and *4*, the joint flights of *Vostok 5* and *6* gave the false impression that the Soviet Union was capable of a maneuvering and rendezvous flight. The Soviet press statements about the flights emphasized that the two craft came within about five kilometers of each other during their first orbit. In fact, the "rendezvous" between *Vostok 5* and *6* was coincidental. Neither craft could maneuver.[2] Their predetermined orbits took the two craft close to one another. Bykovskii landed later than Tereshkova on 19 June 1963 after an almost five-day flight.

The role of this single woman in space merits a separate discussion because it reveals the instrumental role that Valentina Tereshkova played in the intersection of multiple issues over the role of women in the USSR and of Russian women in the Cold War. While the male Red Stuff was frequently a simple display of Soviet power and illusion, the flight of a female cosmonaut was an attempt to satisfy three, often-competing roles of women in the USSR: First, Tereshkova became an exemplar of the quickly changing role of women in post–World War II USSR. Because of the workforce shortage throughout the war and during the early years of rebuilding, women had occupied positions of authority that they never had in the USSR. They took over management roles in heavy industry and had frontline roles in war; some became icons in the partisan fights behind German lines, and female military pilots fought combat missions. The drive for a postwar normalization of everyday life drove Soviet women back to traditional roles. As the first postwar generation matured, women's roles were rapidly redefined to exclude their equal role in the Soviet economy. The expectation that women had to work remained, but they could not have "male" jobs, and they were under the obligation to replenish the war-depleted Soviet population as well.

The second role that Tereshkova's flight played was as a public demonstration of the ease, ubiquity, and reliability of the Soviet technology. The use of staged industrial events to demonstrate the importance and facility of technology was an established Soviet message. This tactic of showing women using technology as an opening to a distant and promising future was not a new one. Tereshkova's flight echoed aviation flights in the late 1930s in so many ways it is almost impossible to believe that the former flight used the latter to remind the public of the glories of the past. This woman's spaceflight contained the added message that Soviet women were becoming modern and urban. As had been the case in the past, the need to make this statement betrayed the contradiction between official Soviet policies and reality. By statute, the issue of gender equality in the USSR had been a settled affair for generations. According to Communist Party doctrine, there was equal opportunity to toil and labor for women in the USSR. The repeated need to demonstrate that equality indicated that reality was far different.

Lastly, Valentina Tereshkova and the flight of *Vostok 6* directly challenged the progress of the American's women's movement during the peak of the Cold War. Some American feminists in the 1960s interpreted Tereshkova's *Vostok 6* flight as an indication that the Soviet Union had not only pulled ahead in the Space Race, but also had done so in equal rights for women. Not only did she become the first woman to fly in space, beating the Americans on yet another first, but also, she defied American stereotypes of the Soviet Woman.

The Role of Women in Post–World War II USSR

World War II was devastating to the Soviet economy. Not only was Soviet industrial capability either damaged or destroyed, but the loss of lives and productivity also distorted the workforce in ways that no other society had felt. Half of the male population of the country was in regular military service and about one-third of those in service died. The only way for the country to survive had been a mass mobilization that brought every able-bodied person and those disabled but capable of work into the workforce. Women had to take on traditional male roles in the workforce, home, and every aspect of Soviet everyday life. The wartime conditions presented an occasion when hard reality forced the USSR to reach for its feminist ideal. Women had to take the leading role in working and managing heavy industry while men fought at war.

When the war ended, thus began the slow movement back to what had been the usual social hierarchy before the war. The process was slow due to the severity of casualties during the war. The official death toll of the war was eleven million soldiers. Demographers surmised that the country might have suffered double that among civilians. After the war there was a dual need to place returning soldiers in positions in civilian industry and to replenish the population of a country that had lost 15–20 percent of its prewar population. These demographic conditions placed the burden of action on women. Not only did they have to step aside from positions of authority in deference to men who were returning from war, but also, they faced the expectation to replenish the population while continuing to work outside of the home to contribute to rebuilding the economy. This situation placed a double burden on Soviet women as working mothers.[3] If that were not enough, women faced added resentment from men as the degradation of the male role model through de-Stalinization and devaluation of national symbols and the substitution of anonymous and symbolic father figures in Stalin's place had diminished the self-confidence of the postwar man in the USSR.[4] The hard life of Soviet women became even harder after the war, and they received few thanks.

On the cultural front, if the new Soviet man presented a new and ambiguous aspirational model for Soviet men in the postwar era, Soviet women faced far more complex and challenging ways to navigate both at work and at home, subject to both romantic and rationalist measures.[5] A woman living in the post–World War II USSR faced insurmountable conflicting forces. While the new postwar male role emerged, women were expected to embrace a new-found femininity, while continuing to contribute to the crippled national economy. They were expected to work full-time, albeit in soft and more traditionally feminine industries, and take care of home and family and fulfill the ideals of womanhood that official popular culture promoted.[6] The traditional women's popular journals of industrial workers and peasants, *Rabotnitsa* and *Krestianka,* respectively, called for "heroic feats of female fulfillment" as the new standard for women.[7] In other words, there were to be no limits to their labor. Unlike men, however, the female achievements were characterized as romantic labor and expressed to serve society.[8] They were not doing this as a choice but as an obligation of citizenship. Women worked and replenished the population in what gender historians describe as a romantic and not a rational ideal.[9] By the end of the Stalin era, the new woman was primarily maternal and domestic, but expected to be ready to adapt, if the need arose.[10] By the 1960s, female labor was essential to both home and the workforce. Soviet leadership had it both ways.[11]

Valentina Tereshkova fit the bill as a new Soviet woman in many ways and evaded the issue in one essential way. In the loss of her father in combat, she had the inescapable link to the Soviet experience in World War II. As an ambitious young woman, she quit her local Yaroslavl classroom school at sixteen to work in a local textile factory and to complete her secondary education via correspondence classes. While working, Tereshkova took civil defense classes in parachuting.[12] She matched the new Soviet female ideal by participating in a "soft" industry and devoting her spare time to patriotic activities. As was true of all ambitious young people of the time, she joined the Young Communist League, Komsomol. All the while, Tereshkova was young enough that she, at the time of selection, was unmarried. She had the potential of bearing the burden of career, femininity, and family, but had not yet faced the complete challenge.

The recruitment of female cosmonaut candidates was far different and less unified than had been the case for the men, both Soviet and American. Personal accounts and archives make clear that there were two opinions on the recruits. Each group had views on what the first woman in space should be and do. This split continued through the year of training that the final five candidates completed. The recruitment for men in 1959 had been a formal and straightforward process. A special cosmonaut selection commission, under the Scientific Research Institute of the Soviet Air Force (NIIVVS), defined selection criteria that would meet the needs of the *Vostok* series of flights. Young men, meeting the under-35 age requirement, accepted invitations for medical screening from among Air Force squadrons across the country. The commission screened those who had passed medical tests for further training. Nevertheless, there were no women in the Soviet Air Force. The lack of a ready candidate pool left open for discussion the qualifications for this flight. On the one hand, a group thought it best to conduct interviews among participants in quasi-civilian parachuting clubs. Others, led by President of the Soviet Academy of Sciences Mstislav Keldysh, preferred to recruit from among the small pool of female aeronautical engineers and students in Moscow.[13] There is a possibility that there was a third and independent avenue for recruitment from among the associated design and construction bureaus that worked in the space program, but by March 1961, there were two groups of female candidates. The first was Tereshkova's group that also included Solov'eva and Tat'iana Kuznetsova. This was the first group to pass the medical testing at the Central Aviation Scientific Research Hospital (Tsentral'nyi nauchno-issledovatel'skii aviatsionnyi gospital', TsNIAG). In March, Ponomareva arrived with nine other recruits from engineering schools to undergo tests that female military doctors supervised.[14] At the beginning of April, two women, Zhanna Dmitrevna Erkina and Valentina Leonidovna Ponomareva, officially joined the first three at the cosmonaut training facility at the Chkalovskii Airbase outside Moscow.[15]

It was not by accident that Tereshkova fit this new model and found her way among the women recruited to become the first female cosmonaut. She possessed precisely the characteristics that those recruiting the female brigade of the Red Stuff sought. From the beginning, the criteria for seeking female candidates for the cosmonaut program were different from doing so for men, even though the recruiters were among the same people who sought out young male pilots. The recruitment teams had to scout many organizations to find their candidates. No women remained in the Soviet Air Forces after the war. Civilian female pilots, some of whom had survived the war as combat pilots and some of whom had gained certification through civil defense programs, had maintained their flight skills under the auspices of emergency civil defense programs. There were women in aviation engineering graduate programs where the male cosmonauts had studied after selection. Other advanced engineering programs harbored women with similar technical qualifications.

During March and April 1962, five women, Valentina Vladimirovna Tereshkova, Zhanna Dmitrevna Erkina, Tat'iana Dmitrevna Kuznetsova, Irina Baianovna Solov'eva, and Valentina Leonidovna Ponomareva entered training to become cosmonaut candidates. The women came from more substantial recruitment of twenty-three individuals.[16] Among the five finalists, three, Tereshkova, Erkina, and Kuznetsova, had been active in parachuting clubs through civil defense training. Only one, Ponomareva, was a licensed pilot who had been too young to fly during the war.[17] All were under 25 years old.[18] Unique among the final group, Ponomareva came from several generations of Moscow residents and was a third-generation engineer. Her father's and grandfather's education had predated the Bolshevik Revolution.

What is significant about the recruitment and training of these first candidates is that debates over who would be the best paralleled deliberations over the role of Soviet women. Valentina Ponomareva writes about the impact of these discussions when she was to stand before the commission, which at this time included Yuri Gagarin, at the end of her medical evaluation. She learned at the time that Gagarin had been against her appointment. He had said that the career of a cosmonaut was "new, hard, unknown and of course, not without risks. It was acceptable to risk the lives of men and pilots, or even conceivably that of an unmarried woman. But it was unacceptable to risk the life of a mother."[19] Ponomareva was the only mother to advance to cosmonaut training.

The fact that Gagarin was so open in his thoughts about the place of mothers and the fact that this line of debate continued through their year of training up until several weeks before the launch indicates that his was not the sole opinion on the matter. Historian Asif Siddiqi has documented that even as the dates for the joint *Vostok 5* and *6* flights had been set, two groups still argued over which of the two Valentinas—Tereshkova or Ponomareva—would be the best candidate.[20] By the beginning of May, two camps emerged. The first, led by Keldysh and Yazdozksii, Director of the Institute of Biomedical Problems, favored Ponomareva. The second, including Korolev, the Director of the Cosmonaut Training Center Karpov, the Commandant of the Cosmonaut Corps Kamanin, and Gagarin supported Tereshkova's candidacy. The commission was not able to resolve the dispute until 21 May.[21] This conflict might easily have resulted from personal styles—Ponomareva had a reputation for being very serious and appearing stern—or interpersonal conflicts within the commission, but if that had been the case there would have been more quantifiable measures on which to base the selection. By her account, Ponomareva did well in the classroom and practical training. An aeronautical engineer with pilot training would expect to do better in these situations than untrained students. In that way, the process favored her. What the commissioners did not factor in was the public appeal that they had found in abundance in Gagarin. The more dour-faced Titov had fulfilled his (and Gagarin's) public duties quite well. The one thing that Ponomareva had done wrong that was irrevocable was that she had fulfilled her role as a new Soviet woman out of sequence. She had had a child and then started her career. To Yuri Gagarin, and by extension to Soviet society, that was an unacceptable example.

Another small obstacle stalled plans for the flight of the first woman. This was due to the delay in the delivery of the "female" spacesuit, the Spacesuit-2 (Skfandr kosmicheskii-2, SK-2). There were three classes of modifications in the first spacesuits required for the *Vostok 5* and *6* flights. The first were engineering improvements that would allow the cosmonaut to remove the suit while in flight, which was to be present for both flights. The second was to modify the cut of the suit to fit the female anatomy. In general, women have narrower shoulders, shorter torsos, and wider hip girth. These changes in proportions dictated that the restraint systems that

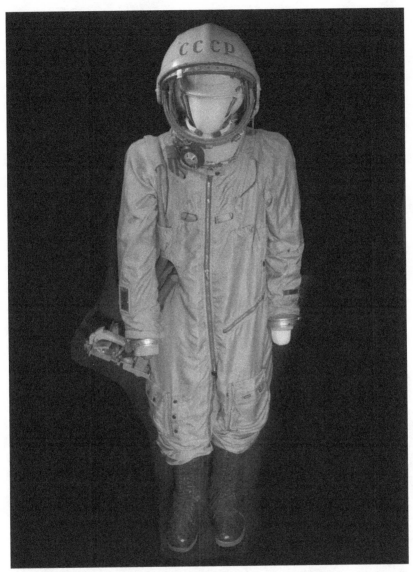

Figure 5. This is the training suit that Yuri Gagarin used to practice before his first orbit of the Earth. All six *Vostok* cosmonauts used similar suits to protect them during their 20,000-foot parachute landing outside the *Vostok* capsule at the end of their flights. It was the modification of this suit to be suitable for Tereshkova that delayed her flight. Courtesy of Emmet Stephenson & Tessa Stephenson Brand.

prevented the suit from ballooning also underwent redesign. It was the third redesign requirement that caused the greatest consternation for the planners of this flight—the urine collection device.[22] Soviet and American men had a tried-and-true device of a rubber hose resembling a heavy-duty condom at one end with a suction flask device at the other to siphon off urine into a container. This was not a suitable solution for a woman. Tereshkova required a design to accommodate her "indoor plumbing" that would not leak into the spacesuit. This engineering challenge caused delays in the SK-2 production. Cosmonaut commandant Nikolai Kamanin complained about these delays and their impact on the flight schedule in his diaries.[23] In the end, the solution to this "pee" problem was hard and uncomfortable and not 100 percent effective. If it had not been messy enough, Tereshkova experienced her menstrual period during the flight. Even though the manufacturer of spacesuits for the USSR, Zvezda, had made the effort to assign a woman, G. I. Viskovsksaya, to the project, the product was not a woman-friendly suit.[24]

At the time the commission recruited the women, the senior parties involved—Nikita Khrushchev, Sergei Korolev, Nikolai Kamanin—and the women in question believed that their careers would follow the well-known pattern of the generation of fliers, male and female, before the war. They thought that the women would train and most likely fly a mission during the program. The accepted wisdom is that Korolev and Kamanin had planned to launch a multimember, all-women crew after the first male multimember crew of *Voskhod 1*.[25] There were also rumors of other all-female missions in the works using *Vostok/Voskhod* hardware. However, as the Soviet space program lost its economic, political, and technological patronage during the latter half of the 1960s, hopes for an extraterrestrial repeat of the heroic female crews of the 1930s faded. The women's cosmonaut brigade remained in training until the early 1970s. Ponomareva's explanation to her grandson as to why she did not fly in space was the most straightforward distillation of the reasons. She said that the reason was that there was not enough spacecraft.[26]

Modeling a new standard for the Soviet woman was, in many ways, Tereshkova's most significant success. Her image exceeded expectations and spread beyond world and Soviet politics. She filled a void in the world of young girls in the Soviet Union. They had few high-profile women heroes and even fewer living ones. Tereshkova drew the attention of young Soviet girls who sought other vocations than repopulating the country as a Mother Heroine, one who had ten or more children. Tereshkova's flight created a spontaneous flow of mail to popular youth publications seeking to find out how they might follow in her footsteps and become a cosmonaut.[27]

The Need to Demonstrate the Reliability of Soviet Hardware

For the last fifty years, people have questioned the quality of Valentina Tereshkova's performance on board *Vostok 6*. Many memoir accounts claimed that Sergei Korolev was unhappy with her completion of basic tasks while in orbit and that he intervened personally to prohibit her from taking manual control of the spacecraft or to make minor changes in altitude, as other cosmonauts had done.[28] Others accuse her of outright failure in her mission due to her experience with space sickness and the fact that her menstrual period began during her flight. It is hard to sort out the innuendo and sexist judgments from the precise story of what transpired during Valentina Tereshkova's flight. According to the stilted official language of her flight plan, Tereshkova's objectives included a comparative analysis of the effect of various spaceflight factors on the male and female organisms, medico-biological research, and further elaboration and improvement of spaceship systems under conditions of joint flight. That meant that the cosmonaut conducted some simple activities to measure basic coordination while ground control recorded her physiological responses. Early in her flight, Tereshkova had complained about space sickness, which had plagued German Titov. It is now known as a common ailment among space travelers that was little understood at the time. When she returned to Earth, Tereshkova had a crack in the visor of her helmet.[29] She maintained that the damage occurred when she removed the film cartridge from the movie camera in the capsule. Tereshkova critics have cited these seemingly insignificant events as the reason for Korolev's disenchantment with women cosmonauts. Of course, any official dissatisfaction with Tereshkova did not find its way into the Soviet press at the time.

As with the five male cosmonauts before her, the Soviet press introduced the USSR and the world to Valentina Tereshkova after her spacecraft was successfully orbiting the Earth.[30] The newspaper *Izvestiia* announced her as the "Daughter of the Land of the Soviets" over her quarter-page portrait on its front page.[31] This introduction seems somewhat modest in comparison to the unreserved announcement of Gagarin's flight in the youth communist paper *Komsomolskaia Pravda* that the "Country Celebrates a

Hero."[32] Gagarin had been an absolute first. Anyone who followed his example would pale in comparison.

The background and differences between Valentina Tereshkova's spaceflight and airplane flights that had preceded her are much more complicated than that, however. Tereshkova was neither the first nor alone in her quest to become this icon of Soviet feminism. There had been female demonstration flights in aircraft a generation before her, and there were other women who trained alongside her for spaceflight. At the time of her flight, there was never an intention that she would be the only Soviet woman in space for almost two decades.[33] Having recruited five women for training, Korolev, Khrushchev, and 1930s aviation hero Nikolai Kamanin anticipated that Tereshkova's flight would initiate an echo of the demonstrations of female Soviet mastery of aviation technology from times before and during World War II.[34] If the rescue flights to the *Cheliuskin* anticipated the celebration of the flights of the first male *Vostok* cosmonauts, the 1938 flight of the *Rodina* anticipated Tereshkova's flight. The parallels are so similar, and Khrushchev's role in both events was so noteworthy, that it is impossible to avoid making comparisons between the 1938 and 1963 events.

On 25 September 1938, the crew of a converted Soviet long-range bomber, an ANT-37 aircraft called the *Rodina*, navigated their airplane to a landing just outside the Siberian village of Kerbi. The landing concluded a 26-hour nonstop flight from Moscow and established the women's long-distance flight record of 5,900 kilometers. The flight not only established a female flight endurance record, but it also established an airmail route that would shave off nearly a week in mail delivery between the two ends of the USSR. Once they returned by train to Moscow, the crewmembers, Valentina Grizodubova, Paulina Osipenko, and Marina Raskova, posed for a photograph at the train station in Moscow. At the center of the photograph was Lazar Kaganovich, the Commissar of Heavy Industry and Commissar of Railways, a member of the Politburo and Stalin's inner circle. Off to the left side in the photograph was the newly appointed Secretary of the Ukrainian Central Committee, Nikita Sergeevich Khrushchev.[35]

Nikita Khrushchev's direct role in the planning for the 1938 flight of the *Rodina* remains imprecise, but during the 1930s, the Soviet Union participated in the international enthusiasm for the new technology of aviation. Aviation was not only an icon for the new technological century, but it was also a potent tool for Soviet officials to overcome the inconvenient geography of the world's largest country. The airplane had the potential to open isolated regions in the north to settlement and exploration and could shrink travel across the eleven time zones of the country to a 24-hour flight. The flight of the *Rodina* established the practicality of establishing airmail service to the Russian Far East. The flight also demonstrated Soviet proficiency in aviation technology to the point where it was routine. The *Rodina's* flight crew had shown that even women could do it in growing anticipation of complete mobilization of the Soviet workforce for war. This subtext somewhat diminished the skills and training of Raskova and Grizodubova, who went on to fly in combat during World War II.

At least the flight was smooth for the well-trained crew of the *Rodina*. They were not just any women. Each had been a trained aviator and participated in the nationwide surge of aviation training that had been a part of the official popularization of aviation. However, none of the three women was among the most highly trained test or military pilots, who were all men and who had ignited popular Soviet fascination with flight at the beginning of the 1930s. All three represented the pinnacle of Stalinist social hierarchy as women. They had attained superior social status through achieving the dream of flight. By the end of the decade, each received the Order of Lenin and a citation as Hero of the Soviet Union, matching the accomplishments of male aviators of the era.[36]

Although the flight of the *Rodina* had its propaganda uses, its crewmembers had legitimate aviation careers both before and after their flight. Pilot Valentina Grizodubova had set the previous women's distance record along with her navigator Marina Raskova in 1937. Grizodubova commanded an all-female bomber squadron during World War II. She also served as Chairman of the Women's Anti-Fascist Committee. She earned the Soviet Hero Star, Soviet Red Star, the Order of the Red Banner, the Order of Lenin, and wore the medal of a member of the Supreme Soviet—all comparable to those of her male counterparts of that era.[37] Navigator Marina Raskova had been the first woman to graduate from the famous Zhukovskii Aviation Academy as a qualified navigator in 1934. During World War II, Raskova headed the all-women combat aviation regiments whose members served as bomber and fighter pilots. Raskova herself died a heroic death while on a resupply mission to the Stalingrad front in January 1943 at the rank of major on a dive-bombing mission.[38] Together, these women represented only the tip of the cadre of pilots who had trained in aviation fields under the auspices of Osoaviakhim.[39] Polina Osipenko and Raskova joined another pilot, Vera Lomako, to set another long-distance record flying east to west in July 1938. Osipenko died during a routine flight in 1939 just before the European war. Only Grizodubova survived the war and the USSR, dying in Moscow in 1993, just short of her ninety-fourth birthday.[40]

There were technical and personnel issues during the *Rodina* flight that did not appear in the Soviet press at the time. Plagued by bad weather, the mission took ten days to accomplish a flight of just under twenty-six-and-a-half hours.[41] The poor design of the ANT-34 did not allow the navigator seat to survive a crash landing. As a result, the navigator, Raskova, bailed out of the craft before it landed in Siberia and struggled to find her way for ten days without supplies. Although she survived, conditions did not allow for the celebratory photo opportunity in Siberia that the planners had hoped. That is likely the reason for the postponed celebration of the flight in Moscow after the crew had returned by rail.

As the legacy of the *Rodina* and the Soviet women pilots of World War II demonstrated, the Soviet government had encouraged women to participate in the aviation movement of the 1930s by offering training and opportunities to fly. This offer of training was as much a necessity as a method to advance the idea of Soviet gender equity. The country was preparing for war in the late 1930s. Everyone had to prepare and participate. Planners for the appearance of a Soviet female with Red Stuff had a more limited view. Spaceflight was never going to be an existential phenomenon for the USSR. Given the way in which the memory of the women fliers of World War II faded, it seems likely that the female Red Stuff was never destined for sustained celebrations. Although the Soviets supported the minimal celebration of the accomplishments of the women pilots of World War II, they remained a part of Soviet public memory.[42] It would have been logical to assume that there would be a similar public showing of women's participation in the space age that went beyond a single mission. Valentina Tereshkova was not the only woman to train to become the first woman in space, but she was not one of thousands as had been the case with the female fliers. A quarter century after the *Rodina* flight, the special committee selected only five women to begin training and testing in early April 1962.

The wide swing from rapid and broad recruitment of female aviators for the war to the small recruitment of female cosmonauts came from the demobilization culture that had shifted women to return to traditional roles after the war. There is an example of the acceptance of this shift and the dislocation that it caused that appeared in popular culture about the time that the special commissions were doing their recruitment. The consequences of discarding women's performance and the contributions of their labor left an impact on Soviet culture. In 1966, a new theatrical film about a woman's experience in World War II, *Wings*, appeared in the USSR. Directed by a recent graduate of the All-Russian State Institute of Cinematography, Larisa Shepitko, the film recounts the depressed state of mind of a World War II female fighter pilot turned construction school principal. While administering a school, Natalia Petrukhina daydreams of her former life of excitement. When given the opportunity to sit in the cockpit of a trainer like the craft that she flew during the war, Natalia starts the engine, to the shock of teasing grounds crew, and takes off to fly one last time. The film had a similar nostalgic tone of other Thaw-era films that expressed a longing for prewar society and people.

The dismal cinematic message about the prospects for women's careers in flying was not uplifting, especially in light of the fact that female crews continued to train for the hypothetical all-female *Voskhod* flight. Korolev and Kamanin had planned to launch a multimember, all-women crew. There were also rumors of other all-male missions in the works using *Vostok/Voskhod* hardware. However, as the Soviet space program lost its political and technological patronage during the latter half of the 1960s, hopes for an extraterrestrial repeat of the heroic flight of the *Rodina* faded as well. The women's cosmonaut brigade remained in official training until the early 1970s, though members had departed earlier. Many explanations of why these missions did not take place emphasized the role that political intrigue played in all decisions related to spaceflight. However, if the simplest explanation is often best, Ponomareva's explanation to her grandson is the most plausible.

Tereshkova, the Cold War, and Women

While Valentina Tereshkova endured harsh judgments against an elusive standard of the modern female ideal and suffered from a double standard for her performance when judged against her male counterparts, there was one area in which she met expectations. That was in the Cold War competition with the United States. Her flight did so by sending a signal to the Western feminist community that that USSR might be ahead of the West in its equality for women and, in doing so, shaking Western stereotypes about Soviet women. Tereshkova's flight caused alarm among some in the women's movement in the West, especially in the United States. Her flight revived for a short time the quiet unresolved debate in the United

States over the role of women in human spaceflight. Much of the debate had centered on the screening of American women pilots at the Lovelace Institute in Albuquerque, New Mexico two years before Tereshkova's flight. The Lovelace Institute had been the same location made famous when the *Mercury* astronaut candidates had received their original physical tests. In summer 1961, twenty-five American women pilots began these same tests, concluding after the second of three stages of tests that the *Mercury* astronauts took.[43] Private funds paid for the women and there was no official authorization or status for the project beyond the grounds of the institute.[44] The thirteen women who passed the New Mexico tests never went on to the proposed further testing in Florida. Of course, many were disappointed personally. The women may have had high expectations, based on their qualifications and performance in Albuquerque, but less well-informed observers held expectations that the program had official sanction. Not all had the clear understanding that the program had been experimental and privately funded and had no official standing with either NASA or the military.[45]

These dueling perspectives led to a public discussion that quickly divided into two opposing positions. One held that only male experimental pilots were equipped to fly in space and the other that it was time for women pilots to join the men. Inside the US, women pilots and their supporters argued that there was no reason to prohibit women who could qualify from competing. In response to the calls for the US to catch up with Soviet feminism, NASA pulled out the heavy guns with *Mercury* astronaut John Glenn and accomplished pilot and former World War II Women's Airforce Service Pilot (WASP) Jackie Cochran making statements doubting the relative preparedness of women for spaceflight. Cochran had paid for the tests at the Lovelace Institute, and most articulately expressed this belief.[46] In her letter to James Webb, Administrator at NASA, Cochran stated:

> Those who are pushing you and your associates at this time about women in space as somewhat of a crash program or because of a present need are, in my opinion, not serving well the long-range interests of women, much less the present overriding national interest. The time for female participation in space flights will come, and I am for it in an orderly way, but it should not be pushed for the present.[47]

She sent this letter almost a year before Tereshkova's flight. At the time, Cochran could not have known that a few months prior to her letter, five

Soviet women had begun training as cosmonaut candidates at the Space-flight Training Center outside Moscow. In addition, she did not know that within a year, Valentina Tereshkova would become the first woman to fly in space.

In immediate response to Tereshkova's flight, another famous American woman wrote an article with much greater urgency than Cochran had expressed. American journalist Clare Booth Luce penned an editorial in *Life* magazine that demanded immediate response to the Soviet "women's challenge" in space. Luce called for American women to participate immediately in spaceflight to match Soviet efforts. In answer to the underlying question of why a Soviet woman was first, Luce said, "The right answer is that Soviet Russia put a woman into space because Communism preaches and, since the Revolution of 1917, has tried to practice the inherent equality of men and women."[48]

Luce was wrong about her impression of the condition of women in the USSR. She had accepted Soviet propaganda and not closely examined their practices. Soviet propaganda always had the intention to confuse myth and reality. Luce was neither the first nor the last to succumb to its influence. Tereshkova's flight had little bearing on the status of women in the Soviet Union. It was a demonstration flight to promote the image of Soviet mastery of space technology. Although the Soviets were the first to launch a woman into space, they made little pretense that Tereshkova's flight demonstrated women's equality with men in the USSR. Equality was not the point of the *Vostok 6* mission. Just as the *Rodina* crew had merely sought to demonstrate that even women could master aviation, Tereshkova's flight did not attempt to demonstrate that women have equal roles in spaceflight. If it had been, Tereshkova would have played an equal role with the male cosmonauts. Her isolation as the only known woman cosmonaut and her professional distinction as the only non-military pilot among the *Vostok* cosmonauts to fly into space emphasized that Tereshkova was not equal, but an exception in both her public and hidden lives. Her argument did not have a prolonged impact on American women's politics of spaceflight in the 1960s, however. Women pilots, fighting for the right to compete with men on equal footing, tended to agree with Cochran's earlier opinion that to hurry through the process to include women would result in a diminution of the perception of the women pilots' qualifications.

Once American consternation over Tereshkova's flight became public, Khrushchev had the potential to relish yet another instance of one-upmanship against the US. Unfortunately, he was not in the position to act on that opportunity. The Politburo removed him from office in October 1964. Unlike after *Sputnik* or Gagarin's flight, Khrushchev neither demanded nor expected a repetition of Tereshkova's flight. He made this decision even though uproar among Americans became yet another unanticipated benefit of the Soviet space program. As historian Susan Bridger stated:

> Though the process of decision-making which had eventually sent a woman into space scarcely reflected any credit onto a state with an avowed commitment to gender equality, this was not to stop its leadership exploiting the international political mileage from Tereshkova's achievement.[49]

Tereshkova, a textile worker and parachutist, served to justify the equality of Soviet science and technology compared to that of the US far more than to justify feminism in the USSR. What the Soviets proved with the launch of *Vostok 6* was that Soviet scientists and engineers (primarily male) could launch even a woman into Earth orbit, and that those Soviet women should strive to appear as modern as Tereshkova did.[50] Her flight did not signify that those Soviet women had broken through any barriers. As an *Izvestiia* cartoon pointedly depicted, Tereshkova had merely changed professions. She had been a textile worker and then became a cosmonaut, performing the same duties in space as she had at the textile mill.

Tereshkova's less noted and probably more widely impactful accomplishment in the West was to shatter American stereotypes about Soviet women. By midcentury, there were two standard American perceptions of Soviet women. Neither was attractive or generous. They both carried the assumption that Soviet women suffered deformation because of communism. The theory went that either Soviet women suffered physically through primitive life in an agrarian or industrial existence or that they suffered deformities because of doing men's work.[51] Rural women suffered the burden of agricultural labor with none of the current benefits of urban conveniences and leisure. An example of this was the perceived matronly appearance of Nina Khrushcheva, who accompanied her husband Nikita to the US in 1959. The second judgment against the consequences of Soviet women participating in heavy industry was even harsher. American perceptions cast Soviet urban women as manly and decidedly unfeminine. Tereshkova broke these American stereotypes about gender in the USSR by being the "blond in space," and maintained a contemporary appearance of femininity.[52] She displayed no deformities of rural or urban Soviet life.

Figure 6. A simple photographic pin of the first woman in space, Valentina Tereshkova. From the collections of the Smithsonian Institution National Air and Space Museum.

Conclusion

Long after the dissolution of the group of Soviet women cosmonauts in 1969, the first public announcement of the existence of the four other women cosmonauts came in the communist women's journal, *Rabotnitsa*, in 1985.[53] This publication was the first time that the five acknowledged their role and met together as a group since 1969. The group disbanded as a concession to what had become evident to many—elevating women's roles in the USSR was never of primary importance but retaining women in the Soviet workforce had been.[54] Because Tereshkova had been the only publicly acknowledged female cosmonaut for the rest of the decade, she was the only one to maintain a career in the limelight. She married cosmonaut number three, Andriian Nikolayev, the only single man among the first cosmonauts. They had a daughter the following year, thus demonstrating the safety of spaceflight much in the same way that the birth of a litter of puppies had demonstrated the same thing in 1961.[55] Tereshkova and Nikolayev soon separated and eventually divorced in 1979.[56] Unlike

Tereshkova, the other women cosmonauts returned to positions related to their original training after release from the program. Erkina graduated from the Air Force Engineering Academy and returned to work as an engineer at the Tsentr podgotovki kosmonavtov [Spaceflight Training Center, TsPK]. Tat'iana Kuznetsova became the Chief of the Geophysical Laboratory of TsPK. Ponomareva joined the faculty of the Institute for the History of Science and Technology of the Academy of Sciences. Irina Solov'eva pursued an advanced degree in psychology and joined the Soviet all-female Antarctic expedition in February 1988, following in the footsteps of a few women dating back to the 1920s.[57]

After Tereshkova's flight, skeptics combined rumors of Korolev's dissatisfaction with her performance during the flight, the American response to the mission, and the absence of another woman cosmonaut or astronaut for nineteen years to draw the conclusion that Valentina Tereshkova's flight had been a stunt with little more justification than publicity against the Americans.[58] Adding to this perception was the fact that the Soviets were poor in delivering the scientific results of their missions to the world.[59] For most of the last half-century, the dominant conclusion was that the flight was a one-time event with the marginal benefit of guiding Soviet women for their new roles in a postwar and eventually prosperous state. In discussing the Red Stuff, separating the sexes of the first cosmonauts might seem to be making an artificial distinction. However, doing so matches the real separation of the sexes that existed during the early 1960s. There was no equal opportunity for women in space in the USSR any more than was true for any other field. The demonstration that women could fly was not an illustration of opportunity but a reassurance to the female population of the USSR that there was a role for them in the workplace, but that they had to fit a precise mold to claim their place. Nor was Tereshkova's flight solely a Cold War challenge.

The assumption among the Soviet population was not that Tereshkova's mission was another publicity stunt. The men who led the Soviet space program had taken a deep interest in the selection and public cultivation of the first female cosmonaut candidates. This care and attention created hopes for follow-up female missions, of which Tereshkova was to be the first. This care echoed the selection of male candidates. The emphasis on selecting a new generation of cosmonaut candidates from among those who had not flown during WWII but still had a legacy had been a sign that the cosmonaut was the Soviet citizen of the future. A female cosmonaut was an analog to the new Soviet man. The female cosmonaut of the 1960s, as personified by Valentina Tereshkova, was a propaganda symbol that transcended the space program. While she had instituted discussion in the United States about the role of women in spaceflight, Tereshkova's flight had created an expectation that the situation for women would change in the USSR. Her accomplishments included success in a feminized version of a male job while maintaining an attractive appearance and fulfilling the reproductive obligations to the state. Tereshkova had become an aspirational model for Soviet women so high that it was a long time before another woman approached the ideal. The role of Tereshkova's success went unfilled for almost another nineteen years, when the next woman flew into space. That time, in 1982, Svetlana Savitskaya flew on board a *Soyuz* to preempt Sally Ride's first American female flight. This flight had been a direct challenge to the Americans. By this time, the presentation of the Soviet woman cosmonaut rang hollower than it had a generation before.

4

New Cultures of the Cosmonaut

Collectibles, Monuments, and Film

While the USSR was overtly competing with the United States in sending humans into Earth orbit and beyond, the state took care to ensure that the domestic audience did not miss the message of Soviet material accomplishments. Newspaper, radio, and other media reports were not sufficient to carry the message of the Red Stuff at home. There had to be a tangible and personal involvement in this new culture to have an enduring and convincing impact. Parades and celebrations were too infrequent for this new idea of Soviet status to build national memories and infuse itself into the national consciousness. The USSR had to invoke new methods to guide the population to absorb the Red Stuff as part of the national consciousness. The Soviet state franchised the expression of the Red Stuff in three ways—through the development of a personal material culture, the erection of monuments and museums in public spaces, and ushering a new era of science-fiction film. First, they all began as state-sanctioned popularization of a vision of the Soviet cosmonaut. Each started with an official message within its medium of an official vision. Collectibles such as stamps and pins relied on official sanction to begin manufacturing and distribution and established themselves as state-sanctioned portrayals of events. Architecture and design remained a politically monitored aesthetic, but one that could dominate skylines and localities. Film in the USSR had been one of the earliest arts—one that Lenin declared to be the "most important." Second, each field distinguished itself in how it began to generate autonomous approaches to the image of the Russian cosmonaut, drawing creative minds to the enterprises. The third distinguishing fact was that each of these sectors later began disseminating its own visions of the Red Stuff beyond the official messages from which it originated. These tools that infiltrated everyday Soviet life—small objects, large architectural projects, films that entertain—fostered the development of personal meaning for an individual. If the meaning, use, and control of these tools remained under tight official interpretational control, they had the potential to form a national bond among the citizens of the USSR.

From Orbit to the Kitchen: Human Spaceflight and Consumerism

Collectibles and spaceflight share historical associations in the Soviet Union. Both emerged during the post-Stalin era of Khrushchev. Each came to symbolize the optimism of the era, and each served as a distraction from the realities of Soviet life. The growth of space-themed collectible consumer goods coincided with the post-Stalinist effort to create a sense of contentment and modernity for the war-weary population. By participating in a culture of leisure activities that had not existed before the war, the Soviet population could consider itself modern. Leisure and spaceflight represented modern living to Soviet citizens as it did for Americans. This illusion of affluence and progress could distract the Soviet population from the lingering sacrifices of the war. These efforts to convince the population of their good fortune extended beyond the small items that the average Soviet citizen could purchase. During the Khrushchev era, propagandists made every effort to identify leisure and recreation activities with cosmonauts. Furthermore, unlike much of Soviet culture, these objects have endured the collapse of the Soviet Union, retaining significance from a brief period of optimism.

The current scholarship of the USSR places emphasis on Khrushchev's attempts to change public expectations of the state as part of his movement away from Stalinism.[1] Some historians, most notably Susan Reid and David Crowley, have turned attention to the material culture of the former Soviet Union, emphasizing discussions of consumerism and aesthetics and how they were used to satisfy the national hunger for private life.[2] Others had pointed out the extent to which these changes touched the day-to-day lives of citizens.[3] Post-Soviet attention to the preservation and conservation of the material culture of the previous era offers the opportunity for a closer examination of shifts in aesthetics and consumerism in the Soviet Union during the 1960s. One aspect of Khrushchev's Thaw was a limited return to the modernist aesthetic that had accompanied the Bolshevik Revolution. Khrushchev's relaxation of Stalin's cultural restrictions did not mean a wholesale return to the prerevolutionary and early Soviet modernist thought. However, it was an opportunity to shed both the aesthetic and sumptuary practices that had symbolized Stalinism.

The material culture of this period of spaceflight came from less tightly restricted circumstances than had previously been allowed. Khrushchev had learned from Stalin that the most effective domestic propaganda involved the promotion of mass celebrations.[4] However, the space program operated in semi-secrecy and did not allow popular participation in planning and staging activities. Mass participation in the space program had to take place after the fact. At its peak in the 1960s, the Soviet human spaceflight program had two missions per year. One way in which the Soviet population could share in the growing momentum of the space program was to read cosmonaut biographies, but reading is a solitary act that did not generate the cooperation of a group activity even when done among school groups and youth organizations. Books did not have the staying power of material goods. The USSR did have a well-established collection of social organizations. Beginning at school age with the Young Pioneers, and through young adulthood in the Komsomol and the Communist Party, information, activities, and materials would appeal to all ages of the Soviet population.[5]

Soviet Collecting

Under Stalin, the Soviet Union prohibited or discouraged individuals from collecting trinkets. In the early 1960s, the Soviet government sought to reverse this policy and encourage personal collecting.[6] Personal collecting was a small concession to consumerism that could be a pressure valve for frustration with economic and material conditions in the country.[7] Small collectible items pervaded Soviet society, diffused through these organizations. The loosening of political mores on personal collecting within the USSR paralleled a relaxing in Soviet society in general. It was threefold. First, the postwar generation had no reason to sacrifice as their parents and grandparents had. They had not personally witnessed war. Secondly, Khrushchev had publicly renounced the overt use of terror to enforce party rule. Finally, Soviet leadership recognized that young people would require some liberalization of government policy to maintain support. The most readily accomplished and least socially disruptive areas of liberalization were the relaxation of laws and rules governing hobbies and allowing contacts with the outside world. By relaxing rules concerning hobbies, the Soviet government sought to encourage the limited acquisition of personal property, including collections. Controlled international contacts with

The more interesting and aesthetically appealing stamps were in high denominations. Higher-denomination stamps were airmail stamps that were destined for foreign destinations.[18] Even when they were more aesthetically pleasing, their messages were strict interpretations of Soviet propaganda. They became "visual statements of the values that the regime espoused and desired to foster among the population. In this light, these virtual representations revealed the regime's conception of how Soviet society should be structured."[19] After Stalin's death, the organization and methods of Soviet philately did not change significantly. The stamps produced through the 1950s were full of propaganda and continued to recap Soviet industrial, technical, and military accomplishments. Instead of depending on symbols and quick slogans, these stamps took on tones that were more ponderous: "In the post-Stalin years, Party platforms continued to occupy a prominent place on Soviet stamps but were presented differently. Gone were the brief heroic slogans of the Stalin era that urged economic mobilization and in their place were rather lengthy excerpts from Party congresses."[20] Despite its best intentions, the Ministry of Post and Telegraphs was not producing stamps whose messages drew attention either at home or abroad.[21]

The design for airmail stamps did not vary much from domestic ones, although they were destined for consumption abroad. Soviet industrial achievements and social and political milestones were the themes that dominated airmail stamps. This trend continued through the 1930s when in 1939 at the New York World's Fair the Soviet Pavilion featured stamp exhibits that recounted Soviet aviation endeavors.[22] Moreover, even in the 1960s stamps continued to include lengthy quotations from party congresses. The resulting stamps left an unsatisfied appetite for aesthetically pleasing and inspirational stamps at the dawn of the space age. Furthermore, they were effective as instruments of propaganda, spreading the message of Soviet accomplishments to mostly capitalist communities that might not learn of these accomplishments otherwise. Around the same time, domestic regulations loosened and stamp collecting became part of an officially sanctioned social organization in the late 1950s and early 1960s. The Twenty-First Party Congress was the first time that the government recognized collecting organizations as independent social groups, receiving official party sanction. Thus, the atmosphere for the domestic collection of stamps was set before the flight of Yuri Gagarin. The first *Vostok* flight provided new imagery for Soviet stamps.

In anticipation of Gagarin's flight on 12 April 1961, the Soviet Ministry of Communications prepared three stamps for distribution in the denomi-

Figure 7. One of the earliest stamps commemorating Yuri Gagarin's flight. Photograph by Eric Long. From the collections of the Smithsonian Institution National Air and Space Museum.

nations of three, six, and ten kopeks.[23] The ministry released these stamps within days of the flight. Youth magazines promoted their sale and collection. For example, the magazine *Pioner* devoted the inside back cover of its August 1961 issue to them.[24] Each of the three stamps was consistent with traditional Soviet approaches to official design and marketing. The three-kopek stamp in a domestic-mail denomination provided only the essential details of Gagarin's flight. The top carries the title "Man from the Country of Soviets in Space." Around Gagarin's portrait are the words "First Cosmonaut in the World." On either side are pictures of a generic rocket and an illustration of the Hero of the Soviet Union medal that Khrushchev had awarded him immediately after his flight. The design could easily replace the front page of *Pravda*. There was no effort at aesthetic innovation.

The six-kopek stamp for international mail followed the post-Stalinist tradition of bearing lengthy quotations from party officials. The two-part stamp illustrates *Vostok*, a ballistic missile, and a launch vehicle rocket flying over the Kremlin with a radar dish on the side on the top portion that carries the postage mark and the same title as the three-kopek stamp. The lower portion carries the quotation from Nikita Khrushchev's early statement about the Gagarin flight: "Our country was the first to lay down the path to socialism. He was the first to enter space and opened the new era in the development of science."[25] The largest denomination in the first Gagarin set was like the other two. The ten-kopek stamp, too, features an image of Gagarin's launch vehicle flying over the Kremlin and the title "Person from the Land of the Soviets in Space." However, this foreign-envelope postage stamp did not have an additional section with a quotation from

Khrushchev because a long quote in Russian was of little value to the international public.

These first stamps honoring Gagarin appeared quickly. It is not surprising that the ministry made little effort to transform the aesthetic approach to stamp design at that time. It had merely adapted the message of human spaceflight to its format of miniaturizing *Pravda* or *Izvestiia* headlines into a stamp format. Each cosmonaut faced in the direction of the rising spacecraft with his or her chin lifted toward the sky. Subsequent stamps that honored the flights of *Vostok 2* through *Vostok 6* were similar in detail. For example, the set of stamps that came out in honor of the dual missions of *Vostok 3* and *Vostok 4*, which carried cosmonauts Andriian Nikolayev and Pavel Popovich, are little different from the stamp issued the year before commemorating German Titov's first full day in space on board *Vostok 2* in August 1961.[26] In all three cases, the stamp featured a portrait of the cosmonaut, his name, the date of the mission, and a stylized illustration of the spacecraft. In all three cases, the stylization of the spacecraft did represent an aesthetic effort, but it did not represent an original design on the part of the stamp designer. The stamp merely copied the fictitious illustrations of *Vostok* that had appeared in the national press. As Soviet officials kept the engineering details about the spacecraft secret until 1967, there was no official representation of the craft, only artists' speculation about how a rocket ship might look.[27]

The stamps from the two-mission *Voskhod* (Sunrise) program with multiple cosmonauts differed little from the *Vostok* stamps. The Ministry of Post issued a set of identical portrait stamps for each member of the 1964 *Voskhod* crew: Komarov, Feoktistov, and Egorov.[28] Once again, the stamps offered little more in innovation and information than had the pages of the official newspapers. These four-kopek, domestic-use stamps offered no new aesthetic enticements. Virtually identical for each of the three cosmonauts, their only appeal was from the information about this latest space mission that appeared to overtake the United States—namely, the fact that this mission involved, for the first time, a multi-passenger crew.

Stamps that commemorated the flight of *Voskhod 2* in March 1965 showed a slightly improved stylization in design.[29] The six-kopek stamp, honoring the commander Pavel Belaev, adapted his official, spacesuit-clad portrait into a slightly abstract version. The stamp that commemorated Aleksei Leonov's spacewalk used the official and inaccurate drawings of the air lock and of the spacecraft he flew. Accuracy notwithstanding, the stamp is dramatic, depicting a free-floating Leonov flying alongside his spacecraft with a motion-picture camera in hand while his commander, Belaev, looks out through the open hatch in the capsule. This stamp was the first attempt to depict action in a space stamp. It is significant that this stamp was the highest-denomination stamp that the ministry issued during the 1960s, for airmail that would more likely find its way to the world philately market via a letter or postcard sent from the USSR to the West.

If previous experience is a guide, the Ministry of Post and Telegraph had designed and printed stamps honoring the flight of *Soyuz 1* with Vladimir Komarov before his launch in April 1967. It had done so in the case of Gagarin's *Vostok* flight and released the stamps almost immediately after the flight.[30] If the ministry staff had followed the same procedures of preprinting stamps in advance, in 1967, they stopped the release of any Komarov/*Soyuz 1* stamps after the disastrous end of his flight on 24 April 1967. Nevertheless, the Soviets continued to create and issue space-themed stamps after that. By one account in 1975, there were more than a hundred of them in circulation. These included ones that celebrated Soviet robotic missions to Venus and the Moon. The space theme came second in numbers only to World War II themes in Soviet philately.[31]

Soviet human spaceflight stamps continued to receive regular attention in collecting journals and in youth publications that encouraged collecting throughout the 1960s.[32] These journals promoted one of the few officially encouraged material culture pastimes in the USSR, and one that encouraged the collaboration of Western participants. Every other issue of the youth magazines *Pioner* and *Semena* featured columns on collecting.[33] Articles noted new stamp issues and made recommendations for completing collections. On occasion, an article would feature a particularly prodigious young collector as an inspiration to others. In all cases, the Red Stuff tightly aligned to the official political situation in the USSR. The cosmonaut is heroic and has become an official emblem of the state.

Znachki

The next step removed from the official portrayal of the Red Stuff was the production of the small, enameled pins known as *znachki*. Znachki measure about two centimeters in diameter. The size of these objects contrasts with the enormous scale of space artifacts that include forty-meter-tall launch vehicles. The small size of znachki, as well as their accessibility, transformed the experience of space exploration into one that was palpable to all Soviet citizens through material consumerism. Comprehensive displays of large-scale spacecraft and engineering artifacts have remained rare even today in the former Soviet Union. During the 1960s, the secrecy and ambiguity that surrounded the space-program hardware made access to such objects nearly impossible for the average Soviet citizen. For those reasons, space-themed znachki offered the complete public image of the Soviet space program.

The small, enameled pins that commemorate Soviet space missions are the material culture of the official historiography of the Soviet space program. They surpassed stamps in this distinction because of their unique conception, manufacture, and distribution that transformed an existing object of limited use into one that symbolized mass participation. The pins are distinct from other forms of collectibles because they have a briefer popular history. They offered the opportunity to unsophisticated individuals to collect items without training in other fields and with only the guidance of the popular press. These small pins illustrated a miniature, idealized chronology of the scientific and technical achievements of the space program. The illustrations presented officially sanctioned and often inaccurate images of the spacecraft. They celebrated the firsts and anniversaries of Soviet accomplishments, and thus through repetition and sheer force of numbers, they reinforced the Soviet propaganda mantra of mastery of spaceflight. The pins were ubiquitous throughout the former Soviet Union and reiterated official Soviet accounts of space activities, embodying Soviet efforts to establish claims of superiority through persistence and repetition.

Znachki in general, no matter what the subject matter, were unique as consumer goods in the history of the Soviet Union. In contrast to the Soviet government's previous efforts to control and manipulate civil society, znachki emerged from a middle layer of Soviet managers, met an emerging demand for consumer goods, and created a minor source of fundraising. It just so happened that while this was occurring, the Soviet Union began to fly in space. This coincidence was quite fortuitous for the manufacturers of znachki. They quickly recognized the new and exciting market in space pins. Their popularity quickly transcended borders. There is no exact English equivalent of the word znachki, which is the plural of the Russian word znachok meaning "badge" or "small mark." The word itself is the diminutive form of the word for a sign, mark, or symbol.[34] Anyone who has visited the Soviet Union, or knows someone who did, immediately recognized the pins as the most common collectible from that country.

The youth movement of the Khrushchev era expanded the associations of znachki workplaces to a broader audience. It just so happened that the youth movement coincided with the public space program. Znachki have developed into the significant objects that they are because of social forces at work within the post-Stalinist Soviet Union and the official government's efforts to meet the demands of a population that had made sacrifices for generations and now held high expectations. The pins became a commodity in a society that was notorious for the absence of consumer goods. In addition, an examination of space-themed znachki offers a unique perspective on the transformation of the 1960s culture of the Soviet Union. The space pins combine the optimism of modernism with the reassuring values of socialist realism. The design of many of the pins hearkens back to the constructivist style of the 1920s and 1930s, while the content bears the reassuring tale of incremental Soviet achievement that characterized socialist realism. Finally, the pins are the material remains of the Soviet effort to appeal to the youth market and control the emerging student movements of the early 1960s. Youth organizations introduced znachki to student groups in the late 1950s and encouraged their collection through their official organs.[35] Toward the end of the next decade, their popularity was so great that they transcended the youth movement, becoming popular souvenirs among tourists.

During the height of the Space Race, znachki collecting gained an enormous following, and the space program was at its zenith in popularity and success. Many collections from that time exist today to demonstrate the peak of its popularity. There are two primary sources for the study of collections in this discussion: the Smithsonian National Air and Space Museum's collection as well as virtual collection resources. The Smithsonian collection has grown over the years through private donations and diplomatic gifts and a much larger private collection that has developed through the advice and recommendations of the international znachki-collecting community. Of the several virtual collection resources, the first is the ever-present market of Russian space memorabilia that is for sale on eBay. These listings frequently offer images of individual znachki that might not be available in an organized collection elsewhere. They also provide insight into the collecting ideas and strategies of Soviet youth during the 1960s, as many of the sellers claim to have amassed their collections in their youth, usually as Pioneers. In addition to auctions on eBay, there are more permanent, and often more comprehensive, websites that attempt to catalog znachki according to the preferences of the owners.[36]

Soviet popular journalism dominated the written documentation on the history of znachki. Among the magazines that focused on contemporary znachki collecting were the popular collecting journals that announced the release of new designs and encouraged collecting. For the most part, these were philatelic and numismatic journals that sought to place znachki within the context of their disciplines. In doing so, they advocated strategies that paralleled stamp collecting, placing greater emphasis on the breadth of collections than on aesthetics or completeness. The official journals of the Pioneer youth organization, *Pioner* and *Semena* (Seed), also encouraged its readers to complete their collections, not surprisingly, as editors of collecting journals also wrote the collecting columns for those youth magazines.[37] Popular science journals, such as *Zemliia i vselennaia*, announced newly available pins according to scientific specialization as well.[38] In recent years, experts have published monographs on specific collecting areas, including space, which paid closer attention to subject-matter grouping, completeness of cataloging, and design sophistication.[39]

Despite an occasional resemblance to other collectibles, znachki are unique due to their ubiquity. Since 1957, they have not been subject to officially controlled production or distribution. It is not the rarity of an individual pin but the ubiquity of pins that enhances its value to collectors and historians. Although this might seem counterintuitive, znachki derive value through commonality. A single pin design owned by many carries its message further than a rare pin owned by a few. The year 1957 was a pivotal year for the popularization of the Red Stuff for two reasons. For space historians, it was a pivotal year because of the *Sputnik* launch and because collecting znachki began in earnest. During the sixth and largest biennial festival of the International Union of Students (held in Moscow), twenty thousand international students—and at least ten thousand Russian students—congregated in the capital. The festival's motto was "For Peace and Friendship," which was a variant of the two elements of all previous and subsequent mottos. However, the primary mission of this festival was to demonstrate Soviet leadership among the growing postwar youth movements. From a political perspective, the festival was a success.[40] From the perspective of the youth participants, long-term success equated with long-lasting relationships among them. An immediate measure of the outcome was the proliferation of the material remnants of the event. By some estimates, there were more than seven hundred distinct types (mainly representing Soviet cities) of znachki at the festival, issued in runs that numbered into the thousands. The assembly of tens of thousands of young people in Moscow began a flurry of trading and exchange among youth.[41]

The official *znachok* (singular) of the festival was notable for its six-color rendition of the official flower symbol of the festival. The use of multiple lacquer colors was a noticeable departure from previous generations of red pins that celebrated the Bolshevik Revolution. A transition in themes soon followed on the heels of the transition in color.

The first indication that space themes would transform znachki collecting into a long-term trend occurred a few years after *Sputnik*. Although the znachki exchange among young people at the youth congress in 1957 had caught on, the attempts to re-create this success were cautious. The first space exhibition in the Soviet Union was small, featuring stamps and znachki; it opened at the Moscow Planetarium three years after *Sputnik*.[42] The exhibit included space-related stamps, postcards, znachki, and commemorative coins. All objects portrayed highly stylized representations of the spacecraft that executed the much-celebrated space firsts of the Soviet Union. None revealed technically accurate details of the space hardware. That was not the intent. These pins were decorative and collectible. The exhibit was the brainchild of planetarium director V. K. Litskii, who encouraged established collectors, at the time primarily adults, to expand their traditional philatelic and numismatic collections to include space subjects. The exhibit also captured the attention of young people, who had been born in a time when collecting did not meet official approval, and thus paved the way for the next generation of officially sanctioned collectors. The placement of znachki alongside stamps and coins was novel and foretold the dominant role that the pins would play in illustrating the Soviet space program.

What makes space znachki most interesting is that the decentralized fabrication of the pins created previously unexplored uses and methods of dissemination. Non-aerospace organizations timed their issues of znachki to capitalize on public attention, issuing pins as soon as the public announcements of program milestones, albeit after the missions occurred.[43] By echoing the official announcements of missions, even these independent distributors reinforced the official historiography of the space program. Like the official infrastructure, they ignored failures and enthusiastically celebrated heroes. Manufacturers chose sequences of successful projects to form a set of collectible pins and thus define the scope of a successful program. For example, *Vostok* pin sets featured the six spacecraft or cosmonauts and were distinct from a *Voskhod* set in either design or theme. Each set would have a distinct style, separating it visually from another program often sold on a presentation card or box.

When examining the thousands of space znachki that exist, it is useful to organize collections according to materials, manufacturer, the purpose of issue, and subject matter. The combination of perspectives helps to organize these objects that are notorious for their ad hoc creation. There were three types of znachki as defined by the purpose of their issue: memorial, jubilee, and souvenir.[44] Memorial znachki honored persons or events (including party congresses, seminars, and scholarly readings), marked anniversaries, or commemorated deaths. The jubilee znachki honored the anniversaries of births and events, usually at ten-year intervals. Souvenir pins came from municipal organizations, museums, sports palaces, and metros to generate revenue and publicity for those places. The sooner a znachok appeared relative to the occurrence of a given event, the more likely that the pin was one-of-a-kind. Sets of pins, sold in a box, or attached to a velour card, usually appeared on the market later, coinciding with an anniversary or after a given program concluded.

The manufacture of znachki was a decentralized endeavor, relying on the issuing organization to commission specific pins on its own instead of going through a central authority. This decentralization was the significant difference between stamps and znachki. The Ministry of Post and Telegraph controlled both the design and distribution of stamps. No single central authority presided over the many organizations that made znachki. When the space program gained popularity, these already decentralized manufacturers expanded their operations to capture their share of the emerging market. The second edition of the first catalog of space znachki published a list of twenty-three known manufacturers that included traditional government pin makers such as the Moscow and Leningrad Mints and surprising enterprises such as the All-Russian Choral Society.[45] Each manufacturer displayed its identifying mark on the back of the pin. Given the range of reporting structures that these twenty-three organizations represented, it was unlikely that they received their content information or design directives from a single source or that any one single body or individual reviewed or approved the designs. They were responding to internal values and cultures that had formed throughout the history of the Soviet Union. In the absence of the central power that stamps had, the prevailing message of the znachki was an indication of the characteristic values of disparate middle managers at various enterprises throughout the USSR.

When enterprises and organizations in the spaceflight industry awarded znachki to distinguished individuals for service and achievement, the pins had no marks indicating their origins.[46] There had been no need to do so because the recipient would likely hold onto this award or remember the circumstances of the award. Souvenirs and collectible pins, however, had a price stamped on the back of the pin—usually a number followed by the letter "K" next to the manufacturer's mark. This number indicated the initial sale price in kopecks and the fact that their distributors anticipated earning money on their sale.[47] Znachki that were issued for more traditional purposes, to honor participation in a project, were products of controlled production and distribution. Because they were not for sale, they did not have the price marking. These issuing organizations ranged from committees within the Academy of Sciences to museums, museum associations, and professional societies. With time, both types have found their way to the collectors' market. The awarding organization usually issued a certificate with the pin and kept track of the recipients.[48]

The first officially sanctioned space znachki were memorial-type pins that the Shcherbinsk Smelting Factory produced.[49] They reflected the limited information available on the satellites, but as news services published illustrations of the first Sputniks, the earliest znachki makers adapted their designs to incorporate small images of the satellites.[50] These pins depicted the first three Sputniks accurately. This accuracy was in sharp contrast to later depictions of Gagarin's Vostok craft that received no accurate public depiction until 1967. However, as early as 1958, eager visitors could see models of the first Sputniks on display at the Brussels World's Fair and the Moscow Exhibition of Economic Achievements that appeared that year, as well as the Soviet National Exhibition in New York in 1959.[51] Rectangular or circular, and straightforward in design and message, these pins matched the somewhat reserved claims that the Soviet press made about the Sputniks. They marked Soviet proficiency in science and technology but made limited claims on Soviet world leadership beyond spaceflight. These claims did not appear until there was a human champion to make them.

Yuri Gagarin's flight around the world on 12 April 1961 led to spontaneous celebrations in Moscow the next day that matched the celebrations of victory in World War II.[52] Moreover, although party officials had authorized the printing of stamps before his mission, their release was contingent on the success of the mission.[53] Although the Ministry of Communications could control the dissemination of stamps, it had no statutory or institutional authority to impose an embargo on znachki. The pins did not fall under strict regulations that governed the production of stamps and coins that had immediate monetary and trade value. Gagarin's flight inspired the first spontaneously produced znachki. During the festivities in Moscow, participants appeared with small (70-by-55-millimeter or 70-by-45-millimeter) paper portraits of Yuri Gagarin on their chests that individuals had taken from newspaper photographs of the cosmonaut.[54] Within days, enterprising producers refined the idea, placing cellophane over a smaller picture (18 millimeters in diameter) of Gagarin. These photographs came from newspaper reports of his flight, as no prelaunch photographs were released.

Three unofficial pins preceded the Shcherbinsk Factory's production of a steel and enamel pin that portrayed Gagarin.[55] Subsequent Gagarin pins from other manufacturers added the detail about his flight as it became public, including approximate launch and landing sites. The illustrations of the launch vehicle and spacecraft were fictional and did not resemble the real objects. Unlike the previous attempts to conceal the design of the *Vostok* spacecraft, this representation reflected the popular conception of a rocket ship.[56]

Nevertheless, this is consistent with all other illustrations of Gagarin's flight in the popular media, including stamps, posters, and cartoons. The earliest pins had to simplify the graphic of his flight in response to the technology of making znachki. The elegant lines of engraving and large pieces of text did not translate into inexpensive metal and enamel. The images became increasingly abstract.

While Gagarin's flight placed the most significant emphasis on the accomplishment of human spaceflight and caught officials unprepared for the popularity of the pins, more numerous, official znachki appeared immediately after *Vostok 2*. On 6 August 1961, German Titov became the first human to orbit Earth for more than a day. The design and complexity of the Titov pins had more detail than did the Gagarin ones. They immediately sought to identify the complexity of Titov's flight, illustrating multiple orbits around the globe. It was impractical to illustrate all seventeen orbits that Titov made around Earth. Four distinct lines around a representation of a globe on one pin made the point that he had made multiple orbits.

The efforts to provide informative yet aesthetically appealing pins continued through the *Vostok* and *Voskhod* programs. The next four *Vostok* flights were paired flights that implied a maneuvering capability that the Soviets had yet to demonstrate. Translating the dual flights of *Vostok 3* and 4 and *Vostok 5* and 6 into pins resulted in similar designs. The Soviet news agency TASS had emphasized the near-simultaneous timing of the flights to the point of insinuating active rendezvous—a technical feat that the Soviets were not capable of. The pins that represented those flights echoed

Figure 8. A portrait pin of Konstantin Feoktistov, *Vostok* and *Voskhod* spacecraft engineer and *Voskhod* cosmonaut. From the collections of the Smithsonian Institution National Air and Space Museum.

this representation. One of the first pins featuring the flights of Nikolayev and Popovich on board *Vostok 3* and *4* show two stylized ships emerging from the tip of a sickle suspended above Earth. The flights of Bykovskii and Tereshkova (*Vostok 5* and *Vostok 6,* respectively) received similar treatment with pins showing two rocket ships orbiting Earth and two ships flying away from Earth in similar trajectories. The Leningrad Mint produced each of these two pins under an official commission.[57]

In the year after the last of the *Vostok* flights, Soviet space designer and manager Sergei Korolev had demanded a redesign of the interior of the spacecraft to accommodate multiple cosmonauts. Soviet engineers built the *Voskhod* spacecraft from the skeleton of the *Vostok* spacecraft but announced it to be an entirely different species. The plan had been to use the *Voskhod* as a challenge to the Americans' maneuverable *Gemini* spacecraft, but the *Voskhod* was little more than a gutted *Vostok* and could not maneuver in space. There were only two *Voskhod* missions, *Voskhod 1* and *Vosk-*

hod 2, which flew in October 1964 and March 1965, respectively.[58] One of the first *Voskhod* pins illustrated the literal meaning of the spacecraft name, using the imagery of a sunrise underneath a soaring spacecraft. Other pins emphasized the multiple crews of this first *Voskhod* with images of three helmeted cosmonauts along with the rocket ship, such as one from the Mytishchinsk Experimental and Souvenir Factory. In fact, the three cosmonauts did not wear helmets or spacesuits during the *Voskhod 1* mission as the cramped room in the spacecraft did not allow it.

The last flight of the *Voskhod* spacecraft provided an opportunity for greater artistic representation because one of the crewmembers, Aleksei Leonov, was himself an artist and drew his impressions of his mission while in orbit. Moreover, his mission reminded the public of ancient dreams of spaceflight. Leonov became the first person to venture outside of a spacecraft and take a walk into space floating alongside his spacecraft. The first znachki to represent Leonov's mission used more abstract images to illustrate the flight than had previous ones. Most depictions focused on the distinguishing aspect of the flight, Leonov's walk in space. In contrast to the grainy and unfocused photographs of Leonov's spacewalk, the pins illustrated a crisp image of a human flying through space, untethered and symmetrical.

These examples of small representations of the Red Stuff made their way into every Soviet household. People could hold, examine, compare, and trade them according to their own individual preferences. At the other extreme of scale, there were movements to create places that reflected the Red Stuff that people could visit and inhabit.

Titanium, Modernist, and Large: Museums and Monuments

By the 1930s, the modernist art, architecture, and design movements of revolutionary Russia had all but ceased. Some proponents of the movements left the country, others gradually burrowed into their respective professional infrastructures, adapting to the changing political aesthetics of the time. Experimentation with the modern idiom continued for some time into the 1930s, but it did not meet with any degree of success. Architects and designers had conceded to official sentiments that pseudo-classicism was closest to the Russian ideal and seemed to abandon the modernist movement in the design and architecture of the 1920s and 1930s.[59] Importantly, in all cases, Stalin's imposition of socialist realist aesthetics in art (including fine art and literature) put an end to the political discussions that had previously intertwined with early concepts of modernism.[60] Stalin's death in 1953 emboldened the dormant modernists. In November 1955, the Soviet Union of Architects renounced "ornamentalism."[61] This time, the proponents were not arguing for an ideological path to communism but argued that modernism was the appropriate style to complement de-Stalinization and the construction of communism.[62] In a strict sense, these new movements in design and architecture, being purely aesthetic, were not the same as the movements of the 1920s and 1930s. The minimalism of these influences was evident in much of everyday life in the USSR in the 1960s. It was apparent in the material culture of the space program. However, the messages that these objects carried were like their socialist realist analogs. This new design strategy had influence over the buildings and exhibits where the public could see the new space achievements of the USSR.

The themes and approaches to space exhibits on the USSR space program also followed distinct geographical tracks: those on display within the USSR attracted the Soviet population, and those in traveling displays were designed for non-Soviet audiences. The best-known examples of each are the Kosmos Pavilion at the VDNKh and the Soviet exhibitions at the 1958 and 1967 World's Fairs in Brussels and Montreal. World-traveling exhibits and the large-scale domestic ones were nearly identical in content, including objects and descriptive components. Ironically, there was greater diversity in content among the domestic displays than in larger urban areas like Moscow. This diversity was in the size of the exhibit and the sources of their displays. Large exhibits, such as the Kosmos Pavilion, relied on the national organizations, such as the USSR Academy of Sciences and Znanie Society for their displays.[63] Smaller exhibits that local organizations built, for example at the Moscow Planetarium, had limited resources for displays. These public displays carefully balanced the needs for secrecy and the celebration of the space program. While they avoided the pitfalls of direct comparisons between the hardware of the US and the USSR, they offered the most significant opportunity for the comparison of each Soviet accomplishment, thus presenting the Soviet Union in a favorable light.

The First Soviet Space Museum

Plans for a large-scale space museum, one that would match the size of what was to become the Smithsonian Institution's planned National Air

Museum, began very early in the Soviet Union, predating American plans, including NASA's earliest exhibits and the Smithsonian agreeing to combine the history of aviation and spaceflight.[64] In 1959, in the newspaper of the Soviet Writers' Union, *Literatura i zhizhn'* [Literature and Life], members of the Soviet Academy of Sciences proposed that the construction of a museum dedicated to the emerging history of spaceflight be located in Kaluga.[65] The three-stage proposal provided the justifications for a museum and the audience that would benefit from it and the location. The authors began their proposal with a tribute to past Soviet accomplishments in science and technology, laying out the Soviet aerospace progress: "The first jet airplane was developed in the USSR in 1942, long before it appeared abroad. Soviet scientists opened spaceflight to humanity by launching the first artificial satellite of the Earth on October 4, 1957."[66] Historically inaccurate Soviet rhetoric aside, the writers continued their argument with the justification that a museum would serve the interests of the working population. The museum would create a physical display that would reward the population's efforts in the form of a tribute to the accomplishments of the state:

> We propose to preserve for the laboring generation all that the Soviet population has done, is doing, and will be accomplished in the future in rocket technology and spaceflight. This is why it is necessary to create a state museum of the history of rocket technology and spaceflight.[67]

Finally, they suggested locating the museum at the same place, Kaluga, where local officials had built a museum memorializing Konstantin Tsiolkovsky inside his house and workshop the year after his death.[68] A new museum would reinforce the legend of Tsiolkovsky and strengthen the claim of the Soviet tradition in rocket technology.

The signatures of prominent individuals over a clear rhetorical argument for establishing a national museum in a low circulation but politically influential newspaper might have meant that that argument had ceased. However, even within the Writers' Union, there was some room for dispute. The Tsiolkovsky House Museum in Kaluga had been a local operation that relied on municipal funds to maintain the site and generous donations from Moscow colleagues to sustain the publications and research activities of the organization.[69] The local community initially balked at the suggestion that an already taxed infrastructure in the small town could take on more burdens. A state museum without state funding would be out of the question, as A. Laskina, a member of the staff of the existing Tsiolkovsky House Museum, wrote:

> In the last year, 15,397 people have visited Kaluga from 356 towns. People have come from all over the world. In ten years, the number of visitors has doubled. Comments from the visitors express their concern about the resources of the museum. Museum Director V. P. Akimov and the research staff at the museum, V. S. Zotov give tours. Money for the exhibits has come from colleagues in Moscow. It is entirely clear that such a severe historical undertaking must be done, not on the regional scale, but on the all-Union [nationwide] scale.[70]

Laskina was keenly aware of the local burdens of the enterprise that the academy members had proposed. What she may not have been aware of was the broad support for the museum within the Soviet space establishment. While his official anonymity prevented publication of his signature on the April letter, Sergei Korolev had elicited the proposal from his better-known colleagues.[71] Korolev played a driving role in establishing the museum in the rocket pioneer's hometown, but he could not publicly argue that the museum was a prize for the people of Kaluga in honor of the Soviet state's accomplishments in spaceflight. Korolev's anonymity dictated that he use surrogates to campaign for the museum. Within two years of the publication of the first *Literatura i zhizn'* article, the project had a public advocate in the person of Yuri Gagarin, who visited Kaluga immediately upon his return from his first goodwill tour abroad, as a tribute to the town's native son.[72] On 13 June 1961, Gagarin laid the cornerstone for the large space museum in the city of Kaluga.

Plans for the museum were subject to the architectural competitions common for the era; constructivist architect Boris S. Barkhin, who initially gained fame as the chief architect of the *Izvestiia* newspaper building during the mid-1920s, led the team. The team designed a building that became a state prize-winning structure. Korolev did not live to see the building open, however. Construction took more than six years, and the building formally opened on 3 October 1967 (the day before the tenth anniversary of the *Sputnik* launch), nearly two years after Korolev's death in January 1966 and close to six months before Gagarin's death.[73] The misfortune of Korolev's premature death might have had implications for the Tsiolkovsky Space Museum in Kaluga. As chief designer of the space program, Korolev retained physical control of the hardware of the space program. Flown ob-

Figure 9. Exterior view of the Tsiolkovsky Museum in Kaluga, Russia. Vyacheslav Erdneev/Alamy.

jects that were national treasures, such as Yuri Gagarin's flown *Vostok* capsule and Valentina Tereshkova's *Vostok* ejection seat, were accumulating in warehouses within the suburban Moscow space complex. Historians with access to Korolev's papers have determined that he was collecting documents as well to support a robust historical record for a museum.[74] None of these items has been on full public display.[75] There remains a possibility that Korolev had intended to make these objects and others part of the permanent collection of the Tsiolkovsky Space Museum. Unfortunately, all that we know of Korolev's intentions was his advocacy for the museum in Kaluga, and we have little understanding of his motivations and rationale.

The geometrically shaped building, featuring parabolas and trapezoids, sits atop the bank of the Oka River near the edge of Kaluga. The museum loosely resembled the constructivist architecture that lost favor during the 1930s and was somewhat out of place among the city's existing architecture of traditional wooden Russian houses and imperial palaces. One can only assume that this contrast made clear that from the nineteenth-century origins of Tsiolkovsky, the modern space age had emerged. Of course, the architecture fit into the growing movement among architects to stamp out the last vestiges of Stalinist pseudo-classicism.

Evolution of a Space Exhibit Icon: The Kosmos Pavilion

During the time of the construction of the Tsiolkovsky Space Museum, other space exhibits in the Soviet Union would become more famous. The 1960s plans for an elaborate, if not centrally located, museum of spaceflight notwithstanding, public exhibitions of Soviet space achievements were located primarily in one place within the Soviet Union during the 1960s. The Vystavka dostizhenii narodnogo khoziaistva [Exhibition for Economic Achievements, VDNKh] became synonymous with space exhibits while the city of Kaluga built the Tsiolkovsky museum. At this site inside Moscow's city limits, the tone and scope of space exhibitions evolved. When a parallel and equally active program of exhibitions began outside of the Soviet Union, those international exhibitions resembled the VDNKh.

The VDNKh in Moscow was the barometer of official Soviet pride in its agricultural, scientific, and technical accomplishments. The exhibition first opened in 1939 as the Vsesoiuznaia selsko-khoziastvenaia vystavka [All-Union Agricultural Exhibition, VSKhV], a celebration of Stalinist collectivization. The purpose of the 1939 exhibition was to demonstrate that there was no famine in the country, only abundance resulting from collectivization and mechanization of agriculture. As a national forum, the exhibition park was a successful attempt to engage the Soviet population in the accomplishments of the early Stalinist state.[76] By coming to the park to reflect on things not available in their everyday lives, the Moscow public was tacitly accepting the lies of abundance.[77] The architectural and artistic themes and architecture of the park date from the high-Stalinist period, featuring exemplars of socialist realism and neo-classicism. Vera Mukhina's sculpture *Rabochi i kolkhoznitsa* [Worker and Woman Collective Farmer] marked the entrance to the park, after representing the USSR at the Paris World's Fair in 1937.[78]

The original displays at the All-Union exhibition betrayed the fact that agriculture dominated the Soviet economy, featuring produce, apple groves, and garden plots in its scientific displays and including halls featuring folk art and artifacts from throughout the country in its buildings. When the occasion demanded it, scientific exhibitions had their place, too. A year before the official opening of VSKhV, the Soviet Arctic Exhibition opened in 1938, celebrating recent accomplishments in exploring and navigating the North Pole. It carried the motto, attributed to Sergei Kirov, "There is no land that Soviet power cannot transform for the good of mankind." The exhibit contained genuine artifacts from the polar expeditions.[79] This pavilion maintained the theme of agriculture and ethnographic displays. It was only after the post–World War II reconstruction of the exhibition halls at VSKhV that industrial accomplishments began to dominate the park, replacing the previous celebrations of agricultural accomplishments and folk culture.[80]

The prehistory of the location of spaceflight exhibits at VDNKh is significant because the place has always displayed pivotal technology in the Soviet Union. Its predecessor, the Mechanization of Agriculture Pavilion at the exhibition, had a long design history.[81] This design was convenient for installing the largest examples of agricultural machines, such as advanced combines and tractors. The building opened in 1939. Soviet monumentalist sculptor Sergei Merkurev designed the statue of Stalin, which stood in front of the building.[82] The Mechanization of Agriculture Pavilion changed over the course of the next two decades. After the evacuation of the exhibits to the Siberian city Chelyabinsk during World War II, the All-Union Agricultural Exhibition reopened in Moscow in August 1954 and experienced several administrative reorganizations in the next decade.[83]

In 1966, outside the pavilion, a *Vostok* rocket took the place left empty by the removal of the statue of Stalin.[84] The launch vehicle, unlike most of the contents of the pavilion, was hardware that the military industry provided to the Academy of Sciences for display. This placement was a direct and straightforward indication that a new icon had replaced the cult of personality. Space was the new focal point of the state. Where Stalin had stood in front of the mechanization of agriculture as a proud accomplishment of collectivization and industrialization, the product of anonymous rocket engineers, the *Vostok* rocket, now presided over the Soviet mastery of Cold War–era technology.

The Space Pavilion (or as it was affectionately known in English, the Kosmos Pavilion, that being the one word that most Americans recognized) did not exist until 1966. However, displays of spaceflight first appeared at VDNKh in 1958, gradually edging out the tractors and combines. That year, models of the first three Soviet *Sputnik*s first appeared in the main entrance hall of VDNKh. The models were moved into a 100-square foot exhibition in the Mechanization of Agriculture Pavilion a year after their original appearance at VDNKh.[85] The display models of *Sputnik, Sputnik 2* (the spacecraft that carried the dog Laika into space), and the heavily instrumented *Sputnik 3* attracted a steady stream of visitors to the main exhibition hall.[86] The models lacked detailed information on the missions and gave no insights into the interior workings of

Figure 10. Exterior of the Kosmos Pavilion at the Exhibition of Economic Achievements, Moscow. Ruslan Kalnitsky/Alamy.

the satellites. Unlike the previous displays that had occupied the pavilion, these exhibits were not working hardware. However, they provided the first public exposure to anything remotely resembling the real objects that had been the focus of press celebration. Not only the Western journalists but also Muscovites were hungry to see the material evidence of the USSR's accomplishments in space. One Western journalist described local interest as a "steady stream of visitors."[87]

A model of *Vostok*, the spacecraft that carried Yuri Gagarin into space, had its first public display on 29 April 1965, in the Mechanization of Agriculture Pavilion, alongside the three *Sputniks*.[88] Instead of revealing information about the history and technology of the first human-crewed spaceflight, the first display of a *Vostok* model represented a carefully crafted effort to conceal details about the human spaceflight program in the Soviet Union by camouflaging details about the design legacy of *Vostok* and its technical properties. The *New York Times* reporter who first wrote about the model interviewed, perhaps unknowingly on his part, the spacecraft designer himself and only received the most cursory description of the spacecraft from that knowledgeable engineer and cosmonaut. At the time of his *New York Times* interview, Konstantin Feoktistov, the chief designer of *Vostok*, was famous as the flight engineer of the first *Voskhod* spacecraft that carried three men into orbit on the first multi-man mission in October 1964. In his recent memoirs, Feoktistov acknowledges that his flight on board the *Voskhod* was, in fact, a reward for redesigning the *Vostok* interior to accommodate three men.[89] Feoktistov did not attempt to describe the operations of the spacecraft.

Another model of the *Vostok* soon appeared also at the twenty-sixth Salon International de L'Aéronautique et de l'Espace at Le Bourget Airport during the Paris Air Show in June 1965.[90] At that time, the Soviet portrayal of the craft was equally cagey regarding the technical details of both Soviet spacecraft. Yuri Gagarin accompanied the *Vostok* model to the Paris Air Show and asserted that the *Vostok* and *Voskhod* craft were "of entirely different design," a lie that the Soviet space establishment would perpetuate for another generation.[91] The *Vostok* at the Paris Air Show served as a decoy, hinting to the world that great technical advances separated the displayed *Vostok* from the still shrouded *Voskhod*. It would be almost a gen-

Figure 11. This is a scale model of the *Vostok* capsule that the Tsiolkovsky Museum first loaned and then donated to the Smithsonian Institution National Air and Space Museum. At this scale, Yuri Gagarin would have to be the size of a newborn to fit into the capsule. From the collections of the Smithsonian Institution National Air and Space Museum.

eration later that the world would read and hear Soviet engineers concede the nearly identical designs of *Vostok* and *Voskhod* and Konstantin Feoktistov would admit the high level of risks taken in refitting a one-person craft to carry three.[92]

These first two sightings of models of the *Vostok* were revelations, albeit minor ones. Until 1965, the few published photographs of the actual *Vostok* were of the protective conical shroud that covered the spacecraft from launch until entry into orbit, revealing no more than the external dimensions of the craft. Before that time drawings had included inaccurate representations of the spacecraft and its functions.[93] This deliberate trail of misinformation served to hide not one, but many secrets about the first human spaceflights. First, the USSR was engaged in a Cold War against the United States and a culture of secrecy prevailed. Even though the Americans were displaying flown *Mercury* spacecraft throughout the world, it surprised no one that the Soviet Union adhered to secrecy at that time. The tradition of secrecy combined with the fact that the space program was an auxiliary part of the ballistic missile program of the Soviet Union, and not housed in an independent, non-military agency like the American NASA. The USSR had no buffer agency to protect even more valuable military and strategic secrets than the design of spacecraft while publicizing its achievements. Beyond the military and political culture of secrecy, the Soviet Union feared the technical comparison with the American *Mercury* and *Gemini* spacecraft. Given the sophistication of Soviet intelligence of the time, Soviet engineers and managers must have understood their hardware would compare poorly in technological sophistication.

The installation of the *Vostok* model in summer 1965 began a content shift at the Mechanization of Agriculture Pavilion into Moscow's first permanent space exhibition.[94] Direct administrative control of the pavilion was under the Soviet Academy of Sciences' Council on Exhibitions, which had directed the content of the scientific, industrial, agricultural, and ethnographic displays at the VDNKh since its rededication in 1959. The Kosmos Pavilion did not devote much of its real estate to displays on Soviet accomplishments in human spaceflight, however. Only the rear, domed portion of the hall featured the activities of humans in space. Most of these exhibits represented scientific activities in spaceflight through high-quality, full-scale models of scientific spacecraft, starting with the 1958 model of the *Sputnik 3* satellite. These displays reflected the scientific interests and expertise of the Academy of Sciences. The pavilion visitors relied on paid staff to explain the displays that provided little interpretation themselves.

The guides were busy with this activity, claiming to complete as many as 150 tours per day.[95]

Memorial Monument Complex

As a form of the Soviet-style recapitulation, another monument to the space program emerged near VDNKh in the Moscow skyline during the early 1960s. On the seventh anniversary of the launch of *Sputnik* in 1964, a 350-foot (107 meters), titanium-clad, stylized rocket and plume emerged just outside the entrance to the park to commemorate the Conquerors of Space.[96] The monument itself was as much a tribute to the revival of unadorned, modernist architecture as it was a tribute to the new Soviet space age. However, the titanium spire was only one component of an architectural complex that comprised the monument to the Conquerors of Space.

To tie the memories of revolutionary explosives experts to the contemporary activities of the engineers and technicians, the monument also included an alley lined with socialist realist tributes to the legends of Soviet rocketry and spaceflight. Soviet sculptor Lev Kerbel, best known for his portrayals of Russian wartime suffering and Lenin,[97] created busts of Russian and Soviet scientists and engineers that lined the walkway leading up to the monument. Engineers and technicians gathered every morning near the base of the monument outside VDNKh to wait for buses that would carry them to work at the space design bureaus in the northeastern suburbs of Moscow where in Podlipki (later Kaliningrad and now Korolev City) much of the space program was based. It was commonly known, even during the period of relative secrecy, that the well-made apartment blocks throughout VDNKh and the botanical gardens had been built to house the growing aerospace community in Moscow. The typical public did not know before 1966 that the apartments had grown around Sergei Korolev's existing single-family home, where he lived until his death.[98] When the Conquerors of Space obelisk opened, the organizers promised that a museum would soon open at its base.[99] Almost seventeen years later, the Memorial Museum of Cosmonautics opened in an excavation underneath the base of the monument in 1981.[100] If the Kosmos Pavilion was established to display models of Soviet spacecraft, the Memorial Museum was to provide an immersive experience. The main gallery of the museum had low ceilings and lighting. The museum itself contained a mixture of models, real artifacts, and artistic interpretations of spaceflight. In many

Figure 12. This is an interior shot of the largely underground museum that sits at the base of the Conquerors of Space monument that features a pioneers' alley and a titanium rocket that dominates the skyline of northern Moscow. This shot shows the stained-glass and metal monument to Gagarin inside the museum. Stanislaw Tokarski/Alamy.

ways, the museum appeared to be a church of spaceflight. The museum experience incorporated stained glass windows and altars featuring images of Yuri Gagarin and offering the visitors a ritualized experience culminating in a light show that flashed images of spaceflight to the soundtrack of contemporary popular music.

The aerospace engineers who left for work each morning at the base of the titanium monument did their jobs among a very different set of museum displays, exhibits that never drew anywhere near the public crowds that the Kosmos Pavilion claimed. These museums only rarely had outside visitors—usually foreign dignitaries and selected school groups. These isolated, private museums housed the remnants of the actual spacecraft and equipment that had flown or could support human life in space. Each design bureau and enterprise jealously guarded its collection of objects that represented the material legacy of its contribution to spaceflight. The resulting displays presented fragmented views of the space program. At the institution that Sergei Korolev had established were the remnants and components of rocket engines and spacecraft. A few miles away at the design and testing facility that developed spacesuits and other life-support equipment, a basement museum stored a complete collection of spacesuits dating from the ones that engineers had designed for dogs to wear during rocket experiments.[101]

The creation of these private museums resulted from the circumstances of the space program more than it reflected a deliberate attempt to isolate them from the public. The Soviet space program operated on a shoestring budget that did not produce secondary backup hardware. Engineers were using all hardware that was available to them to continue testing and further refinements of their projects. Recovered hardware was precious to them. After much of the flown hardware did not survive the destructive post-flight testing, what remained rarely left the factory of origin. Under the supervision of an individual who would collect and arrange the artifacts for the edification of his colleagues, a legacy display developed. The purpose of the legacy display was both to reassure old-timers of their accomplishments and to educate new workers about the heritage of their mission.

It was not until the 1980s when Gorbachev demanded financial accountability of state-run organizations that these closed museums promoted visitation from the public. At that time, they had to justify their existence through standards of accounting that no longer allowed manufacturing enterprises to support these pet projects financially solely for internal use. After the collapse of the Soviet Union, these museums took on the role of educational institutions and then began measuring their outside visitation as an indication of public service.[102] Between the world of the Kosmos Pavilion and the private enterprise museums, there was a world of small museums that sprouted up during this period, each filling a specific demand from an audience or a patron. For example, the display that ultimately became the Gagarin Spaceflight Training Center Museum in Star City, just outside of Moscow, had started under the tutelage of Yuri Gagarin himself in 1967.[103] He initiated collecting for the museum as a repository for the gifts that cosmonauts received over the years from local and foreign admirers. He donated the first object to the museum's collection, a small, metal cast figure of a smelter (reflecting his early training as a smelter) that the people of Czechoslovakia had given to him during his post-flight tour.[104] The museum took on a somber and personal tone when Gagarin died in a training flight accident in 1968. At that time, the Commandant of the Cosmonaut Corps, Nikolai Kamanin, decreed that everything associated with Gagarin at the center become part of a memorial museum.[105]

Kamanin oversaw the re-creation of Gagarin's office on the site of the museum in Star City and the official opening of the Gagarin Museum at the Spaceflight Training Center. This museum served more like a memorial than a museum, where objects provided little or no context and little information about their background. In Gagarin's memorial office, the clock permanently displays the time when Gagarin died.

The Red Stuff in Film: From Instructive Present to a Usable Past

Film directors' attempts to portray the Red Stuff operated under the regulations of two distinct artistic guidelines, those of science-fiction writing and those of the film industry. Even after the death of Stalin, Khrushchev maintained only slightly relaxed restrictions on science fiction. The first Congress of Science Fiction Writers in 1958 made these limitations public. Under these guidelines, science fiction should not violate the dictates of scientific socialism. That translated into a restriction that in written or filmed science fiction, the technology could not drive society to the next social level but could only be a tool with which society carried on with its already established path into the future. Even the most celebrated science-fiction writers of the time adhered to this message. Ivan Efremov set his 1957 novel *Andromeda Nebula* in the distant future. He told a tale of two advanced civilizations existing under communist utopias that meet peacefully. As the first to venture into the science-fiction genre after the death of Stalin, Efremov was understandably cautious in his writing, so much so that some of his peers thought his writing dry.[106] Those writers who diverged from this habit and alluded to criticism of the current state, or implied that social improvement was possible, were expressing dissent. Moreover, there was a third category of writers who deliberately navigated between what was and was not acceptable to create semi-subversive science-fiction tales that undercut the optimistic tone of the early 1960s. The Strugatsky brothers, Akady and Boris, were the prime example of this even though their writing implicitly veiled criticism of the status quo. The underlying rationale of the mandate from the science-fiction writers' organization was that there was no need for social progress, revolution, or war in science fiction because, without the brutality of Stalin's rule, no adjustments were necessary.

If navigating the new rules of science fiction were not enough, filmmakers also had to navigate the new rules of Soviet cinema under Khrushchev, which, though promising, were far less clear. As film historian Josephine Woll has shown in her discussion of 1960s Soviet cinema, the films of

Khrushchev's Thaw had their own guidelines that directed them toward introspection and away from potential subversion of the Soviet experience. The bulk of the popularly and critically accepted films made some mention of the war experience and a baseline from which to measure contemporary conditions. Even those that were introspective and poignant carried high expectations for the USSR during the early 1960s, relying on the memories of World War II.[107] Three of the best-known examples of these films are *The Cranes Are Flying, Ballad of a Soldier,* and *Ivanovo detstvo (My Name Is Ivan)*. Each of these films maintained a close connection to the Soviet war experience. All, in subtle ways, challenged the official Soviet memory of the war by pointing to injustice, cruelty, and sacrifice. In each case, these Thaw films established a path to safe and subtle criticism of the USSR while honoring the memories of the Soviet people.

Three theatrical films produced in the USSR in the earliest years of the human spaceflight era relied on an interpretation of the Red Stuff to carry their message. Each attracted positive popular response from the Soviet public, though some skirted official Soviet film policy. The public had been starved of science-fiction movies since Stalin. This release of a new type of film coincided with the relaxing of theatrical filmmaking rules and a new, postwar generation of directors that precipitated the rush of Thaw films of the Khrushchev era. The first came from a frustrated engineer who became a filmmaker, the second was an animated rekindling of an early twentieth-century Futurist poet, and the third was a socialist realist attempt at science fiction.

Pavel Klushantsev's Films

It is somewhat ironic that the most accurate depiction of human space-flight in the USSR came a from a film director who ultimately suffered denouncements for his politically disruptive depiction of Soviet cosmo-nauts. Pavel Vladimirovich Klushantsev was born on 25 February 1910, in St. Petersburg, Russia. The timing and location of his birth proved fateful for the rest of his life. Born to an aristocratic family, he grew up during the cultural and intellectual whirlwind that defined Russia for the first two de-cades of the twentieth century. According to his surviving family members in the early twenty-first century, his first desire had been to study engi-neering in Leningrad.[108] Unfortunately, Klushantsev's coming of age co-incided with Stalin's purge of science and engineering and the consequent subordination of everything to state authority. The sons and daughters of middle-class engineers had encouragement to enter engineering schools to build the Soviet future. The sons and daughters of the aristocracy, even if their families had supported the Bolshevik Revolution, found themselves behind the offspring of workers and peasants when it came to educational opportunities.

Pavel Klushantsev never became an engineer, but he spent most of his working life translating the work of other Soviet engineers to the public in film and later in books. In 1930, Klushantsev attended the Leningrad Fototekhnikum (Leningrad Photography Technical School), a leading motion-picture production company that existed primarily to strengthen the USSR's "most important art." He went on to work for four years in Minsk at Belgoskino, the National Belarus Cinema, as a cinematographer. In 1934, Klushantsev returned to the city of his birth at Lenfilm/Lennauchfilm (Leningrad Film and Leningrad Scientific Technical Film), the leading film production company in Leningrad, as a director and producer.

Lennauchfilm is probably the most famous unknown film studio in the world. The production company, founded on 11 February 1933, was an offshoot of the Leningrad Fototekhnikum. The production company's initial mandate was the film documentation of the historical and cultural legacy of European Russia and gained an international reputation for producing films on the art collections of the USSR in Leningrad and Moscow. Like most aspects of state-induced activities in the 1930s USSR, Lennauchfilm redirected its mission toward military and civil defense training films, reflecting the general mobilization of society. By the mid-1930s, Lennauchfilm was a leading producer of military training films as part of the national mobilization toward war. During the war, the film studio's mission turned toward the film preservation of cultural heritage. Seven camera operators and staff filmed under the extreme conditions of the blockade, creating films of the front lines, warships, and area aircraft. The studio was the sole source of movie footage from blockaded Leningrad. Footage from Lennauchfilm appeared in Jeremy Isaacs's acclaimed multipart World War II documentary, *The World at War.*

After the war, Lennauchfilm began work on the genre of educational science films. This turn in direction was part of the plan to return to their original mandate to document Russian and Soviet culture and knowledge. The studio had several thematic directions elaborated for decades: physiology of higher nervous activity, fine arts, Russian history, and outer space. The last theme was the one to which Klushantsev devoted the rest of his film career.

During the 1950s and early 1960s, he directed a series of six spaceflight-related documentary films that reiterated the realistic style of science-fiction movies that emerged in Russian fantasy films in the 1930s. Using a combination of scientific lectures and advanced special effects to portray the near future, the filmmaker offered Soviet viewers a style of film that Western viewers saw in the televised collaborations between Wernher von Braun and Walt Disney in the *Man in Space* series. Klushantsev produced his films for a Russian audience, who viewed his unique combination of science education and science-fiction films in movie theaters. He did not have the advantage of a ready-made theme park or television distribution that Disneyland had. Klushantsev articulated a direct connection between the current state of Soviet science and engineering and proposed a near future of routine human spaceflight to the Moon, Mars, and Venus. His films represent a significant component of the public culture of spaceflight during the 1950s and early 1960s in the Soviet Union. These movies assumed a technical sophistication among the viewers by presenting basic principles of physics, astronomy, and earth sciences as background to the tales of exploration. In Klushantsev's films, the future was imminent, and technology was about to fulfill its promise. Technological innovations were inevitable, but there was no need for further social improvement. These Klushantsev films offer a peek into what the average Soviet citizen thought the space age was to bring.

In 1957, the young director translated Russian mathematician and rocketry theorist Konstantin Tsiolkovsky's book *The Road to Space* into a film. The resulting film combined historical reenactments, scientific demonstrations, and science fiction projecting what space travel would be like soon. This format became the director's style for the next decade in portraying spaceflight. Klushantsev's style of popular scientific movies that combined documentary with scientific fantasy became known as *nauchpop*, a combination of the Russian words for science and popular.[109]

Although popular cinema never achieved the propaganda effect that Lenin and Lunacharsky had predicted during the Revolution, it did remain a popular diversion from everyday life. Despite ideological mandates and the international popularity of modernist film directors such as Eisenshtein and Vertov, Russian audiences preferred a coherent storyline and rational adventure. As a result, the earliest Soviet science-fiction movies subordinated the ideological aspects of space travel to the fantasies about the appearance of other worlds. Subsequent films attributed a nation's ability to fly in space to personal loyalty and party discipline. This trend continued in the new Soviet science-fiction films at the dawn of the space age and would continue in films and the portrayal of cosmonauts through the collapse of the USSR.

Klushantsev made no effort to create the feeling of introspection about his visions of the near future in contrast to the Thaw films of the time. His films were hopeful, but not too much so. There was no technological drive toward a better future. The personal cost of space exploration, if there was any, was the loss of those weaknesses of pride and sentiment that the Komsomol discouraged. The reenactments of Tsiolkovsky's life in *The Road to Space* featured a historical demonstration of how Tsiolkovsky came upon the concept of staging rocket engines. The old but not elderly Tsiolkovsky was trying to explain the concept to a young student while rowing in a rowboat. To make his thoughts clear, the old man began to throw objects from the boat (buckets, baskets, and finally the oars) over the bow of the boat. As the boat moved forward with each object thrown overboard, the film character Tsiolkovsky explained to his student that each time the force of throwing an object overboard propelled the boat forward, it did so with increasing distance per mass of force because the boat grew lighter as it expended energy.

Theatrical science-fiction film disappeared in the Soviet Union under Stalin, as had science-fiction literature. The 1934 production of *Kosmicheskii reis* [Cosmic Voyage] was the last theatrical space science-fiction film screened in Moscow public theaters until 1960. The film recounted a thinly veiled adventure of a scientist who closely resembled Konstantin Tsiolkovsky and his companions on a mission to the Moon. In 1960 the East German film based on the Stanislav Lem science-fiction story *Der schweigende stern* [The Silent Star] opened in theaters in the Soviet Union under the Russian title *Bezmolvnaia Zvezda*.[110] This new infusion of science-fiction films, beginning with the prescient Soviet film *Nebo zovet* [The Sky Calls], followed later in 1960. It predicted a Space Race between the United States and the Soviet Union to Mars.[111] Like *Kosmicheskii reis* before it, *Nebo zovet* took pains in demonstrating the effects of spaceflight through special effects and set design.

Two years after the release of *Nebo zovet*, Klushantsev presented his fictional interplanetary tale *Planeta bur'* [Planet of Storms].[112] The movie began with a crash. A meteor crashed into one of three spacecraft en route to the planet Venus. The crews of the surviving spacecraft had to decide to explore the planet while waiting for a third craft to join them. They decided jointly that in the name of the party, humanity, and the Soviet Union, they would go ahead with the risky exploration, leaving alone a crewmate and the only woman, Masha, in orbit. With them, the men took the robot and its designer, an American, to the surface. In this movie, the cosmonauts found themselves separated from their spacecraft on the planet Venus, fighting prehistoric animals and surviving erupting volcanoes en route to their spacecraft. The men survived the mission by maintaining Komsomol discipline. The robot was not capable of sacrificing himself for the good of the collective; therefore, the cosmonauts abandoned the robot in a river of molten lava. Meanwhile, on board the sole remaining spacecraft orbiting the planet, Masha struggled to maintain discipline over her desire to commit a pointless act of heroism and remain in orbit. In the end, Masha overcame her emotions, obeyed orders, and aborted her rescue attempt to the planet's surface.

Planeta bur' resembled *Kosmicheskii reis* in content and values. Both films relied heavily on the principles of science education for content, although the latter used wild fantasy in the Venus segments of the movie. The former presented an image of cosmonauts that announced the new age of human spaceflight to the world. *Planeta bur'* demonstrated that cosmonauts took the values of party discipline with them as they explored the solar system. Despite his bold venture, the film met derision at home for its political and social context. Even with his scientific realism, Klushantsev had ventured too far in *Planeta bur'*. His vision of the near future combined both a Komsomol reality and international cooperation with the United States. It is not very clear whether he was at all aware of the emerging Moon race or even President Kennedy's May 1961 declaration that the United States was sending men to the Moon before the end of the decade. In any event, his mere anticipation of US-Soviet cooperation in space violated the clearly stated guidelines for science fiction.

Planeta bur' gained the most fame of Klushantsev's films in the West because it was edited and expanded by producer Roger Corman for American distribution as *Voyage to the Prehistoric Planet* (1965) by Curtis Harrington and as *Voyage to the Planet of Prehistoric Women* (1968) by director Peter Bogdanovich. The original scenes drew the most acclaim. A self-taught special effects engineer far ahead of his time, Klushantsev devised many effects and techniques used by major motion pictures for decades to come. He still carries a reputation for his meticulous design and creation of "John the Robot" featured in *Planeta bur'*, with over forty-two points of articulation on its major body joints, one of the most technically elaborate robot costumes of its time.

Despite the international interest in the movie, *Planeta bur'* did nothing to improve Klushantsev's lot within the Soviet film industry. Before World War II, he had committed the crime of having been born to an aristocratic family. During the Cold War, Klushantsev had committed two equally career-killing crimes—internationalism and sentimentalism. It is easy to forget that this film had its theatrical release during the hottest years of the Cold War. The Soviets had shot down a U-2 spy plane and its pilot Francis Gary Powers; Nikita Khrushchev had pounded his shoe on the desk at the UN in 1960. US president Kennedy had announced his country's plan to send men to the Moon, and the Berlin Crisis had culminated in the erection of the Berlin Wall in August 1961. During 1962, the Soviets deployed nuclear missiles in Cuba. This was no time for sentimental science fiction about multinational crews exploring the solar system. If that had not been enough formal accusation to prevent Klushantsev from continuing in theatrical films, unbeknownst to him, five Soviet women had just begun training to become the first woman in space. Klushantsev had portrayed his female cosmonaut Masha as weak and emotional, but technically competent.

The political and social message of *Planeta bur'* notwithstanding, Klushantsev demonstrated outstanding special effects talents in this movie. He had his artistic crew create scale models for shooting the external scenes for the film. He staged as many scenes as possible outdoors, using illustrated backgrounds only on rare occasions. Details in Klushantsev's spacesuit models are meticulous. The suits bear little resemblance to the suits that Yuri Gagarin and German Titov wore on their *Vostok* flights the year before, but more closely resemble models that Konstantin Tsiolkovsky had predicted thirty years prior. The Venus suits for *Planeta Bur'* were somewhat reminiscent of Tsiolkovsky's modified diving suits from *Kosmicheskii reis*. Spacesuits in subsequent movies evolved in function. The suits from *The Moon* had full articulation from between the shoulders and wrists and hips and ankles. Perhaps the most impressive scene in all of Klushantsev's movies was Masha's weightless ballet that the lone female cosmonaut performed while alone in the spacecraft.

After the lone theatrical venture, Klushantsev continued his work with a 1960s trilogy that included movies *The Moon, Mars,* and *I See Earth*. All three films adhered to his previous style combining science education with realistic portrayals of science fiction, even though the latter two were hybrids of documentary and theatrical film, switching from scientific lectures

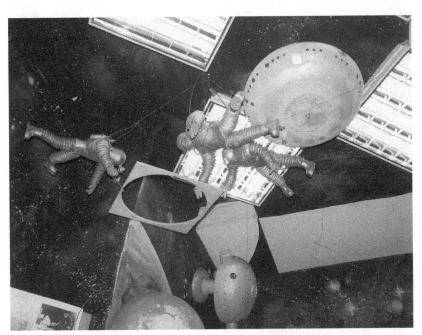

Figure 13. Film models from Pavel Klushantsev's movie *Planeta Bur'*. Even though politics cut short his filmmaking career in the 1960s, Pavel Klushantsev remained a celebrated filmmaker in his hometown of Leningrad (St. Petersburg) until his death at the end of the last century. The intricately designed props and models remain on display inside the museum of Lennauchfilm Studio (2009).

and interviews to dramatic demonstrations of scientific principles. This new genre reinforced the cosmonaut message during the 1960s. Through party discipline, the Soviet Union was leading the way into space.

At the end of the 1970s, Pavel Klushantsev left Lennauchfilm, while remaining in his native Leningrad. For the next twenty years, he dedicated himself to writing children's popular science books. He utilized his artistic skills that had stood out in his films to illustrate scientific and technical principles to children. Although he incorporated American as well as Soviet space achievements in his books, they never received the outcry against internationalism that *Planeta bur'* had.

The Flying Proletariat

The second film was *The Flying Proletariat*. This film was a short, animated film based loosely on parts of the 1925 poem of the same name that Russian Futurist poet Vladimir Mayakovski wrote.[113] The poet, who died by suicide in April 1930, had long been an ambiguous figure in Soviet history in both the subject and style of his writings, yet he never received state denunciation. His poem was as an ode to the promise of aviation to the newly formed state. The film itself purposely conflates the promises of aviation with those of spaceflight, presenting one as a seamless expansion of the other. Relying on Mayakovski's original prologue and the second part of the poem, "The Future of Life," the film contained a tribute to Mayakovski written by Aleksandr Galich in between, along with verses from other Mayakovski poems. The film opens with Mayakovski's denunciation of the militarization of the world and continues with his hopes for the future. The animation presents the fantasies of the poet and includes no more hardship and a much lighter industrial infrastructure that includes no need for roads. The future of 2325 will rely on flight, and the mechanization of life, including farming and weather.

One telling line in the film, "The sky was inspected both inside and outside. No gods, no angels were found," is a quote from the Mayakovski poem. This significance of this line is that throughout the 1960s, the press and popular culture have attributed this line to Yuri Gagarin and German Titov without any reference to the poet. When the narrator speaks these words, the animation shows children throwing away their religious icons in response. This was likely an intentional linking of the 1920s Futurist anti-religious movement with Khrushchev's contemporary anti-religious campaign. Putting the poet's words into the mouths of cosmonauts bolstered a tie between the Red Stuff and the Russian cultural past.

The modern scenes in the film that drew on times after Mayakovski's death added more modern global disputes, including World War II and fascism and the long-running caricature of the atomic weapon-carrying capitalist, into the worldview that technology would solve all problems. The movie also included sentimentality. There are love scenes featuring a protagonist who can soar into the skies almost at his own volition seeking a quiet interlude with his love on another world. Moreover, in the conclusion the film makes a firm link between Mayakovski and Khrushchev through a montage of images.

A Dream Come True

A Dream Come True had its premiere the following year from the Odessa film studio.[114] The directors Mikhail Karzhukov and Otar Koberidze created another film on spaceflight that debuted prior to the advent of human spaceflight. *The Sky Calls* was another film that found its way to the American market via Roger Corman.[115] The premise of the earlier film was that there was a non-cooperative race to Mars. *A Dream Come True* echoed some of the themes that were in *Planet of Storms;* a multinational mission travels to another world. In this case, cosmonauts are called to rescue an alien ship that has crashed on the Martian satellite Phobus. The alien ship had heard the singing of a Russian communications operator. Music has become the universal means of communications. Two confident and experienced cosmonauts take the assignment to retrieve that lost spacecraft. The film makes it clear that these two are unquestionably the best qualified for the job. They are among an army of ready cosmonauts, technicians, and journalists who live at the fictional cosmonaut training center. In addition, to fill the third seat on the craft, an American named Laughton makes the crew international. The film does not miss an opportunity to distinguish between the USSR and the United States. While the cosmonauts are brash and heroic, the American fears the unknown and hesitates to rescue an alien visitor to the solar system. The unsympathetic third man on the flight distinguishes the honorable characteristics of the Russians' Red Stuff even more.

These three films represent the narrow range of acceptable and independent portrayal of the Red Stuff in movies. Klushantsev veered beyond acceptable limits in his positive or even heroic representation of the American scientist and the scenes in which Masha was not merely sentimental but was weak. The animated film based on Mayakovski served another purpose in defining the Red Stuff. *The Flying Proletarian* formalized the link between early revolutionary Russian culture and the contemporary state. *A Dream Come True* acknowledged the internationalization of human spaceflight but placed the Red Stuff at the forefront of humankind's move beyond Earth.

Conclusion

These three disciplines—collectibles, monuments, and film—served the Soviet state well in spreading the message of the Red Stuff during the 1960s. Each in its way created neural networks into civilian Soviet minds that promulgated the message that the new breed of space explorers was uniquely Soviet and offered promise for the future. The USSR relied on these new and renewed instruments of conviction to sustain the celebration of Soviet accomplishments in space beyond the press coverage of the events. There was nothing particularly innovative about the use of these tropes. All had long histories that went as far back as Imperial Russia and transcended national borders. As authentic and effective as these disciplines were, each offered an entrée to dissent that few would have anticipated in 1961.

For the first time since Stalin had imposed a *fantastiki* ban, science fiction made a comeback in the USSR during the 1950s and 1960s. The coming of the space age gave a filmmaker who had longed to be an engineer the opportunity to revive his love of technology through films. Unlike fantasy creators from a generation before, his ambitions began with understanding science and technology. Pavel Klushantsev did not directly probe social issues like the American television show *Star Trek,* nor did he broach the extreme artistic individuality of Andrei Tarkovsky and others in the class of Thaw filmmakers of the 1960s. His point of exploration began with the contemporary chronological setting and character development. His characters are strong Komsomol representatives and were little different from party portrayals at the time. He focused on the technological work behind the scenes, explored ways to expand on scientific theories and experiments, and emphasized science literacy and everyday work of technicians. Klushantsev produced self-affirming work for engineers—there was an immediate payoff for sacrifices, but with little regard to strictures on gender and internationalism. Masha cried, violating the decorum of a Komsomol'ka. His crew was multinational, even including an American during one of the hottest periods of the Cold War. These transgressions ended Klushantsev's theatrical film career. Subsequent films during that early period seemed to stay within the loosely articulated structures of presenting science-fiction versions of the Red Stuff. The other Soviet science-fiction films from that time conformed either to the ideology of the era, atheism, or the politically accepted style of socialist realism.

In the absence of systematic exhibits to promote the space program, znachki took on the role of telling the tale of Soviet spaceflight. Children and students learned the lessons of Soviet spaceflight through Pioneer and youth organizations that encouraged collecting via routine articles and columns that announced new issues. Znachki are also significant because they represent a significant departure from previous public culture movements. They shifted public commemoration of national accomplishments from solely mass events to a personal scale. No central authority dictated the content and message on all pins. However, as there remained only a single source of information on the space program, pin makers shared the same content as other memorabilia makers. The opportunity for innovation was through design. That was the basis of distinction among znachki manufacturers. Finally, the pins are significant for their endurance. Large collections remain intact, and like modern American baseball cards, they have taken on a following of their own beyond the subject that they illustrated. By allowing the independent design and production of these objects, Khrushchev had unwittingly allowed an unofficial portrayal of human spaceflight to develop. Space-themed stamps, znachki, and other collectibles survived the Khrushchev and Brezhnev eras and even survived the USSR.

Stamps and znachki offer the opportunity for comparison of two styles of celebrating national accomplishments. Both embraced the spaceflight subject matter, and each responded to the increased demand for collectibles in the Soviet Union. The Ministry of Post and Telegraph had joined with the rest of the country to relax scrutiny on domestic collecting. Spaceflight prompted manufacturers to expand their production of znachki, which had only recently established them in a souvenir role. There were also differences between the two. Although there were more than a hundred space stamps during this time, there were thousands of individual space znachki. Each pin came from autonomous producers who demonstrated no hesitation about flooding the market. The Ministry of Post, with no competitors in the market, had no reason to attempt innovative designs but conceded public interest through its attention to the new space age subject matter.

When it came to exhibition design and museum and monument construction, Soviet designers and architects were willing to adopt any technique available to them to celebrate the Soviet space age. That included using any small remnant of the space program, adapting Stalin-era buildings, and constructing new buildings that vaguely echoed the powerful constructivist designs of the 1930s. Their approaches voiced the opinion that new and modern would be nice, but anything would do if the display or building amplified the message that was predominant in the Soviet press: the USSR was on the road to conquering space. In contrast to the United States, Soviet space hardware played a small initial role in public exhibits. Engineers and managers held onto their flown hardware for testing and secrecy's sake. High-fidelity exhibition models had to satisfy the public demand to see the hardware that had accomplished the heralded Soviet space firsts. They were not relics of Soviet technical prowess but were images of notable accomplishments. Demand remained high inside the USSR and abroad, and these models served their intended purpose throughout the world.

The slowest official response to the public demand for information, images, and the excitement of the space program came in the forms of formal museums and monuments to spaceflight. Buildings took longer to plan and construct and required buy-in on the part of the local community before they could appear in public. The Academy of Sciences of the USSR was first to respond by adapting the Stalinist monument to collectivization and industrialization, the Mechanization of Agriculture Pavilion, into a monument to Khrushchev's achievements in spaceflight. Later, architects seized on the reemergence of modernist styles and late twentieth-century military materials to create the permanent buildings and monuments to the space program. These buildings and monuments would satisfy an expectation of what spaceflight should look like and remain near-permanent markers of the official expectation of human spaceflight during the 1960s.

5

A Removal and Three Deaths

The Declining Official Need for Heroes

By the mid-1960s, all the significant components of the Red Stuff were in place, and their public presentation was familiar. Brief versions of cosmonaut biographies appeared in Soviet periodicals almost immediately after their flights. More elaborate and officially sanctioned versions appeared soon after as books. The Academy of Sciences transformed Stalin's Mechanization of Agriculture Pavilion into the Kosmos Pavilion, presenting a broad narrative of Soviet mastery of space. The titanium-clad Monument to the Conquerors of Space towered above the Moscow skyline. Nationwide, the Pioneer organization encouraged young children to collect cosmonaut stamps and more importantly, znachki. In each case, the public culture told a tightly controlled story of the Red Stuff and the men and one woman who flew in space. These were stories of success. More importantly, they supported the Soviet state's official optimism for the future using well-honed methods refined for a generation. It appeared that the arrangement of historical narrative and visual and material culture were immutable. They were not. Four events over a period of a few years shook national confidence in the Red Stuff. And these events began to loosen the grip that the Soviet state had in the interpretation of the Red Stuff. By 1969, the United States had landed men on the Moon and brought them back safely to Earth. The simplified version of the race that the USSR had initiated in 1957 was at an end. Nevertheless, both sides continued to demonstrate their national prestige in space. It was not only because the USSR had lost its Moon race that Soviet control over the Red Stuff was faltering. There were domestic reasons, too. The tight-knit and secretive organization that had orchestrated the space program had begun to crumble in full view of the world.

The six *Vostok* and two *Voskhod* spacecraft had been domestic and international public relations successes. Between 1961 and 1965, each flight seemed to upstage or preempt announced American plans for human space missions. The final *Vostok* flight had reasserted the role of women in the Soviet population. Behind the propaganda, Soviet engineers knew the technological and fiscal limitations of the Soviet program and that it would be only a matter of time before they could no longer upstage the US. In late 1965, the Soviet Council of Ministers reorganized the research institutes and design bureaus that were involved in spaceflight to redirect their efforts toward a direct challenge to the American *Apollo* program.[1] The space program's focus was shifting from opportunistic one-upmanship to an attempt to match American technological efforts. This action would inevitably draw off resources from existing activities. What no one anticipated was the rapid public unraveling of the program that occurred between autumn 1964 and spring 1968 while these administrative changes were taking place. During this period, engineers and planners worked on their technological challenge to the American *Apollo* program. However, from the perspective of a Soviet audience that had grown accustomed to bettering the Americans on a regular basis, the end of the golden age of Soviet spaceflight came about through a sequence of stumbles.

Four events rocked public perception of the Soviet space establishment between October 1964 and March 1968. They also had effects on the engineers and technicians who had made the Space Race possible. The first event was the ouster of Nikita Khrushchev in October 1964. Both a bane and a supporter of the space program, Khrushchev had encouraged the firsts in human spaceflight, as he had demanded an immediate follow-up to *Sputnik* in 1957.[2] A little over a year later, the genius of the space program, Sergei Korolev, died and his obituary revealed his identity as the anonymous "chief designer." In April 1967, the first spaceflight fatality occurred when Vladimir Komarov died in the maiden voyage of the *Soyuz* spacecraft. Less than a year after Komarov's death, the first man in space, Yuri Gagarin, died in an aircraft training crash. The last successful spaceflight mission conceived under the Khrushchev-Korolev alliance occurred in March 1965. The next partially successful Soviet spaceflight did not occur until the Americans were within two months of orbiting the Moon and were less than a year from their goal of landing astronauts on it.[3] By then, there was no mistaking that the USSR had fallen behind the US in the propaganda war of the Space Race.

This chapter focuses on a period of three and a half years in the Soviet space program (October 1964-March 1968) that marked the end of the public perception of easy victories. First rumors and then the memoirs and diaries of rocket engineers Nikolai Kamanin and Boris Chertok, and cosmonaut candidate Valentina Ponomareva and the Energiia Corporation's published documents and observations of events contrasted with the press reports of the time. In some cases, technical defeats became public because of internal and technical turmoil that remained shrouded, and there was no official justification of the circumstances. That left an environment that was fertile ground for rumors and freelance speculation. In all cases, the infrastructure that had created the Soviet cosmonaut culture found itself ill-prepared to respond forthrightly to these public stumbles as the interests of the space program diverged from its initial source of public success.

The most significant tests of the Soviet fascination with the Red Stuff occurred at the same time as the greatest setbacks of the program. This challenge occurred just at the point when the Soviet space program had played out its repertoire of existing hardware. It had essentially repeated the same mission without any real refinement of the hardware. Even though extensive work continued for the next step of the space program, those activities were secret. As far as the Soviet population knew, beginning in the mid-1960s, the Soviet space program had stumbled, and the missteps continued to the end of the decade. Weeks after the American *Apollo 11* mission completed the first successful Moon landing, Soviet rocket engineer Boris Chertok hailed a cab in Moscow on his way home to his apartment. Chertok titled his introduction, "Voice of the People," very cannily in a volume of his memoirs on the Moon race between the United States and the Soviet Union:[4]

> In August 1969, I got into a taxi and told the driver my address, Ulitsa akademika Koroleva. During the drive, the taxi driver let it be known that he knew who lived in the "Korolev" apartments. It was evident that he wanted to tell me what the "People" thought, "We no longer have Korolev, and the Americans have landed on the moon first. Does that mean that we could not find another brain among us?"
>
> Until 1964, Khrushchev displayed such involvement that people linked our triumphant space battles with him. In January 1966 the world found out that our successes were first and foremost the result of the creative activities of Academician Korolev. And after Korolev, everyone understood that everything was being done "under the wise leadership of the Central Committee of the CPSU."[5]

Over the generations, the Soviet population had grown used to the idea of secretive industrial and military programs housed in remote towns and villages known only by a post box number. This had been a Soviet reality dating back to Stalin's first purge of engineers in the late 1920s. No one spoke the truth that not all arrested during the purges died by execution. References to the anonymous chief designer evoked visions of yet another valuable individual who had conflicted with the authorities. The Soviet population might not have known Korolev's name, but they understood his story. The American accomplishment was an indication that things were not operating smoothly in the secret Soviet space program. Chertok understood that the Soviet political leadership had let its people down. He knew that they were failing technologically at that point.[6] He also recognized that they had failed to explain adequately why the Americans had achieved their goal and the once dominant space power seemed far behind. The cab driver had his own theory, shared by many that the leadership of the CPSU had inhibited the growth of another genius of the magnitude of Korolev. Through his continued work on ballistic missiles, Chertok had more excellent knowledge of the inner workings of the Soviet space program than the average person did, and Chertok shared these sentiments. The system had failed to live up to its promise.

When discussing the space program of the Soviet Union, it is useful to examine the program from three interrelated perspectives. First is the actual hardware and technology of the program, including the technical and scientific expertise. Second is the policy that defined the direction and missions of the program and included its supporting infrastructure. Last is the public value of the program that it represented. Comparing references from engineering memoirs with media accounts that were the primary source of information to the Soviet public at the time offers an understanding of the differences between what happened and what the public understood. Often one of the three perspectives lagged, but the other two remained particularly vigorous, giving the impression that the program was going well. During the first years of human spaceflight in the USSR, between the formation of cosmonaut recruiting commissions in late 1959 and the death of Yuri Gagarin in March 1968, the intricate interplay of these three perspectives of the space program contributed to the vitality of the program. That dynamic interplay gave the space program an appearance of vibrancy when the hardware and missions were frequently technical repetitions of previous flights. With the death of Gagarin, the last illusion of vigor in the space program faded away and revealed a program in turmoil that had extended beyond its technical capabilities.

If these terrible losses had not been enough, on 27 March 1968, the most celebrated hero of the Soviet space program, Yuri Gagarin, died in an airplane crash. The era of instant public successes in space was over. Chertok and his taxi driver agreed on the state of the Soviet space program in 1969. They had drawn the same conclusions based on different and, to some extent, overlapping sets of information. The engineer understood the technological deficiencies with which they had managed to create an illusion of achievement during the beginning of the decade. The cab driver probably had no engineering insights but understood that after years of certainty about the Soviet Union's role in spaceflight, suddenly his country was no longer in the lead. Both men were disappointed in the situation.

Public Loss of Leadership

Nikita Khrushchev was the chief author of the Red Stuff. One of the keys to Nikita Khrushchev's rise to power was his ability to articulate the frustrated hopes and aspirations of the Soviet population. Beginning with the secret speech of 1956 in which he denounced Stalin's crimes, he initiated the fleeting political Thaw that was to characterize his reign.[7] The 1956 speech seemed to encourage dissenting literary publications and artistic expressions. The speech proved to be an illusion of encouragement. Promises of agricultural gains that exceeded what the country's agronomists could accomplish, even with the army and Communist Party labor at their disposal, characterized his Virgin Lands campaign. It was Khrushchev's challenge to the United States in the Cold War that best characterizes the nature of the illusions that he constructed. Khrushchev had reasoned that as the Soviet Union had been the ally of the United States in World War II and had faced the full brunt of the German invasion, it followed that the country should be equal to the US after the war. Although arguably true in political terms, there was no way for this to be true in technological terms. Khrushchev's canard played out most clearly in the space program. By preempting planned American missions, Khrushchev had been able to maintain the perception that the Soviet Union was ahead of the United States in mastering the technology of spaceflight.

However, as the Thaw gave way to repression, the Virgin Lands campaign produced increasingly disappointing harvests, and direct military confrontation with the United States threatened nuclear war over Cuba, the illusion of a space challenge began to evaporate. The Soviet space infrastructure could not maintain challenges to the US. Unlike Khrushchev's other failed illusions, the decline of the golden age of the Red Stuff took place after his fall from power and did not contribute to his fall. Khrushchev's political fall was only one of the factors contributing to the decline of the golden age of Soviet spaceflight. Khrushchev had been a master manipulator of public sentiments. The loss of his agility in reinterpretation would have an impact over the course of the next few years.

Autumn marked the time when the party measured and published the results of the harvests. The harvests in 1964 were Khrushchev's final failure, on top of foreign policy failures that undercut his authority in the Politburo. October 1964 was his time to pay for his dangerous international political adventures and unsuccessful ventures in overhauling the domestic economy. The 1964 harvest had been abysmal. Even with three cosmonauts wedged in a spacecraft in orbit, preempting the American Gemini two-man program, his confrontational foreign policy had granted no substantial benefits to the USSR and provided him with insufficient political capital. Khrushchev was forced out of office before the *Voskhod* crew returned to Earth. With his departure, the Politburo had lost its most active advocate for the use of the space program as an instrument of propaganda and national prestige, a strategy that lent itself to the repeated recasting of existing hardware into new missions, neglecting a long-term, reasoned program.

Loss of an Anonymous Manager

The style and organization of funerals in the midcentury USSR grew from traditions that developed after the Bolshevik Revolution. Historian Catherine Merridale has cited the proximity of death to the Bolsheviks' foundation myth, thus connecting death and revolution in the public culture.[8] The deaths of 238 people in Moscow during the Revolution afforded the Bolsheviks the opportunity to prescribe the location and ritual for all "revolutionary" deaths after that. The Bolsheviks selected the ideally secular location along the Kremlin Wall for its ideologically consecrated location as well as for its proximity to the headquarters of the new power.[9] They also used the shift in funeral locations to encourage cremation over prolonged religious ceremonies centering on the corpse.[10]

Without Khrushchev's enthusiastic embrace, the Soviet space program was on its own to demonstrate its utility to the state in the form of national prestige. Sergei Korolev recognized that he would not survive competition with his peers on the merits of his previous accomplishments alone and immediately set out to demonstrate to the Politburo the tangible and propaganda benefits of a space program. He offered them spy satellites, military communications satellites, and a crewed lunar mission.[11] His program was a combination of strategic hardware and a renewed vision of the Red Stuff. Of these, the public would only learn the full extent of the Moon program, the continuation of the Red Stuff competition with the US, after the collapse of the USSR. His appeal to the Politburo emphasized his role in creating the military hardware under the direction of Khrushchev. In describing his past performance, Korolev had demonstrated that he could be a ruthless taskmaster in his efforts to extract results from his engineering staff.[12] In embarking on a Moon mission and the other projects, Korolev would have to motivate his staff to produce additional hardware, both spacecraft and rockets, and manage a mission that was far more complex than the missions that had satisfied Khrushchev through 1964.[13]

To take the Soviet Union to the Moon, Sergei Korolev's General Design Bureau, No. 1, OKB-1, would have to utilize new launch vehicles, spacecraft, life support, and navigation. It would have to produce a new launch vehicle that would be capable of carrying the weight of two men and life support to the Moon. An entirely new, maneuverable spacecraft was necessary. Korolev would also be responsible for commissioning vehicles that would assure that a cosmonaut could return safely to Earth. Moreover, finally, he would have to overcome the weak Soviet record in navigating spacecraft beyond low-Earth orbit.[14] We may never know whether Korolev believed himself and his staff capable of meeting these demands. We do know that the pressure on him, as the projects that he managed were slow to materialize, built dramatically during 1965, which could have only contributed to the rapid deterioration of his health.[15]

Korolev died on 14 January 1966, at age 59, during what began as a routine surgery. The surgeons quickly discovered a large cancerous mass; efforts to excise it caused substantial bleeding and subsequent cardiac arrest of an already weakened heart.[16] Korolev's death came at a particularly bad time for the public space program. If anyone in the Soviet Union could, by sheer will, turn *Vostok/Voskhod* into a successful lunar program on top of producing advanced military space infrastructure, it was Korolev. He had already demonstrated his capabilities by succeeding in multiple uses of the *Vostok* hardware. He was undoubtedly under a great deal of stress at the time, though probably no more stress than he had been under at the Kolyma gold mines, where he served his term after his arrest during Stalin's purges.[17] However, the Soviet public was not aware of any of this. It was only upon the time of his death that Korolev's role in Soviet spaceflight became public knowledge.[18]

The day that he received the news that his boss was dead, OKB-1 Deputy Director Vasilli Mishin wrote an obituary for Sergei Korolev. When he completed it, he called Party General Secretary Leonid Brezhnev's office for permission to release the essay to TASS. Mishin sent a draft copy of the death announcement to the Central Committee via Boris Chertok. Brezhnev responded that they could strengthen the notice.[19] What Brezhnev meant was that the Korolev's death notice would not appear as a small note on the death of an obscure engineer, but that the notice would announce officially the role that Korolev had played in the Soviet space program. The resulting death notice appeared on page four (the traditional location for obituaries) in the newspapers *Pravda* and *Izvestiia,* on 16 January 1966.[20]

The death announcements were identical in all venues. Although Vasilli Mishin claimed authorship, the obituary appeared with the signatures of Leonid Brezhnev, followed by the rest of the Politburo, leading Soviet scientists, and engineers. The obituary itself was matter of fact in its references to Korolev's significance to the space program. It was the first public acknowledgment of his role but outlined his life through a litany of familiar prizes and honors, including full membership in the Academy of Sciences, party membership, and receipt of the Lenin Prize. The obituary listed Korolev's accomplishments in rocketry and space exploration. Although the death notice was a breakthrough in acknowledging Korolev, it did not deviate from accepted formulas of the culture of the Red Stuff. The author repeated the legend of Korolev's acquaintance with Konstantin Tsiolkovsky in person and mind. The obituary also cited his leadership of the nameless and faceless army of scientists, engineers, technicians, and workers who comprised the Soviet space program. At no point did it broach the subject of why Korolev's identity had been a secret during his lifetime. There was a tacit assumption that all understood why this had been.[21] Secrets were commonplace in the USSR.

Korolev's death notice was part of an ensemble in Soviet newspapers. Unsigned notes from the USSR Academy of Sciences and the funeral organizing commission appeared beneath the official announcement. The Academy of Sciences' note repeated the list of honors from the official obituary.[22] The funeral commission provided new information on the viewing hours at the Hall of Columns of the House of Unions and the schedule for the funeral at the Kremlin Wall.[23] The only surprising information was a coroner's report that bore the signatures of the Minister of Health and several members of the faculty of the Academy of Medical Sciences.[24] The report declared that Korolev died from cardiac insufficiency that surgery for a rectal sarcoma had aggravated. There is no way to measure the public reaction to this level of intimacy at Korolev's first public introduction.

The departure of Khrushchev from the Kremlin and the death of Korolev in the following year marked the end of a management team devoted to maximizing the public effects of spaceflight on national and international popular cultures. The two had made a successful match by using the existing hardware and technology and maximizing the political impact of each launch. A single orbit, followed by missions of increasing length and simultaneous numbers, and co-orbits that hinted at the possibility of maneuverability had kept the Soviet space program in the world's attention. This currency granted Korolev the opportunity to develop and execute a new plan to challenge the American lunar program. The engineers and technicians who worked in the program knew that their capability was going to have to stretch. They were aware that the Americans were pulling away from them technologically. Between 1964 and 1966 the average Soviet citizen had no reason to think that Khrushchev's removal and Korolev's death would have any impact on the Soviet Union's continuing conquest of space. After the *Apollo 11* Moon landing, the events grew in retrospective significance.

Collapse of the Hardware Gamble

Soviet cosmonaut Vladimir Mikhailovich Komarov was the pilot of the October 1964 *Voskhod* mission that appeared to preempt the American *Gemini* program by taking three men into orbit in the same spacecraft. He had been the first post-*Vostok* hero of the Red Stuff even though the truth of his mission had been less technologically dramatic. Three men had packed themselves into the modified *Vostok* spacecraft without the safety precaution of spacesuits or ejection seats. The single significant modification to the craft had been an improved braking system that allowed the crew to land inside the spacecraft for the first time. Landing inside the craft was a necessity; there was no way to fit three ejection seats into a one-man craft, and even if it had been possible, no one would survive ejection at 20,000 feet without pressure suits carrying an oxygen supply. The *Voskhod* mission was yet another example of the Khrushchev-Korolev strategy of recasting old hardware for new missions. If Khrushchev had been shortsighted enough to believe that the celebratory firsts could continue, Korolev was not. Korolev focused on one of the essential pieces of hardware for challenging the Americans: a maneuverable spacecraft capable of changing orbits called the *Soyuz*. He was preoccupied with this project at the end of his life. His successors carried on his plans after his death.

On 23 April 1967, Komarov took off in an entirely new spacecraft from the Baikonur Cosmodrome in the Kazakh SSR. In retrospect, most of the engineers and technicians involved in the launch agreed that it was not ready for a human-occupied test flight.[25] However, in the words of engineer Boris Chertok, no one at the time felt prepared to declare to the authorities that the spacecraft was unready.[26] The OKB-1 engineers felt acute pressure to deliver remarkable feats on the level achieved during the Khrushchev-Korolev alliance. Another design bureau, Lavochkin, had taken on the tasks of robotic lunar exploration, which was going well. A successful *Soyuz* launch with a human inside was the only option to continue OKB-1's reputation. The public had read no enthusiastic proclamations of space successes for over two years. In what was most likely an effort to buoy the morale of the Korolev engineers, Minister of Defense Industries Dmitrii Ustinov allowed Yuri Gagarin to be the backup pilot to Komarov on the *Soyuz 1* mission.[27] According to Chertok, only Gagarin's (and possibly Komarov's) mood benefited from this act.[28]

The first crewed flight of the *Soyuz* suffered problems from beginning to end. During the first orbit, the left solar panel that powered the broadcast antenna failed to deploy; the backup antenna that was supposed to supply telemetry failed similarly.[29] According to the quick notes that Vasilli Mishin, Korolev's former deputy and replacement at the design bureau, wrote at the time, no one could determine what was going on in the craft and arguments broke out within minutes of the spacecraft attaining its orbit.[30] By the second orbit, when Komarov first successfully contacted ground control, the men realized that the craft did not orient correctly and did not respond as expected under automated or manual control.[31] During the first few hours of the maiden *Soyuz* flight, engineers canceled the launch of a second *Soyuz* and planned to rendezvous with *Soyuz 1* as they turned to the problem of landing Komarov's spacecraft. During the mission, Mishin quickly scribbled down the times at which each new orbit would begin, reflecting the scramble to optimize the landing time and site commensurate with the available electrical charge from the partially functioning solar panels.[32]

The engineers estimated that the optimal time for an emergency landing would be after the seventeenth orbit when the craft would repeat its ground track from its first orbit.[33] As the pilot and spacecraft reentered the Earth's atmosphere, the ground crews thought that they had managed to save the mission, but the *Soyuz* main parachute did not deploy properly, tangled in the backup parachute, and failed to slow the craft's descent. If in the unlikely event that Komarov survived the high accelerations of an improper reentry, he died upon impact at an estimated 145 km/hour.[34] As a final blow, the solid fuel in the soft-landing engines of the spacecraft, which should have softened touchdown, ignited on impact and the spacecraft burned before rescue crews could approach it and recover the cosmonaut's body.[35]

In his notes from that day, Mishin began preparing an investigation team immediately after his staff briefed him on the landing plan for Komarov and the *Soyuz* craft. Mishin listed in his notebook twenty candidates for the safety commission to investigate the flight.[36] As had been the tradition under Korolev, the candidates were heads of design bureaus answerable to OKB-1, including Guy Severin (the chief designer at Zvezda, the manufacturer of spacesuits and other cosmonaut personal equipment), Nikolai Kamanin, and engineers who worked for the Korolev bureau.[37] Korolev had done the same in the past when missions had failed. He had maintained close control of his projects, created and staffed his investigative teams with the tacit approval of the Ministry of Defense Industries, Ministry of Defense, and the Politburo.[38] Vasilli Mishin was no Sergei Korolev, and the Politburo did not bestow instant approval as it had under Khrushchev. Komarov's death was the first time that an astronaut or cosmonaut had died on a spaceflight mission. (Three *Apollo* astronauts had died in a launch pad accident three months before.) Mishin's efforts to imitate his mentor's management style ended abruptly. Within two days, he was before a special commission on the catastrophe at the same time Moscow was preparing to bury the cosmonaut in the Kremlin Wall.[39] The commission consisted of Vice President of the Council of Ministers (and former Minister of Defense Industries) Dmitrii Ustinov, president of the Soviet Academy of Sciences

Mstislav Keldysh, Marshal of the Air Forces Konstantin Vershinin, and Yuri Gagarin. Mishin and OKB-1 would no longer bury hardware failures in an internal investigation.

The initial press reports on the *Soyuz* flight and Komarov's status appeared on 24 April 1967, the day he died.[40] The reports carried enthusiastic and optimistic accounts of his mission. The subtitle of the headline in *Pravda* declared the "successful completion of the preliminary program of research."[41] These blatant falsehoods appeared while engineers were debating the likelihood of returning the cosmonaut from orbit in his malfunctioning craft.[42] The next day, death announcements for Vladimir Komarov appeared in newspapers in the editions where journalists had planned continued celebrations of his flight. His portrait appeared on the front of *Pravda* with the traditional black frame indicating his death.[43] The *Pravda* obituary alluded only to the parachute failure at the end of Komarov's flight. It made no mention of the other complications of his flight. In fact, the paper declared Komarov's flight victorious until the parachute failed, echoing the previous day's reports.[44] His remains took their place in the Kremlin Wall, alongside earlier patriotic and military heroes. Even as his death was one of the routine sacrifices of modern society, it influenced the space establishment that would later have a public impact in Soviet society. In her biography of her son, published just before her death in 1983,[45] Anna Gagarin wrote about how the sequence of deaths in 1966 and 1967 affected her son:

> After Korolev's death, something changed in Yuri. He became the backup to Komarov in January 1967. Komarov died in April. Yura [diminutive form of Yuri] turned dark with grief. We asked him how it happened, why the flight was completed. It found itself unexpectedly over the Earth twisting in the stirrups of the parachute. "The unknown is always unpredictable, and treacherous," answered Yura. The explanation did not completely drown out the pain.[46]

Gagarin could not dismiss this first cosmonaut death. Neither could the Soviet public. The Red Stuff remained heroic, but it was no longer invincible.

The official announcements of Komarov's death simplified the cause of death. They made no mention of troubles in orbit. The *Pravda* cover story blamed failed parachute lines for the cosmonaut's death. Also unmentioned were the abbreviated flight plans including the canceled *Soyuz 2* mission and the planned spacecraft rendezvous. Control of the narrative of *Soyuz 1* closely resembled the control of the narrative of the *Vostok 1* flight, where only the barest details became public. Embarrassing details that might challenge the heroic and utopian image of the cosmonaut did not support that cause.

Death of an Icon

Yuri Gagarin's death was the most public and profound loss to the Soviet space program. For years after his mission, Gagarin had played an essential role in promoting Khrushchev's messages to the Soviet population and the world. The Soviet public seemed to accept, as had the Americans about John Glenn, that their first man in orbit would not return to space. Given the paucity of flight hardware and the relatively large cosmonaut corps, there had been no reason to risk another flight for Gagarin. Many trained cosmonauts had not yet flown and might not have the opportunity to do so. It was not until after Khrushchev's departure and Korolev's death when the Soviet space program was trying to challenge the Americans that Gagarin's assets as a trained, experienced cosmonaut began to outweigh his value as a living icon. It was at this time that Gagarin petitioned the cosmonaut and Air Force hierarchy for a return to flight status. Yuri Gagarin died on 27 March 1968, in a jet-skills training flight, along with his instructor, when their craft crashed near the town of Kirzhach, less than one hundred miles southeast of Moscow.

The Soviet system responded as it traditionally had. Gagarin's death generated the standard repertoire of ritual, memorialization, and monuments. The material and public culture that emerged after his death notwithstanding, the space community was never able to pronounce a definitive reason for his death. There was never a coroner's report. For forty years, the cause of the plane crash that killed Gagarin was a subject of debate, first in secret and internal to the Soviet defense industry and then, for the last twenty years, very public. Unanswered questions had left cracks in the cosmonaut image.

As we have seen, Yuri Gagarin cut the ideal figure for the role of first Soviet man in space. His life as described in official portraits and biographies followed the classic path of success in Soviet society. His genuine good nature made him an effective ambassador to the world after the flight. The way in which the Soviet public celebrated Gagarin's flight indicated that it was not subject to simple manipulation from above. By all accounts, the celebrations that took place after his return to Moscow on 14 April

1961 were spontaneous.[47] After his 1961 flight, Gagarin began a dual life between the two sides of the space program. His immediate, public post-flight duties were diplomatic and managerial.[48] The diplomatic duties had him giving speeches and traveling throughout the world. The managerial duties had no public profile. On the other side of the steel gates, he was working behind the scenes among still-secret cosmonauts and a still anonymous space workforce. Almost seven years after his historic *Vostok* flight, Yuri Gagarin began training for the spaceflight of *Soyuz 3*, the redesigned and tested version of Komarov's spacecraft. The one possible salvation for the public space program that remained within the government's grasp would be to return cosmonauts into space to its triumphant levels of the past by launching the first man to orbit the Earth into space once again. Gagarin's return to space would signal to the world, but most importantly to the Soviet people, that despite setbacks the inner strength and potential glory of the Red Stuff remained. The political risk of placing a living icon back in the hazardous business of spaceflight was less than the public relations gain of a successful flight. Yuri Gagarin died during a training flight to regain his flight certification in March 1968.

There are few details available today of the actual decision-making process of returning Gagarin to space. What is intriguing about the situation is the absence of a detailed and precise protocol for determining his flight status. No one had ever written that he would not fly again, but there was little incentive for allowing him to train for another spaceflight. There was enough coterie of cosmonauts yet to fly in space, and the skills for flying a *Vostok* spacecraft and experience did little to distinguish Gagarin. The consensus among Gagarin's contemporaries was that some official associated with the space program, either Kamanin, Korolev, Ustinov, the Politburo, or Khrushchev, decided that Gagarin should not fly in space again after his 12 April 1961 flight.[49] When and who made this decision is up for speculation. What is clear is that by 1967 all of those who were in the inner circles of piloted flight in the USSR knew that for Gagarin to return to flight status, someone at the highest levels of government would have to approve his status. Female cosmonaut candidate Valentina Ponomareva made this assertion, as did Boris Chertok in his memoirs and Nikolai Kamanin in his published diaries.[50] Even though the fact that the approval process was underway seemed to be common knowledge among insiders at the time, it is also unclear what document or event triggered Gagarin's return to flight status. As far as the public was concerned, this decision process was completely unknown. Those involved only recently revealed the complicated return to flight process through the sale and publication of some of the documenting correspondence.[51]

One of the requirements for Gagarin to return to flight status was that he refresh his piloting skills to the level that they had been at the beginning of the decade when he had been a young lieutenant in the Air Force. To do so, he was assigned the chief flight instructor for the cosmonaut program, Vladimir Seregin. Seregin had been a decorated fighter pilot during World War II and the chief test pilot for the Air Forces for the MiG-15 Uchebnyi trenirovochnyi istrebitel' (pursuit training craft, UTI). The MiG-15 UTI sacrificed speed and maneuverability for a second seat for training.

As is true with many episodes of disasters in Soviet space history, the memoirs of all agree on the details of the hours before the flight. However, the accounts of the moments immediately before the crash vary. Everyone seems to agree that on the morning of 27 March 1968, Gagarin rose and walked from his apartment to his office in Star City. Once there, he made a few telephone calls, including arranging an interview with a magazine reporter and speaking with his wife who had been on duty at the hospital for the previous month.[52] From his office, he joined several comrades on the bus from the cosmonaut training center to Chkalovskii Air Base, where they joked among themselves while en route.[53] It was there that Gagarin met Seregin to perform his final flight test before being certified again to fly jets solo. The two pilots were scheduled to fly a MiG-15 UTI.

Gagarin and Seregin may have received an unreliable weather forecast on the 27 March flight. The wind was unusually gusty as they flew their MiG-15 at high speed between two cloud layers at relatively low altitude.[54] There are also questions about the responsiveness of the aircraft. The MiG-15 carried two extra fuel tanks that made it slower and perhaps even less responsive in adverse weather conditions.[55] There might also have been a nearly averted head-on collision with another craft or a weather balloon. Marks on a pressure gauge showed the cockpit had depressurized before the MiG-15 hit the ground, meaning that either it collided with something, or the pilots tried to eject before the crash.[56] In addition to the questionable flight conditions, a minute after Gagarin and Seregin took off in their MiG-15, a pair of faster MiG-21 jets took off and overtook the smaller plane. They were followed a minute later by a second MiG-15 whose pilot reported that he did not see the MiG-15 carrying Gagarin and Seregin. According to the flight plans, the planes (Gagarin's MiG-15 UTI and the MiG-21s) came "in dangerous proximity to each other," less than 1,640 feet apart, according to a team of investigators in 1987.[57] On that day, clouds probably prevented the pilots from seeing the horizon. If Gagarin and Seregin had maneuvered to prevent a collision or avoid a stall, the lower-than-expected cloud ceiling might not have allowed them enough time to prevent the crash.

The Official Investigation into Gagarin's Death

Given the highest level of involvement by the Politburo, it was clear that the Central Committee saw the loss of Gagarin as potentially more publicly catastrophic than the unexplained failure of a critical piece of Moon mission hardware a year previously. Komarov's death in the *Soyuz* remained isolated from the secret Moon program in the public's mind. Yuri Gagarin had been the living icon of the Red Stuff. It was not possible to wall off Gagarin's death from the heart of the program. It demanded immediate and high-level attention. The Central Committee immediately convened a panel to determine the cause of the crash. This Central Committee investigation was the first in a series of inquiries that started at the highest level almost immediately after the crash, followed by periodic reexaminations, the most recent published in 1998.[58] The original report remains classified, perhaps in perpetuity.[59] Although the depth and the scale of the investigation into Gagarin's crash were unprecedented in the history of spaceflight and rocketry, there exists in Russia a particular public interest in the autopsy of rulers and leading figures in history. Early, prominent examples include the public details about the deaths of Tsar Alexander III in 1894, Lenin in 1924, and even Stalin in 1953.[60] The circumstances of the crash spared the population the publication of the specific autopsy results. There was little in the way of human remains to autopsy. Instead of an autopsy, the investigation team examined the crash site and flight records of the craft flying within proximity of Gagarin that morning.

The first formal investigation of Gagarin's death began on 28 March, when the Central Committee of the party clarified the organization and leadership of the investigation.[61] This was a high-level commission with external appointees who had every incentive to conduct a complete investigation. Vice President of the Council of Ministers Ustinov, to whom the space industry reported, was the president of the commission.[62] The investigation commission's work was divided and staffed in four areas or sub-commissions. The first area was to investigate flight operations at the Air Force base and Gagarin's and Seregin's individual flight preparations on the day of the accident. The second examined the flight history and wreckage of the UTI MiG-15 No. 612739, the one that Gagarin and Seregin flew. The third reviewed flight preparations of cosmonauts at the Space-flight Training Center (TsPK). The fourth had a purely administrative role of preparing conclusions and assembling the final report for the Central Committee.[63] Each sub-commission had a staff of 12–15 people, as well as a representative of the KGB and the Central Committee, resulting in a population of close to one hundred people who had intimate knowledge of the investigation process. That included cosmonauts and representatives of the Air Force who participated in each, as well as representatives from civilian organizations. The first three sub-commissions began work on 1 April, within one week of the accident.[64]

All the investigations into Gagarin's death searched for a discrete cause of the crash or a clear vindication of hardware or personnel. The lack of definitive conclusions about the cause of the crash from this initial investigation and subsequent ones lead one to suspect that the MiG-15 probably crashed for many reasons, not one. For the last three decades, the cause has been the focus of vigorous internal debates within the space establishment, with many interests willing to blame others for its cause.[65] A computer simulation of the crash by an official review commission reenacted the last stage of flight. It concluded, "The plane went into a spin characterized by maximal energy loss. Subsequently, it recovered from the spin after which the aircraft plowed into the ground."[66] Either Gagarin's MiG-15 fell into the vortex wake of another aircraft, or else it banked sharply to avoid hitting another plane or an instrument probe, the computer simulation showed.[67] The commission determined the method of death, but not the means. The public, however, was unaware of these disputes. The government has never published an official report on Gagarin's death. Individuals have published unofficial reports that disagree on the most significant points.

A subsequent 1987 investigation uncovered air safety violations in the 1968 exercise, but, in 1989, *Novosti* said, "The commission's findings unequivocally showed that careless flying or lack of discipline on the part of the crew, as well as a ground-control negligent attitude or someone's malicious intent, were ruled out as possible causes for the disaster."[68] Offering no proof of its assertion, the translation of the *Novosti* account of the 1987 report was the opposite of its literal meaning. The doublespeak that was the official public language of the time had pointed out the causes that it had tried to conceal.

The Soviet government has never released an official report from any of the inquiries.[69] Some participants have claimed that the report remains classified in perpetuity. In the absence of an explanation, alternative views of the results emerged rapidly. Moreover, recent personal accounts have revealed another rift within the space community over the investigation. As one reporter stated, the investigation quickly deteriorated into an assignment of blame, obscuring the cause of the technical reasons for the crash.[70]

The Funeral and Mourning

Yuri Gagarin's death was announced almost immediately after it happened. TASS announced it on the radio during the evening of the same day, 27 March 1968, before authorities had the opportunity to notify the families. Gagarin's mother reported that she learned the news of her son's death from a radio announcement that she heard in her home in Gzhatsk. She heard only the end of the broadcast ("In completing a training flight with Gagarin there was a tragic death"), before the regional party secretary arrived and confirmed the news.[71] The full text of the announcement read:

> It is with deep sorrow that the Central Committee of the CPSU, the Presidium of the Supreme Soviet of the USSR and the Council of Ministers announces that on 27 March 1968, as a result of a catastrophe that occurred during the course of a training flight, the first conqueror of space, the famous pilot-cosmonaut of the USSR, member of the CPSU, delegate to the Supreme Soviet, Colonel Yuri Alekseevich Gagarin died tragically. In this aviation catastrophe also died commander of the aviation group, member of the CPSU, Hero of the Soviet Union, Engineer-Colonel Vladimir Sergeevich Seregin.
>
> The Central Committee of the CPSU, the Presidium of the Supreme Soviet of the USSR and the Council of Ministers of the USSR express their deepest sympathies to the families and relatives of the deceased comrades.[72]

The same text appeared in the print press inside the black box that indicated an obituary. It did little more than identify the two pilots and name the date of the crash while repeating the identity of the signatures three times. Among the personal condolences to Gagarin's and Seregin's wives and children was a brief article whose title was almost as long as the article itself. This article, entitled "From the government Commission for the explanation of the circumstance of the deaths of pilot-cosmonaut of the USSR, hero of the Soviet Union, Col. Y. A. Gagarin, and Hero of the Soviet Union Engineer-Colonel V. S. Seregin," provided the most information about the crash, including the location and the identity of the aircraft.[73] The official obituary had neglected these details.

One Western obsession with the Soviet space program focuses on the idea that there have been mysterious deaths among the cosmonauts.[74] The underlying assumption is that deaths were hidden, denied, and a shameful part of the secrecy-shrouded program. It is evident through the study of the culture of death in the Soviet Union that premature and unanticipated deaths are not surprising in Russian culture and the press reports on them, even when they go unexplained. For example, the deaths of World War II received commemoration at the time and remained a fixture in Russian culture. The unexpected deaths of aviation heroes, such as Valerii Chkalov, the long-distance record pilot from the 1930s, occasioned solemn and public commemoration of their lives.[75] The example of Chkalov's death does much to disprove the existence of a prevailing culture of secrecy over death itself. Secrecy often covered the causes of death—Chkalov died unexpectedly in a crash during a test flight. While not attempting to cover up his death, his eulogizer, People's Commissar of Defense Kliment Voroshilov, felt free to embellish the manner of his death. He said, "To the very last minute Chkalov, the best citizen of the Soviet Union, gripped the controls of his fighter, to the very last he honorably served his homeland, which will never forget his compelling image as a true Bolshevik and people's hero."[76] Voroshilov's account of Chkalov's death was fictional. There was no way that anyone could have known how he spent his final minutes. This fabricated account reinforced the heroic image of Chkalov. The same had been true in the case of the lives lost at Baikonur in the Nedelin disaster, when a rocket that failed to launch unexpectedly exploded, killing seventy-eight men. When the public learned of Gagarin's and Seregin's deaths, the precise way they died remained a secret. In Gagarin's case, his death provided an opportunity to enhance the monuments to the Red Stuff, but it did not burnish his image. Komarov's death had made the death of cosmonauts imaginable. The Soviet public had learned that even space heroes could die, in the same way that aviation heroes had died before them. They, too, knew that the Americans had suffered losses in their space program.[77] Gagarin's death held open the possibility that a cosmonaut's death could be meaningless.

There had been no effort to conceal the event as the broadcast moved more quickly than the party secretary could. The plans for his funeral began as soon as they were able to collect his remains. Cremation of the remains took place at 9:15 p.m. Moscow Standard Time (MSK) that evening. Gagarin's and Seregin's relatives who lived near or around Moscow, all the cosmonauts, high-level Kremlin and military officials, including Ustinov,

Verhshinin, Moroz, Kamanin, and other generals and officers, attended the cremation. Urns with ashes were placed in the Krasnoznamnii Hall of the Central House of the Soviet Army, TsDSA, the traditional place of national mourning. The reception began on 29 March at 9 a.m. MSK. Forty thousand people visited on that day.[78]

Despite the solidly Bolshevik culture that prevailed, Gagarin's funeral referred to religious observances as well. His funeral services took place on 30 March 1968, exactly three days after his death, conforming to the Orthodox tradition that dictates that the body funeral services take place as the spirit leaves it on the third day. The funeral was a concession to the Orthodox sensibilities of his mother, much in the way that funeral ceremonies had been since the early "red funerals" of pre-Bolshevik Russia incorporating new and old rituals.[79] Although the ceremony was devoid of the overt Orthodox content of the traditional funeral, the forms and procedures of interment adapted Orthodox ceremony to the secular, revolutionary content of the Soviet regime.[80]

Brezhnev and other members of the Politburo orchestrated and attended Gagarin's funeral.[81] The funeral activities began early morning on 30 March, with the arrival of Gagarin's widow, Valentina Ivanovna, and other relatives of both Gagarin and Seregin at the Central House of the Soviet Army. The doors had opened at 8:30, ahead of schedule, as there had been a long line waiting to enter since early that morning. There were 200 official mourners included in the procession, in addition to the 130 invited relatives, cosmonauts, pilots, and closest friends.[82] The proceedings began promptly at 1 p.m., when an honor guard carrying the urns containing the remains entered the hall. From there the entire mourning party walked to the Kremlin Wall for interment. In the front were the invited attendees. Behind them were the thousands of un-ticketed spectators.[83]

The final resting place for the remains of Gagarin and Seregin was adjacent to other recently deceased national heroes. Among their neighbors were Rodion Yakolevich Malinovskii, the Minister of Defense from 1957 until his death in 1967, ill-fated *Soyuz* cosmonaut Vladimir Komarov, Nikolai Nikolaevich Voronov, Marshal of Artillery and World War II hero who had commanded the artillery during the defense of Stalingrad, and, of course, Sergei Korolev.[84]

The Martyr Remembered

After the quick succession of Gagarin's death, funeral, and the initial investigation into his crash, attention turned to the shaping of the public memory of this Soviet hero. Planning for the first wave of commemoration began almost immediately, under the direction of Kamanin. He began his work on 1 April, with decisions about allocating pensions for the Gagarin and Seregin families and the erection of memorials for each. By the end of the workday, he had begun a preliminary plan for construction of a memorial to Gagarin at the Spaceflight Training Center, TsPK. According to his diaries, he made the initial decision to preserve everything tied to Gagarin's life at TsPK, including personal items. Kamanin promoted a continuation of Gagarin's efforts to support a museum at TsPK. He went further by establishing the Gagarin office-museum at the center. Kamanin solicited the support of S. P. Pavlov of the Central Committee of the Communist Party, who agreed to assist in the project and include the efforts of the Komsomol.[85] This decision was the foundation of the memorial office that would open a year later at the Gagarin Spaceflight Training Center Museum.[86]

By the evening of 1 April, Kamanin had also met with Aviation Marshal Vershinin and other representatives from the Air Force. In addition, present at the meeting were prominent engineers from the OKB-1. The team gathered to compile a list of commemorative projects for the Politburo to review and approve. The list followed the expected course of namings and renamings of places and monument commissions that followed the Soviet habit of renaming things in honor of deceased or living political leaders and heroes. This list echoed the systematic renamings that had occurred in the aftermath of Stalin-era pilot Chkalov's death.[87] Gagarin, an international hero, known outside the world of specialized enthusiasts, merited higher profile namings than had been bestowed upon Chkalov. The working group made the following recommendations, all of which the Central Committee approved:

1. The Center for Spaceflight Training, TsPK, would be renamed in honor of Gagarin
2. The Monino Aviation Academy became the Gagarin Academy
3. The town of Gagarin's childhood, Gzhatsk, would be renamed Gagarin in his honor
4. The square at Kaluskaia Street in Moscow was renamed as Gagarin Square and a monument erected in his honor
5. Obelisks or monuments were to be built where Gagarin landed in 1961 (outside of Engels); where he died; in Gzhatsk (soon to be Gagarin); the village of Kyshino; and at the TsPK

6. A state prize was to be established in honor of Gagarin and an award for the greatest achievements in human spaceflight
7. Three fellowships were to be established in honor of Gagarin at Monino and Zhukovskii Aviation Academies
8. The 70th Regiment in the Air Forces (VVS) was to be renamed the Seregin Regiment
9. Pegokskii Street in Moscow was renamed for Seregin.[88]

Kamanin wrote in his diary that he also recommended that they establish a fund in the name of Gagarin to raise money for the project to memorialize the cosmonaut. He proposed that the donations come from individuals, domestic organizations, and other voluntary organizations abroad. Epishev and Zakharov objected to this plan and recommended that they forward the ten proposals (including the pension provisions) without mentioning any source of funding. They argued that the USSR was not a developing country so the state should not face the embarrassment of asking the public for funds.[89] Kamanin's proposal for public fundraising never had consideration at the Central Committee, although Pavlov encouraged him to pursue his ideas for voluntary contributions with the Komsomol.[90] On 2 April, Marshal Yakbovskii presented the formal proposal for the pensions and memorialization to the Central Committee.[91] It approved them all.

With a turnaround between suggestions and approval of less than one week, one might think that the proposed memorialization would come about in quick action. However, the statue of Gagarin at the former Kaluskaia Square in Moscow did not appear until July 1980, over twelve years later. The head of the Moscow City Committee of the CPSU, V. V. Grishin, and USSR Academy of Sciences President A. P. Aleksandrov made speeches at the public ceremony that declared the formal opening of the new titanium-clad monument. Aleksandrov declared that the monument represented the contributions of the "scientists, workers, designers, engineers and technicians" who brought the Soviet Union to the stage of exploring space, repeating the mantra of the anonymous masses that were responsible for all previous space achievements.[92] Forty meters tall above the long-renamed Gagarin Square, the memorial was the product of the sculptor Pavel I. Bondarenko and the architectural team of Ia. B. Belopol'ski and F. M. Gazhevskii. The speakers, as well as the spectators, tied the opening of the monument to the 26th Party Congress, Brezhnev's last, which was occurring at the same time as though the monument and the congress

Figure 14. This titanium monument to Gagarin, the first man in space, was planned almost immediately after his untimely death in 1968 but took twelve years to complete. Erected in the southern part of the city, near the Presidium of the USSR Academy of Sciences, the monument dominates the skyline among the sprawling development that has taken place in Moscow since the collapse of the USSR. Felix Lipov/Alamy.

shared a symbolic meaning.[93] There was no explanation for the twelve-year delay between the approval of the monument, the other renamings, and its opening. Once again, the public had to draw its conclusions.

Kamanin's original proposals, drawn from his experiences from the 1930s and 1940s, established the basis for the ways of honoring Gagarin's memory throughout the former Soviet Union. Gagarin's honors took the form of namings and monuments. However, specific underlying memorial tones have emerged over the last thirty-five years that Kamanin would not have proposed. One remarkable characteristic is the increasing importance of the celebration of his mother in the Orthodox tradition. Gagarin's mother began to take on this role after his death. The celebration of Gagarin's mother in his birthplace culminated with a heroic-scale bronze sculpture donated to the town by the Hungarian government (officially the Hungarian people). She sits alone on at the end of the bench, apparently waiting for her son to take his place. The elements of religion are close to the surface. In each of the three Gagarin houses in and near Gagarin, guides note the traditional icon corner common to all peasant houses. The noting of the corners is an acknowledgment that the Gagarin family maintained some connections to Orthodoxy throughout their lives.[94] Gagarin's death had set in place the notion that the Red Stuff could freeze in time.

Rumors and Speculation

Despite a well-orchestrated effort to manage Gagarin's memory through the public culture of the Soviet Union, rumors about the circumstances surrounding his death emerged in the atmosphere of secrecy. There had been little attempt to manage the public message about his death. Although the crash investigation commission completed its work by 1969, reports on the results did not become public until 1988.[95] These reports were not part of an official publication but emerged as individual accounts of the process. The most significant consequence of the official silence on the causes of the crash was the spread of rumors to satisfy the public's curiosity. With few official details to work from, rumors and conspiracy theories arose immediately in both the Soviet Union and the West. The first and most widely circulated rumor was that an inebriated Gagarin had died in a drunken car crash and not an aircraft accident after all.[96] This explanation was plausible, as many that were knowledgeable of the social conditions in the Soviet Union at the time were aware of the persistent problem of alcoholism in the country. Gagarin had a reputation for enjoying drink, even if there were no indications that he had a drinking problem. The rumor served as a warning that no one would escape the tragedy of alcoholism in the USSR. Variations on the alcoholism rumor included ones that claimed that either or both were drunk while piloting their MiG-15 UTI or that one of the other three pilots flying in the vicinity had been drinking and ultimately caused Gagarin's plane to crash.[97]

A secondary set of rumors also pointed to pilot error. In those, a sober Gagarin was hot-rodding in his aircraft, attempting prohibited acrobatic maneuvers, or he was conducting illegal activities while in flight. One example of this genre purports that Gagarin was trying to hunt bear from an open cockpit. This portrayal of Gagarin's recklessness after his death grew from some of the legends of Gagarin's womanizing and otherwise reckless behavior.[98] In contrast to Tom Wolfe's celebration of the purportedly high-spirited NASA astronauts, these rumors assigned punishment for this behavior.

The final set of rumors centered on the prevalence of conspiracy theories to account for unexplained deaths throughout the Soviet period. There is a Stalin-era saying, "No one dies without reason in Moscow."[99] The rumor that sinister forces had executed Gagarin and the rumor that Gagarin was not dead, but hidden away in a psychiatric institution as punishment for his indiscretions gained currency.[100] In the more recent variations applied to Gagarin's death, stories held that the Kremlin killed Gagarin for one of the following three reasons: to avoid embarrassment over his womanizing; to remove completely all reminders of Khrushchev's adventurism; or to stop the emergence of a new cult of personality of either Khrushchev or Gagarin.[101] Of course, the ends and the means did not match. There were undoubtedly simpler ways to stop Gagarin's womanizing without killing two national heroes. The best way to eliminate reminders of Khrushchev's adventurism would have been to shut down all favorite aspects of the then-publicly inactive space program. The Soviet Union had vast experience with political assassinations, murders, and executions. There was no better way to strengthen a cult of personality than to create a martyr. Under Stalin, the prosecution of villains quickly followed martyrdom to leave no doubt who was at fault for the first death. Brezhnev's Kremlin never assigned blame for Gagarin's death. As a martyr, Gagarin would have served no purpose.

The development of rumors in the popular culture begs the question of why the government was so vague about the causes of Gagarin's death for so long. The public portrayal of his life had been molded to fit post–World War II Soviet expectations. In contrast, his manner of death defied explanation, especially when considering the image of his life. If his boisterous womanizing was a source of shame, so too was his manner of death. To admit that he had died due to poor management and maintenance of the aerospace industry would be an open admission of failure in a field that carried much of the USSR's reputation. The truth could have put the illusion at risk. However, the absence of an official explanation left reasonable people with precisely the idea that the space program had been a sham.

Although Gagarin's death was the last in a series of events that altered public perception of the Soviet space program, it did not mark the end of the Soviet human spaceflight. Work on the human-crewed lunar mission continued in secret after a break in the wake of Komarov's death. Vasil-li Mishin continued the massive launch vehicle project, the *N-1*, which would rival the power of the American *Saturn V*. As before, all this work occurred in secret with no bold proclamations of their intent. On 3 July 1969, the first test-launch of the *N-1* failed.[102] An American spy satellite took photographs of the aftermath of the *N-1* explosion, but neither Soviets nor Americans learned of the failure or the photograph until the late 1990s.[103] Mishin continued three more attempts to test the rocket over the next five years.[104] The Soviet public did not know that the program existed when it ended in 1974.[105] The true extent of Soviet lunar aspirations did not gain public attention until the participants began to publish their memoirs.[106] As far as the public was concerned, the Soviet Union had no successes and no plans for human spaceflight between 1965 and 1968. The public silence from the Soviet cosmonauts made the events of July 1969 even more resounding.

Early in the morning on 21 July 1969, engineers, politicians, scientists, and cosmonauts assembled at NII-88 to watch the broadcast of Neil Armstrong taking the first step on the Moon. The United States was broadcasting the images worldwide, but China and the USSR declined to relay the signal on their national broadcast channels. The central Soviet television station provided live closed-circuit relays over cable lines to special, privileged groups of Soviet officials who watched Armstrong walk on the Moon.[107] Of course, it would not have been possible to embargo the news of *Apollo* entirely. Soviet news agencies did announce the *Apollo 11* landing after it happened. The Soviet print media published stories about *Apollo 11* in the internal pages, among other foreign and science stories.[108] On the day that Armstrong stepped on the Moon, *Pravda* carried a front-page block announcing *Luna 15*'s simultaneous presence on Earth's satellite.[109] The probe had crashed and failed to send a robotic return sample to the

Earth before the Americans could successfully return their astronauts. Four days later, on 22 July 1969, the grainy photographs of the American Lunar Excursion Module and an American astronaut appeared on page four of *Pravda*.[110]

Neil Armstrong stepped on to the Moon one year and four months after Gagarin's death and at the time, few people in the Soviet Union knew how seriously the Soviet space establishment had worked to send their men to the Moon. What the Soviet public did know was that the Americans had done as Kennedy had promised. The decade that had started with the opinion of the USSR in the person of Gagarin leading the world into space ended with an entirely different reality. Khrushchev, the official cheerleader for spaceflight, was gone. The chief designer, Korolev, to whom was attributed the genius of the early years of the program, was dead. A cosmonaut had been the first space mission fatality. Gagarin, too, was dead. Moreover, the Americans had stepped on the Moon. That was the knowledge that Boris Chertok and his cab driver shared.

Conclusion

Between October 1964 and March 1968, the public culture of the Red Stuff suffered dramatic losses that shook it to its foundations. The human space-flight program no longer had the political and managerial support that had masterminded the appearance of a wildly successful program. The removal of Khrushchev and Korolev's death deprived the program of its ability to camouflage technical failures as routine tests and spin routine flights into great technical achievements. Finally, it lost its most recognized symbol of the pinnacle of its accomplishments and with it the last-ditch effort to regain the prestige of the past. The program did continue, however, salvaging what remained of the space program in subsequent years and continuing to launch humans. Ustinov maintained a commitment to space activities, with an eye to mimicking American accomplishments, as he did with other military technologies. Work continued on the human-crewed lunar mission, albeit without the public enthusiasm, confidence, and successes of the beginning of the decade. Therefore, Yuri Gagarin became an icon of the glorious years of Soviet power with monuments and museums in every republican capital and regional library. Beyond this effort on behalf of Gagarin's memory, there was relatively little attempt to create a new, fictional universe in which the Soviet space program was at the center.

There are three features of Gagarin's death and burial and the events that preceded it that are important to the history of the Red Stuff. First, it signified the resounding end of an era of the easy space firsts. The Soviet space program lost its political patronage (Nikita Khrushchev), administrative genius (Sergei Korolev), and finally public hero (Yuri Gagarin) all in a little over three years. The program had no immediate source from which to draw quick distractions as the Americans approached their goal of landing a man on the Moon. Moreover, to add further insult, Soviet foreign policy was under stress from dissent among its allies. Soviet tanks rolled into Prague in August 1968 to suppress Czechoslovakian liberalization.

Second, the official commemorations of Gagarin's death reveal much about the extent to which the postwar Soviet state was flexible in reaching out to the sensibilities of the Soviet—mostly Russian—population to assure the preservation of Gagarin's memory and by association its legitimacy. The managers of his funeral took few risks. The rituals and customs used during his burial and subsequent memorialization resembled ancient peasant, Russian, and more recent Stalinist traditions that continued to hold sway, even among the recently urbanized population of the USSR. The methods of his memorialization paved the way for his memory to endure through the collapse of the USSR.

Third, efforts to decipher the reasons for Gagarin's crash frustrated the minds of the Soviet population as well as those of the aerospace community. To this day, there is neither popular nor scholarly agreement on the reasons for his crash, even after a long and painstaking examination of the crash site and all available avionics evidence. Neither an official explanation nor the full and ambiguous report of Gagarin's death reached the public. In response to the absence of an official cause of death, Soviet popular culture generated rumors to account for it. Unlike previous mysteries in the USSR, the precise cause of Gagarin's death remains a mystery despite thorough and deliberate inquiry. The broader, more general conclusions about his death bring forth an indictment against the organization of one of the most esteemed parts of the Soviet state. Gagarin, who had been a Hero of the Soviet Union, became its most famous victim.

Gagarin's death marked a point at which the layers of the international space competition came away from the official stories of the Soviet and Russian space program. It marked the end of the post-Stalinist, post–World War II optimism that the Soviet Union could match or overtake the United States in spaceflight. After the reorganization, the public space program turned to a lower key, more attainable program than the one that Korolev and Khrushchev had begun. Once again, Soviet space missions methodically repeated themselves with reliable and well-tested hardware. This time, however, the program flourished without the flamboyance of a Khrushchev and the direct challenge to the United States. Moreover, the Soviet public could no longer turn to the space program as a reliable fairy tale of their relative status in the world. In a few short years, the Red Stuff went from a reliable narrative of Soviet existence to one that was beginning to generate questions and skepticism about that reality.

6

Outpost in the Near Frontier

Orbiting Space Stations during the Era of Stagnation

The 1970s marked a visible slowdown in the public culture of the Red Stuff. One can attribute part of the slowing to an overstretched government struggling to meet the expectations of a postwar generation, and part was due to a lessening in resolve and faith in human spaceflight being the assurance of popular support. Despite the deep resolve to commemorate the lost heroes of the 1960s Red Stuff, there was little to show for it in the subsequent decade. It took nearly twelve years to complete the monument to Yuri Gagarin in Moscow near Kaliuskaia Square that his memorial commission had proposed soon after his death. The final monument, a 42.5 meter (139 ft.), titanium alloy-clad modernist sculpture took its place at the renamed Gagarin Square on Lenin Avenue, near the Presidium of the Academy of Sciences, just in time for the Moscow-hosted Olympics in 1980. The technical requirements to complete the monument were enormous. Construction of the collaborative artistic and architectural design required technical contributions from the All-Russian Institute of Aviation Materials to develop the technique to cast ingots of the aluminum-titanium alloy in a low-oxygen environment.[1] After engineers refined the technique on the scale that the monument required, it took less than a year to complete it at the Balashikha Foundry and Mechanical Plant. The figure of Gagarin stands erect on top of the pedestal that resembles the plumes of a rocket's exhaust, with his shoulders swept back and arms firmly at his side, almost as though he was soaring into space on his own power. The Gagarin monument is the world's largest-scale monument made of a titanium alloy, the head alone weighing 300 kg (660 lb.). At the base of the statue is a stylized model of the *Vostok* descent module that bears the following inscription: "On 12 April 1961, the Soviet spaceship *Vostok* with a man on board flew around the globe. The first person to penetrate space is a citizen of the Union of Soviet Socialist Republics, Yuri Gagarin." A generation had been born and attained adulthood in twenty years between the launch of Gagarin and the dedication of the monument. Many of the Olympic athletes passing the Gagarin Monument in Moscow in 1980 had no personal memory of the origins of the Red Stuff. A new generation had replaced the post–World War II youth generation.

A decade after the Moscow Olympics and thirty years after the flight of Yuri Gagarin, signs of the collapse of the USSR began to show through growth in independent and skeptical domestic assessments of it. Criticism of the USSR was no longer relegated to whispered conversations among friends in kitchens. It had become a national pastime. This skepticism about promises and experiences of the past had an impact on the portrayal of the Red Stuff as well as all aspects of Soviet life. Doubt and reexamination of the mainstays of Soviet life came in two diametrically opposite points of view. These opposing views briefly took the form of two groups that dedicated themselves to assessing the Stalinist legacy. These two similarly named groups emerged to take charge of dictating the legacy of the Soviet Union and its meaning to Russian culture. The first, and most conservative, was called Pamiat' (Memory) that began in the 1970s to preserve Soviet-era monuments and landmarks as part of a Russian historical legacy. It developed into a Russian nationalist movement by the 1990s only to fade into relative obscurity in the twenty-first-century Russian Federation.[2] Their sentiment that the Soviet period was a positive period in Russian history returned after Vladimir Putin became President of the Russian Federation. The second group was Memorial. Soviet dissident Andrei Sakharov led that group. The organization began its work by making petitions to the party congresses and holding conferences at the end of the 1980s. Their underlying goal was to demand the acknowledgment of the crimes of the Soviet state. Memorial's assessment of Soviet history paralleled that of Western historians, emphasizing its tragic consequences.[3] Memorial counted collectivization, forced industrialization, the purges, and the USSR's inadequate preparation for World War II as the tragic events in Soviet history, and the lost generation under Brezhnev was counted among the casualties.

This bifurcation of interpretation of the Soviet experience reflects the difficulty that the Soviet population had in rationalizing the 74-year Bolshevik experiment. This split had its origins in the 1970s, a time when open dissent over conditions in the USSR had begun to emerge. Memorial and Pamiat' represent two extremes of interpretation of the Soviet experience, each of which began in the final years of the state. There was other, more subtle unraveling of the official Soviet narrative going on at the time. Every aspect of the USSR would come under scrutiny from both perspectives.

The Red Stuff was not immune to this harsh assessment. Each side did not limit its argument to the precise role that the Red Stuff played in society. By the close of the Soviet period, the Red Stuff stretched between its role in official Soviet society and its portrayal in the emerging and independent Russian culture. Cosmonauts became foils for discussing the human condition in the USSR. There was no rapid collapse of centralized authority over the public culture of human spaceflight. It happened very gradually. Piece by piece, the ownership of the program dissipated over the years under Brezhnev. Each appropriation did not neglect the previous but left the public image of the space program that no longer had a single focus or regnant myths. Even the times at which the state invoked its right to define the Red Stuff, the parameters were narrow, fitting only a small range of Soviet public culture. The tug of war between unofficial and official cultures had a long-term effect on the public image of the Red Stuff after the collapse of the USSR.

The unraveling of Soviet control over reality began early. There is a Russian phrase describing the period between the ouster of Khrushchev and the dissolution of the USSR: *epokha zastoia*—the epoch of stagnation. Historians date the beginning as early as the first turnabouts of Khrushchev's social reforms, pointing to the Daniel and Sinyavsky trials in which government lawyers prosecuted the writers for the words of their characters. Others point to the international oil crisis of the 1970s as declining economic growth rates resulted from increased investments in military technology. Of course, it is much easier to describe its end. In 1989 the Berlin Wall fell, and the USSR dissolved in December 1991.

The Soviet public responded to the return to repression in a deeply subversive way. During the 1980s, the Soviet joke flourished. Known in Russian as *anekdoty,* jokes were the favorite insubordinate expression of dissatisfaction with the Soviet state. One space-themed joke unfavorably compares Brezhnev to his three predecessors:

> Lenin, Stalin, Khrushchev, and Brezhnev are all traveling together in a railway carriage. Unexpectedly the train stops. After sitting for a while, each leader takes his turn to cope with the situation.

Lenin suggests: "Perhaps, we should call a *subbotnik* [a voluntary salary workday] so that workers and peasants fix the problem." Despite that effort, the train does not move.

Stalin puts his head out of the window and shouts, "If the train does not start moving, the driver will be shot!" However, the train does not start moving this time, either.

Khrushchev then shouts, "Let's take the rails behind the train and use them to construct the tracks in the front." However, it still doesn't move.

Brezhnev then calmly proposes an alternative, "Let's close the shades and pretend that we are on a rocket orbiting the Earth." The train still does not move, but the passengers are satisfied.[4]

The joke relates the public's attitude toward Brezhnev and his disposition to ignore domestic issues that were festering under his rule. The content implies that Brezhnev was content to ignore all dysfunction that was beyond his control with a series of space stations orbiting no further than Gagarin's flight in 1961.[5] While Brezhnev chose to close the curtains, others began to create their narratives of Soviet human spaceflight. These creations began small and subtle and had little effect at the beginning, but they established the framework for a post-Soviet public history of human spaceflight that would emerge by the beginning of the 1990s.

When Brezhnev replaced Khrushchev in 1964, the public veneer of Khrushchev's space program was beginning to dwindle, the consequence of multiple conditions. Khrushchev had driven space engineers to perform beyond the narrow margins of their capability, defying their ordinarily conservative sensibilities. The series of losses and failures had a negative impact on both the material and propaganda parts of the space program. By 1969, the Moon race was over, and the USSR had failed to send men there. The loss to the Americans in the Moon race forced the Soviet Union to redirect its human spaceflight program toward more reasonable goals. Hardware that engineers had developed as part of the secret Soviet effort to send men to the Moon were debuted to the public as components of new, orbiting space stations.

This change in direction necessitated that the Soviets reexamine the necessity of their regnant myths of Soviet firsts. Meanwhile, avoidance of taboo topics, such as the deaths of Komarov and Gagarin, continued. As Brezhnev approached his death, however, the role of the Red Stuff for the state had shrunk to fit specific circumstances. Soviet space stations remained in near-Earth orbit, like that of Gagarin. In 1978, The USSR began to launch cosmonauts from socialist and allied countries as a demonstration of their political allegiance to the USSR. The Red Stuff was now a key link to Soviet magnanimity to their allies. The human spaceflight program had undergone a political transformation, turning from Cold War heroism to foreign policy currency. The Red Stuff was no longer trying to upstage the Americans, but solely maintaining national prestige.

While the state began to rely less on the Red Stuff for legitimacy, artists and cosmonauts began to stake their claim on the public culture and presented their interpretations of the experience. Unlike those who followed the narratives established in official Soviet circles, these groups created their myth and mapped out their taboo subjects, creating narratives that were not always consistent with the official Soviet public culture of human spaceflight. As the state became less reliant on the image, other forces began to increase their dominance. Fields that had previously conformed to severe public restrictions on form and content began to use the Red Stuff as a foil for criticizing the Soviet state. Artists who had once submitted to strict definitions of creativity learned to evade those rules. Their creations incorporated the Red Stuff in their art along with political and social subterfuge or outright dissent. The line between official and protest music had blurred. Moreover, even the cosmonauts themselves exercised more autonomy in their own identities. While dissent was permeating the Red Stuff, the USSR under Leonid Brezhnev sought to expand the appeal of the Red Stuff beyond its own borders to allied and affiliated countries and renewed the challenge against the United States on women's rights in space. All of this took place during the decline and sudden collapse of the USSR.

Literature and Arts

This protracted period of the loosening of the official Soviet grip on the Red Stuff coincided with a flourishing of independent expression in the arts. Beginning in the 1960s, the Soviet arts came to life after decades of repression. Before the Soviet Cultural Revolution in the 1920s, Russian artists had led the world in new and innovative expression. Russian artists established new trends in literature, film, fine arts, and design. Lenin's consolidation of political power and Stalin's repression of civil society curtailed free artistic representation beginning in the 1920s and lasting through Stalin's lifetime. The creative minds that had led the Revolution were left with few palatable alternatives beyond perfecting the masking of their statements and beliefs through veiled language, and many artists fled the newly formed USSR, were arrested and/or murdered, or held their most political creations in secret until after Stalin's death.

Even after Nikita Khrushchev signaled a relaxation of some of the strictures against free expression in the arts, ambiguous and situational limitations remained. For example, under Khrushchev, officials tolerated pseudo-modernism in form, which imitated the modernism of revolutionary Russia without the political challenges to authority. Modernism in thought was not permitted. Acceptable forms of expression had to reinforce the Soviet message and did not challenge it. Stamps, znachki, and buildings could borrow from the styles of the avant-garde artists of the 1920s, but only if their messages supported the accepted state-sponsored narrative and did not venture into taboo topics.

Khrushchev's fall from power did not trigger a snap back to the Stalinist norm. In the field of literature, under Brezhnev, the editors of political literature sought to improve the acceptance of their work by cooperating with the Union of Writers. Their goal was not solely to improve their writing, but also to broaden their appeal to a more discerning and larger Soviet reading audience. But this was not to be a neat merging of missions. The policies that took effect brought well-trained fiction writers in to write propagandistic biographies of revolutionary figures. This precipitated an increase in ideological tensions between the pull of literary quality and the push against political control negotiations.[6] These tensions likely had replicas throughout the Soviet artistic world. During the efforts to meld the political and artistic worlds, the results might have been improved and produced more edifying work in the long run, but the immediate effect was to create stark contrast between the quality of the two groups (political and artistic).

After a quarter century of navigating the narrow interpretation of what constituted the proper balance of form and content, artists took the initiative and created their narration of human spaceflight. Two artists, one fine artist and one writer, took on the Red Stuff during the Brezhnev era. The first began operating well within the confines of literary socialist realism. The second, in response to his exclusion from the arts world, used his interpretation of the space program to break the current aesthetic, social, and political boundaries of fine arts. Both men broke with the regnant myths and political and technical taboo subjects of the Soviet space program. Both legacies outlasted the USSR.

The Day Lasts Longer than a Hundred Years

The biography of Chingiz Aitmatov is very important to his story. It points to the weaknesses underlying the concept of a multinational Soviet identity. As the glue that kept the USSR together began to give way, so did adherence to the forms and beliefs that held it together. Chingiz Aitmatov was a Soviet and Kyrgyz author who wrote in both Russian and Kyrgyz. He is the best-known figure in Kyrgyzstan's literature and among the best in all Turkic literature. Aitmatov was born in the Soviet Republic of Kyrgyzstan in 1928 to a Kyrgyz father and a Tatar mother. As was true for millions of Soviet children of that generation, his father was arrested and executed in 1937–38. Surviving the war, the younger Aitmatov began studying at the Animal Husbandry Division of the Kyrgyz Agricultural Institute in Frunze in 1946. He later switched to literary studies and transferred to the Maxim Gorky Literature Institute in Moscow, where he lived from 1956 to 1958. For the next eight years, he worked for the Soviet newspaper *Pravda*. His first two publications, short stories, appeared in 1952 in Russian: "The Newspaper Boy Dziuio" and "Ashim." His first work published in Kyrgyz was "Ak Zhaan" [White Rain] in 1954, and his well-known story "Jamilya" [Jamila] appeared in 1958.

His writing advanced him in the Soviet literary world. Aitmatov published his first book in 1980 while a member of the editorial board of the *Literaturnaia gazeta,* the premier journal of the Soviet Union's literary community. Aitmatov did not have a reputation as a non-conformist author after working at two of the most politically vetted publications in the USSR. His next work, a serialized book, originally came out in the popular though frequently dissenting Russian journal *Novyi mir* in 1980 and was titled *I dol'she veka dlistsa den'* [The Day Lasts Longer than a Hundred Years].[7] The series was republished as a book and renamed *Burannyi polustanok* [Burannyi Waystation] in its Russian-language edition. The title comes from the location of the opening and closing scenes of the novel. The English translation adopted the original *Novyi mir* title. The first title was an overt reference to Alexander Solzhenitsyn's *One Day in the Life of Ivan Denisovich,* the novella that recounted a single day in a 1950s Soviet labor camp.[8] Unlike Solzhenitsyn, Aitmatov's day is written in a classic socialist realist style. *The Day Lasts Longer than a Hundred Years* places its hero on a physical and spiritual journey. In this case, the hero is a Central Asian and his quest does not lead to an affirmation of Soviet life. Threaded throughout the novel is a parallel science-fiction subplot that plays out in space that makes overt criticism of the Soviet state. In this story, the Red Stuff represents neither heroism nor villainy, but its personifications can be unwitting victims of the Soviet hierarchy.

The Day Lasts Longer than a Hundred Years begins and ends near the village of Boranly-Burannyi in Kazakhstan.[9] Edigei Zhangel'din, or Burannyi Edigei, as his friends know him, learns that his old friend Kazangap has died.[10] The two had settled together at the railroad junction forty-four years before. Edigei decides to honor his friend by taking him and burying him at the legendary Kazakh cemetery at Ana-Beiit with full Kazakh rituals.[11] He takes a daylong journey with his friend's corpse as they travel along the rail line. His memories of their youth fill his thoughts throughout the journey, and he dreams of traditional Central Asian folklore as he naps on the way.[12] When Edigei arrives at the cemetery, a Soviet official informs him that he cannot bury his friend there, because the cemetery is about to be destroyed.[13] Discouraged, Edigei finds another burial site and leads the burial rituals himself to the best of his abilities, drawing on his faded memories of his culture and language. Dissatisfied with the outcome of his journey, he returns to the guard to argue over the destruction of the cemetery. The resolution of Edigei's story is decidedly anti-Soviet.

The human spaceflight subplot in this story takes place on board an orbiting space station, but it connects to Edigei's story because the launch facility is nearby. At the hour that Edigei learns of his friend's death, he also witnesses a launch at the cosmodrome Sary-Ozek in the distance.[14] The rocket launch is an unscheduled mission. Its purpose is unannounced, secret, and frightening for the local population, who have grown to expect advanced warning of explosions. As the story unfolds, the reader learns that the Soviets have launched a replacement crew to the orbiting international space station *Parity*. The launch coincides with the near-simultaneous launch of a crew from the United States to the same station. This scenario borrows from the events of the *Apollo-Soyuz* test project that took place in 1975. It is an emergency because the unnamed current cosmonaut and astronaut (Soviet and American) at the station have not responded to messages from ground control.[15] As Edigei's ill-fated mission unfolds, the reader learns the secrets of the cosmonauts 1–2 and 2–1, as they are known in the book. Fear stems from the knowledge that the crew has contacted an advanced civilization from the planet Lesnaia grud' [Forest Stand]. Despite bilateral orders to the contrary, the space station crew continues their communications with the planet. In addition, they ignore the signals from the two spacecraft as they approach the station. As the official announcement of the situation explains:

> *Parity* had not acknowledged signals from the joint control center on Convention for more than twelve hours now and was not reacting to the signals from the two vessels approaching it. They had to find out what had happened to the crew of the *Parity* space station.

When the also unnamed rescue spacefarers arrive at the station, they find a letter from their predecessors explaining their decision to accompany the inhabitants of the advanced planet back to their home.[16] They have abandoned their station.

Press reports of increasingly alarmist orders from the ground controllers to the errant cosmonauts punctuate Edigei's journey. The orders begin with pleas to return home and to cease contacts with the *lesnogrudtsy* (inhabitants of the planet Lesnaia grud'). The final ground control orders concerning cosmonauts 1–2 and 2–1 declare that the new inhabitants of the space station should make no contact with the previous occupants of the station on the assumption that alien culture would contaminate them through their contact with the foreign species.[17] This xenophobic nature of Soviet rule is the deep message of the book. Aitmatov drives this message home through his narrative of the conflict on the ground and in orbit. As far as the Soviets are concerned, Edigei is as contaminated as cosmonauts 1–2 and 2–1. Edigei expresses his contamination through his attempts to bury Kazangap at the traditional gravesite with traditional methods. The twist is that Edigei's alien culture is his own native culture that he is attempting to revive.

The novel ends as the Americans and Soviets launch a protective satellite system that will form a barrier against all contact with the people of Lesnaia grud' to prevent future contamination from alien worlds. Launches from the Sary-Ozek Cosmodrome begin and end Edigei's story. As he rides to confront the officials about the cemetery, the rockets' launch startles Edigei, and he decides to return home to his wife.[18] The planetary shield also serves as a cultural shield against the indigenous traditions on Earth. Edigei suffers the fate that the cosmonauts 1–2 and 2–1 elude. The state thwarts his and the world's contact with an alien culture and consequent enlightenment.

The anonymity of the space station cosmonauts is significant in Aitmatov's story. He calls them cosmonauts 1–2 and 2–1, in stark contrast to the heroism and celebrations that defined the Red Stuff in in his contemporary USSR. Aitmatov is calling out the fake association between traveling in space and the government. According to him, it did not matter who the cosmonauts were. It only mattered that they obey the restrictions against contact with foreign cultures, making the point that the Americans were no longer considered alien, but any culture that was not officially allied could potentially contaminate all who came into contact with it.

The Man Who Flew into Space from His Apartment

This dissociation between the state and cosmonauts is a common feature of late-Soviet portrayals of the Red Stuff. Simultaneously, a little-known Soviet child's book illustrator was planning a subversive art installation that would deconstruct the official imagery of the Red Stuff. Here, too, the artist's biography provides hints as to the ways in which he would choose to illustrate the Red Stuff. Ilya Iosifovich Kabakov was born in Dnipropetrovsk in Ukraine on 30 September 1933. Ilya was evacuated during World War II to Samarkand with his mother. There he started attending the school of the Leningrad Academy of Art that also had been evacuated to Samarkand. His classmates included the painter Mikhail Turovsky. From 1945 to 1951, he studied at the Art School, Moscow; in 1957, he graduated from V. I. Surikov State Art Institute, Moscow, where he specialized in graphic design and book illustrations.

Because of his graduation from the Surikov Institute and subsequent productivity, Kabakov became a corresponding member of the Union of Artists in 1959 and a full member in 1965. Membership in the Union of Artists was a prestigious position in the USSR and brought with it substantial material benefits and perks. It secured him a studio, steady work as an illustrator, and a relatively healthy income by Soviet standards. In general, Kabakov illustrated children's books for three to six months each year and then spent the remainder of his time on his projects. For the first years of Khrushchev's rule, membership in artists' and writers' unions granted members income and the freedom to create as they pleased in their spare time. The situation began to change after six years. In 1962, there was an exhibition of the Moscow Artists' Union at Manezh Hall. This exhibit incurred Nikita Khrushchev's infamous attack on modern art that firmly stated his limits to cultural reform in the USSR. The incident marked the end of Khrushchev's Thaw Era that had begun in 1956. It did not mark the end of modern art in the USSR, however. As had happened in previous eras, the modern went underground. That same year Kabakov produced several series of "absurd drawings" in his personal time. He continued to develop his style unofficially as he worked and prospered as an illustrator.

Even from the beginning, his private projects had a subversive flavor. In the original "Shower" series from 1965, Kabakov depicts a man standing under a shower without water. Kabakov interpreted the work as a universal but straightforward metaphor about the individual who is always waiting for something, but never receives anything—the life of the average Soviet citizen, who was known to queue up reflexively without knowing what was for sale at the head of the line. Kabakov was not explicit in the meaning of his work, and neither were his Moscow colleagues. It was the Italian art critics and critics of communism who overtly interpreted the work as criticizing Soviet culture and its lack of material reward. The minor publicity Kabakov received prevented him from getting work as an illustrator for four years, forcing him to work under someone else's name. The use of an alter ego would become a standard tool in Kabakov's unofficial artwork for the rest of his life in Russia.

Kabakov continued to stretch artistic boundaries in the USSR when he allied himself in the 1970s with the Moscow Conceptualists. This group was an unofficial group of artists that was famous for its preoccupation with Soviet power.[19] The group developed out of the Stretensky Boulevard Group, named for the street on which the Surikov Art Institute was located and known as the Sots group. It is problematic to determine who was a member of the group, as the term is fluid, broadly encompassing the Sots artists and the Collective Actions group, which both were influential in the construction of Russian Conceptualist art.

Before creating installations and architectural models, Kabakov created fictional albums. He had created fifty. Each album was a story about one character who is often able to overcome the banality of everyday existence, or "of a small man, possessed by big ideas." The first ten albums are a series called Ten Characters (1972–75). A man, attempting to write his autobiography, realizes that nothing much ever happened to him, and most of his life amounted to impressions of people, places, and objects. Therefore, he creates ten different characters to explain his perception of the world. In 1985, Ilya Kabakov first displayed the ninth of his Ten Characters in his installation Chelovek uletevshii v kosmos iz svoei komnaty (The Man Who Flew into Space from His Apartment).[20] The installation was unlike previous pieces of Russian modern art. The viewer could not enter the room, only peer into it through slots in the wall. The purported cosmonaut, "The Man," was not present, either. All indications in the display led the visitor

Figure 15. A closeup view of Kabakov's maket (artist's model) *The Man Who Flew into Space from his Apartment,* providing the details of the life inside this one-room apartment. The power of the propaganda contents that line the apartment have propelled the occupant into space. Malcolm Park editorial/Alamy.

to believe that the man had succeeded in his quest and flown into space. The ceiling had a hole in it, and all that remained in the room was a spring-loaded contraption, a table, a bed, a pair of men's shoes and the posters and space memorabilia that decorated the walls of the space traveler's apartment.

These posters consisted of classic Soviet propaganda including, but not exclusively, spaceflight propaganda. The implication was that the posters were the source of energy that propelled the man into orbit. Art historian Boris Groys has asserted that the posters supplied the energy necessary to send Kabakov's cosmonaut into space:

> So, you could say that the hero of Kabakov's installation was able to fly out of his room because he had accumulated the energy inherent in the posters, which in turn tap into the collective energy of the Soviet people.[21]

Kabakov's cosmonaut was able to absorb the cosmic dream of the Red Stuff from the communist ideology by appropriating the energy from the posters to carry him into space. "The Man" has taken the Soviet material and visual culture, curated it for himself, and thereby generated enough energy to launch himself. This act, like the artist's work, was subversive. The occupant of the apartment had appropriated the power of utopian aspirations of spaceflight to fulfill an individual dream. In the 1960s that dream had belonged to the anonymous collective of scientists, engineers, technicians, and workers and the state. In the words of art historian Boris Groys:

> So, you could say that by his misuse of the collective cosmic dream, Kabakov's hero has liberated it from its misappropriation by the ideological apparatus of the Soviets.[22]

Reclaiming the revolutionary energies of spaceflight that the party had claimed during the 1960s was a way to demonstrate that the bond between state power and utopian energy was indeed soluble. Kabakov had begun the process of disconnecting human spaceflight from an increasingly unpopular and unstable political system. In one way, he had come to the same conclusion that cosmonauts would on their own. The lack of direct attention from the state had liberated symbols for individuals to curate and create meanings of their own. Cosmonauts sought their own version of the Red Stuff.

Cosmonauts Refining Red Stuff

While artists were creating their own version of the Red Stuff, its meaning among the men and woman who flew into space was changing as well. The Soviet cosmonaut corps, isolated at their training center far outside Moscow, was a small and cohesive group. Before Yuri Gagarin's death, Soviet cosmonauts had begun to establish their rituals leading up to launch. For example, the primary and backup crews were isolated and monitored together the night before launch to assure their health. They passed the time together in isolation. Other traditions stemmed from the rituals and superstitions of pilots. For example, except for Valentina Tereshkova all urinated on the wheel of the bus that took them to the launch site. Gagarin had done this as a matter of necessity before his own flight. Others repeated the act. The last remaining members of Stalin's Falcons left control of the program to younger officers who spent more time in the cosmonaut corps than they did in the Air Force. They built on their own experiences to determine what brought good luck or tempted fate. As externally enforced rituals dwindled, cosmonaut-generated ones took their place. For example, visiting Gagarin's office turned memorial was added to the stations of reflection before a flight.[23] The list of ritualistic steps before and after spaceflight grew to twenty-nine in the post-Soviet years, incorporating blessings from Russian Orthodox priests as the Church nurtured a close allegiance with the state. These rituals remained an open secret among the cosmonauts and the technicians and engineers that surrounded them and became official through the addition of International Space Station crews' incorporation into flight preparations since the decommissioning of the US space shuttle in 2011. Each step emerged as the cosmonauts established their own customs over the years.

These customs paired with official and semi-official prohibitions. After the deaths of Vladislav Volkov, Georgi Dobrovolski, and Viktor Patsayev during their entry in the *Soyuz 11* return capsule after a three-week stay on board the Salyut 1 space station in 1971, a list of forbidden and encouraged steps grew. Prohibition against launches on 24 October originated from a non-human spaceflight disaster but applied to all launches since 1960. In response to the *Soyuz 11* mission, never again would a cosmonaut pre-autograph and post-date his signature, as Patsayev had done just before the ill-fated *Soyuz 11*. He had autographed a doll for one of his colleagues' daughters and post-dated the signature for after his planned return. His colleague later sold the doll at auction.[24] As a result, cosmonauts decided for themselves to never again tempt fate. Cosmonauts will not sign or post-date memorabilia before a flight. On the technical side, the Soviets decided to never again risk human spaceflight without the safety precaution of wearing emergency spacesuits during launch and entry.

Along the lines of less critical missions or fate-tempting activities, the cosmonauts had more leeway on their activities and rituals. Nevertheless, in instances where the additions seemed innocuous, a subversive element could be present. It might have been only an unrelated footnote to Soviet culture in the Brezhnev period that the first Soviet "Eastern" (imitation of the American Western genre) film was released. During the 1960s, relaxation of cultural import restrictions facilitated an unnoticed permeation of American culture into the USSR. Music, most notably that of the Beatles, flooded youth culture, but there was also an awareness of American and Western film culture. Soviet citizens were profoundly aware of the cultural impact of the American Western, which was full of American hero legends and wilderness imagery. Soviet film producers were proud of the creation of their version of the Western, dubbed the "Eastern" to reflect the geographic direction of the Russian frontier. It was not merely a transplant of

American culture to Soviet Central Asia. It combined Russian and Soviet frontier and imperial culture and mythology in an entirely new genre.[25]

Due to its novelty, *White Sun of the Desert* became one of the most popular Russian films of all time. The movie blended action, comedy, music, and drama and has achieved the status of a cult film in Soviet and Russian culture. The story follows the adventures of a hero as he tries to go home after the civil war. His dream of returning to his Russian bride is interrupted by a series of improbable events that challenge him along the way. His main obstacle takes the form of corrupt local officials, the disquieting practices of the indigenous people, and a harem, which he has rescued. The film dialogue has contributed many popular expressions to the Russian language. It features an ironic theme song, "Your Honor Lady Luck." Soviet dissident poet and lyricist Bulat Okudzhava composed the song in 1967, long before he made any official recordings. He gained early fame for penning lyrics in support of playwrights Andrei Sinyavsky and Yuli Daniel, whom the Soviet state prosecuted for the words of their characters. The song speaks pessimistically of loneliness, hope, luck, and love. Director Vladimir Motyl made a canny decision to sanction Okudzhava as the composer for this excellent Mosfilm production, adding a layer of irony to what was otherwise a fantasy adventure film. By that act, he had sanctioned Okudzhava's music as a cultural artifact that the official Soviet government had previously tried to ignore. This decision was fateful, as cosmonauts incorporated the film into their culture.[26]

The launch of the *Soyuz 12* mission in September 1973 marked the return to human spaceflight in the USSR after the tragic conclusion of the *Soyuz 11*/Salyut 1 mission two years prior. Engineers redesigned the spacecraft, correcting the fatal valve flaw that had cost the previous crew their lives. The reimposed requirement that the crew wear pressure suits during launch and reentry forced a reduced crew size, from three to two—Commander Vasily Lazarev and Flight Engineer Oleg Makarov. Unsuccessful attempts to replace Salyut 1 with a civilian or a military space station curtailed the *Soyuz 12* mission. Lazarev and Makarov had only to demonstrate that the new version of the *Soyuz* could successfully launch, orbit, and return to Earth with no loss of life.

The modesty of this mission meant that cosmonauts who attended the launch did not have time to return to Moscow between launch and landing. They would have to wait in Baikonur with little to amuse themselves. During their wait, Major Nekrasov in the Strategic Rocket Forces made a recommendation for a movie to watch to pass the time, *White Sun of*

Figure 16. The cover of the Soviet "Eastern" film *White Sun of the Desert*, which became a ritual viewing piece for Soviet cosmonauts in the 1970s. The tradition continues to this day—cosmonauts and all who launch from the Baikonur Cosmodrome watch the film the night before launch.

the Desert. There was no copy available in Baikonur, but the cosmonauts were able to use their resources to have a copy of the film brought to them via the Kazakh capital Alma-Ata. They watched and enjoyed the movie and the crew landed safely. That copy of the film remained in Baikonur, and the cosmonauts incorporated a viewing of the film into their preflight good luck rituals. With the addition of international spaceflights, *White Sun of the Desert* and its soundtrack gained immediate identification with the cosmonauts and human spaceflight.

A more important question to ask about the movie is why it remains a part of cosmonaut culture. At some level, the film had a cultural resonance among the cosmonaut corps, like the Russian SSR. There are several reasons why it has appeal to that group. At the basest level, *White Sun of the Desert* belittles other nationalities of the Soviet Union, mocking the nomadic Islamic cultures, which at one time inhabited the region where the cosmodrome sits. The film also hearkens back to the tradition of the Russian frontier and empire literature of Lermontov and Gogol. This sensibility remains in the language of referring to the launch trip to Kazakhstan as an expedition. Thirdly, the movie is famous for its collection of one-liners, some of which have a double-entendre resonance among cosmonauts. The expression "the East is a delicate matter" refers to the East, *Vostok* in Russian, that is easily misinterpreted as a reference to the spacecraft of the same name. The saying "Are there questions? No, there aren't!" can be interpreted as a sardonic reference to the implausibility of objecting to any direct commands from superiors. The expression, "Mahmud, set the fire!" is a punchline for anyone about to be launched into space on top of hundreds of tons of rocket fuel. Finally, the setting, in the desert of the East, makes the movie's backdrop of the coast of Turkmenistan indistinguishable from western Kazakhstan.

The Brezhnev Doctrine in Space

Even as artists and cosmonauts were making their independent tweaks to the Red Stuff, the Soviet state had not completely abandoned the image of humans in space. There was still a (somewhat diminished) role for cosmonauts to play in national politics. Interkosmos was an international space program out of the Soviet space program, designed to give nations on friendly terms with the Soviet Union access to space. The program began in April 1967 as a multinational agreement among Warsaw Pact countries to conduct robotic research satellite missions. Interkosmos counterbalanced the American NASA efforts to encourage aerospace development among their allies. After a few successful scientific satellite missions, the Interkosmos program expanded to include human spaceflight. The first human-crewed mission occurred in February 1978. The Interkosmos missions enabled fourteen non-Soviet cosmonauts to participate in *Soyuz* spaceflights between 1978 and 1988. These missions made human spaceflight become international as the USSR recruited its allies to send guest cosmonauts to visit its orbiting cosmonauts. The Red Stuff was now a cordial host to other nations.

The human spaceflight Interkosmos program was one of the few times that the Brezhnev regime directly dictated the substance of the space program. The Interkosmos was a tool of the Brezhnev Doctrine of foreign policy. The Brezhnev Doctrine called for the USSR to intervene by any means necessary wherever socialist rule was threatened by internal or external forces. The Soviet invasion of Czechoslovakia was the first expression of the doctrine.[27] Leonid Brezhnev reiterated it at the Fifth Congress of the Polish United Workers' Party on 13 November 1968: "When forces that are hostile to socialism try to turn the development of some socialist country towards capitalism, it becomes not only a problem of the country concerned but a common problem and concern of all socialist countries." It followed logically that all socialist countries would benefit from remaining united. The Interkosmos program made a clear statement of the hierarchy of obedient nations.

The first Interkosmos cosmonaut to fly was Vladimír Remek, a pilot from Czechoslovakia on a mission to the Salyut 6 space station in 1978. Cosmonauts from Poland and the German Democratic Republic followed him later that year. Flights with nationals from Bulgaria and Hungary followed in 1979 and 1980.[28] The program introduced non-Warsaw Pact cosmonauts from Vietnam, Cuba, and Mongolia later that year, leaving Romania as the last of the Warsaw Pact nations to send a cosmonaut to a space station. Charles de Gaulle interrupted the sequence in 1968 when he became the first non-aligned head of state to visit the Baikonur Cosmodrome.[29] Franco-Soviet agreements that included cooperation in space led to Jean-Loup Chrétien's *Soyuz T-6* mission to Salyut 7 in 1982. Chrétien's mission paved the way for other non-treaty-aligned countries to make their way to Soviet space stations, including pilots from India, Syria, and Afghanistan, before the program ended in the late 1980s. The public face of the Interkosmos program made a clear message to the international community about the status and currency of the Soviet human spaceflight program. Alliances, treaties, and cooperation were the currency that converted into placing pilots into orbit. The addition of an international corps of cosmonauts made a domestic case for an increasing profile of the Soviet space program in the world. The Soviet Union was no longer solely trying to outpace the US in public coverage of the activities on board the Salyut series of the space stations from the mid-1970s. In absence of pressing goals for firsts, they could create the impression that the USSR was the only human spacefaring nation. Between the years 1975 and 1981, that was the reality, especially if one ignored the preparation for the American shuttle program, which the Soviets did. The American introduction of the first class of shuttle astronauts began to change the image of spacefarers, however, and rekindled old issues within the Soviet human spaceflight community.

The Return of the Women's Issue

The 1980s offered the USSR another and unanticipated opportunity to redefine the Red Stuff in matters concerning women and to relitigate their rivalry with the United States. In January 1978, the National Aeronautics and Space Administration announced its first new astronaut corps selection in nine years. The NASA Astronaut Group 8 broke diversity barriers that had existed since the creation of the space agency. Included in the thirty-five candidates, six were white women, and three were black men. NASA astronaut recruitment had changed significantly since the *Apollo* program. Planning for the American shuttle program had led to a redefinition of the role of astronauts. NASA did not define a women's role in the space program, but its redefinition of an astronaut opened opportunities for accomplished women and men to fly in space who were not former military test pilots. The shuttle corps featured pilots, engineers, and mission specialists who would contribute to engineering activities, payload deployment, and scientific research in space.[30] As internally motivated as this change had been, the perception that the Americans had returned to social issues left unresolved in 1963 caught those responsible for the Soviet cosmonaut corps off guard. The first Soviet women's brigade left the cosmonaut training center in 1969. There was no thought of reconstituting it, and no apparent thought of returning to Khrushchev's gender experiment.[31] There were no women in the Soviet Air Forces, and female engineering graduates from the prestigious Moscow Aviation Institute and the Baumann Higher Technical School were rare. The only established pipeline to the cosmonaut corps that could potentially nominate women was the Institute of Biomedical Problems, the medical institute that had produced Dr. Boris Yegorov, the third in the 1964 *Voskhod* flight. The goal was to find a woman to challenge the American Class of 1978 and rise above the controversies that had swirled around Tereshkova's mission in 1963.

The careful selection of Svetlana Yegenyevna Savitskaya as a counterpoint to the American women astronauts to preempt their missions in space was, in retrospect, an obvious choice. Savitskaya was the daughter of a World War II fighter ace, who had twenty-two individual and two group victories. During the 1970s Evgeny Savitsky was the Commanding General of Aviation of the Air Defense Forces of the Country and Marshal of Aviation just before his daughter's selection. In contrast to her male counterparts in the USSR, Savitskaya was neither a military pilot nor an engineer. She was a test and sports pilot who earned eighteen world records, starting from 1974, in MiG aircraft and she possessed three records in team parachute jumping. Her fame extended to the international arena, including first place at the 6th Fédération Aéronautique Internationale World Aerobatic Championship in 1970. Twenty years after Tereshkova's flight, a female cosmonaut joined the corps based on her aeronautical flight experience. Savitskaya had not been a military pilot. As an aerobatic pilot, she was on par with other American and European pilots who competed in the international arena. With her piloting background, there would be no public doubt as to her flight qualifications.

Svetlana Savitskaya flew in space twice in less than two years. Her first mission being on *Soyuz T-7* on 19 August 1982, she became the second woman in space, preempting Sally Ride's STS-7 already-announced mission on the space shuttle *Challenger* by ten months. Savitskaya's second flight was again in the third seat of a visiting mission to the Salyut 7 space station on board *Soyuz T-12*. During that July 1984 mission, Savitskaya became the first woman to walk in space, preempting Kathryn D. Sullivan's spacewalk during the STS-41-G on 11 October 1984. NASA had announced this mission in November 1983.[32]

It may never be documented that Savitskaya's selection for the cosmonaut corps was an attempt to upstage the United States in demonstrations of women's equality, but the timing indicates that it was more than coincidence. The Soviets sent her into space twice in as many years, thus abandoning all pretense of having more than one woman cosmonaut trainee or a bench of women trainees. The Soviets had long abandoned the domestic pretense that women cosmonauts would encourage future generations of women in engineering when they directed Valentina Tereshkova into a political career.[33] On the domestic front in the 1980s, there was less incentive to make the case that Savitskaya was a symbol for Soviet women. The women's movement in the USSR was moribund. There was no need to create cultural incentives for women to remain in the workforce in traditionally women's fields. Women participated in the workforce at high levels out of necessity. The only plausible motivation to select Savitskaya was the desire to create an illusion of equality that paralleled the US program that set the Soviet program up as an alternative for what the United States was doing. The Brezhnev Doctrine had reinserted party politics into the space program for a brief instance. Even though it did begin to accept women for the cosmonaut corps after Savitskaya, they would all be engineers and physicians. Only one would fly, ten years later, after the collapse of the USSR. At that time, Elena Kondakova's selection, too, would be tinged by accusations of nepotism.[34]

Mir Means Peace, World, and Commune

One of the greatest lost opportunities to enhance the impression of the Red Stuff did not result from new hardware and programs to the Soviet space program but was the rescue of the failing piece of orbiting hardware. In June 1985, two cosmonauts made a dramatic rescue of the Salyut 7 space station. Soviet secrecy prevented the public from learning about this event in any detail until well into the twenty-first century.[35] The story betrays the impact of lingering Soviet self-doubt that caused them to shun public exposure. This fear that their inadequacies would face public examination restrained official promotion of the Red Stuff in the final years of the USSR and the early years of the Russian Federation. Had other instincts prevailed, the Soviet and Russian cosmonauts might have been subject to levels of adoration that approached that of Yuri Gagarin and his contemporaries.

In February 1985 the no longer occupied Salyut 7 space station repeatedly failed to respond to its regular status check signals from Soviet flight control. Its flight patterns as seen from ground observatories seemed odd. Course correction orders from ground computers produced no results. The Soviets faced three options. First, they could allow the dead ship to continue tumbling until the station reentered Earth's orbit. This option had the likely result of embarrassing the USSR when large pieces of the station returned to Earth, potentially over the most populous areas of the world. The second option was to seek American assistance in either controlling the reentry or reviving the station. The Soviets chose a third option of sending their best pilots and engineers to dock with an unresponsive station that was exhibiting no attitude control, reviving the station, and preparing it for additional crews so that eventually the ground controllers could send it onto a safe course into the Pacific Ocean. *Soyuz T-13* launched on 6 June 1985, with commander Vladimir Dzhanibekov and flight engineer Viktor

Savinykh. Against the odds, the pair successfully docked with the station, revived the life-support system, restored the electrical system, and repaired the solar array sensor that allowed the station to maintain its power. Western observers, once they understood the stakes, described the mission as "one of the most impressive feats of in space repair in history."[36] These accolades emerged well after the fact and despite Soviet efforts to shroud the mission into the mundane. Contemporary Soviet press had declared Dzhanibekov and Savinykh's mission was a routine visit to revive the station for future occupation. In retrospect, they had let an opportunity to celebrate magnificent feats pass them by. By the closing years of the Soviet space program, this culture of caution dominated the Red Stuff. In hindsight from the twenty-first century, their caution was well-placed.

Throughout the Brezhnev period, the human spaceflight program of the USSR sought its own direction without relentless political mandates. The series of Salyut space station missions in the 1970s set long-term orbiting missions as their strength. On 20 February 1986, TASS announced the launch of Mir. The new space station was to be a five-year replacement for the Salyut 7, which had survived for the expected four years. The launch of Mir also coincided with a successful Soviet scientific mission to Venus and the anticipation of the subsequent encounter with Comet Halley.[37] The transition from one space station to another did not proceed as smoothly as planned. In June 1985, cosmonauts Viktor Savinykh and Vladimir Dzhanibekov were dispatched to rescue the Salyut 7 outpost once it became unresponsive to ground commands. Their mission had been to revive the station's communications and navigation systems to allow the completion of the transfer of the equipment and a controlled reentry into the Earth's atmosphere.[38] This seriousness and the emergency nature of the mission was not public at the time. The following year two experienced cosmonauts became the first Mir crew to visit the retiring Salyut 7 space station. Leonid Kizim and Vladimir Solovev first docked with the Mir space station on 15 March 1986. On 6 May, the pair were the last visitors to the station. News of the nuclear meltdown in Chernobyl broke in the last week of April 1986, briefly eclipsing the bright, promising future of the USSR in space.

Almost as a postscript, the adventures of Kizim and Solovev on board Salyut 7 in 1985 became a feature film in 2017.[39] An American-trained Russian director, Klim Shipenko, created reenactment of the retrieval of the space station with the close assistance of the Russian Space Agency and the Russian Space Corporation Energiia. The film not only re-created the dramatic events of the mission but delved into the emotional and philosophical makeup of the two cosmonauts. Unlike the cosmonauts of previous films from the gilded age of human spaceflight, these men were self-possessed and capable of overcoming a stagnant bureaucracy through skill alone. This film was not immune to the nationalistic sensibilities that pervade twenty-first-century Russia. The contemporary Soviet concern over the US space shuttle and its potential to retrieve spacecraft, including Salyut 7 in orbit, is played up in the film to the point of appearing as parody to the Western mind. And sexism is featured in the portrayals of a Savitskaya-based character, who freezes at a key moment in her mission, and a hypervigilant female flight doctor whose exaggerated caution threatens the *Soyuz T-13* mission. For Shipenko, the Red Stuff had been born again.

Despite their chronological coincidence, Mir did not suffer the dramatic failure like the reactors at Chernobyl. However, it did not entirely fulfill its expectations. It was a modular space station, adding modules well after the central section had served past its projected lifetime. The plan was for Mir 2 to replace it after five years. Instead, it remained in orbit for fifteen years. Over 100 cosmonauts and astronauts from various countries visited and lived on the station; it was continuously occupied from 1989 to 1999 and had made more than 85,605 orbits by the time controllers sent it to crash in the South Pacific on 23 March 2001. Even surviving for as long as it did, Mir never lived up to expectations of becoming the base structure of what was to become a permanent Soviet-Russian outpost in orbit. The hardware survived the economic collapse of the USSR, but the program did not. After coaxing from the US and other International Space Station partners, Mir's deorbiting went flawlessly, and what remained of the station came to rest on the seabed of the South Pacific.

Mir endured a few mishaps during its time in orbit, most notably an onboard fire and a wayward *Progress* supply ship poking a hole in the Spektr module. Its accident-prone service history made the analogy of an old car well past its prime clear. Mir was no longer cost-effective to repair. Boris Yeltsin's final blow to the official heroic narrative of the Russian cosmonaut was signing the agreement with George H. W. Bush for joint space station operations in June 1992. This multilateral agreement marked the end of the geopolitical competition that had spawned the Space Race. Yeltsin's abandonment of the heroic space narrative removed it as a public display of Russian might, ironic given that the Soviet program became the venue for US-Soviet cooperation. This imagined future had been the topic of disparagement in official circles in the 1960s. The fantasy of *The Silent*

Star and Klushantsev's *Planeta Bur'* had become a reality. Internationalism notwithstanding, however, Mir remained a Soviet station and added the impression of the Red Stuff as a long-enduring mastery of near-continuous occupation in low-Earth orbit.

The Last Soviet Red Stuff

The enduring impression of the Red Stuff's identity from Mir took on the embodiment of the man known as the "last Soviet cosmonaut." Sergei Konstantinovich Krikalev was born almost a year after the launch of *Sputnik* and was not yet three years old when Yuri Gagarin flew in space. The son of a naval engineer and a schoolteacher, he followed his father's example and graduated from the machine-building faculty of the Leningrad Military Mechanical Institute with a specialty in the production of aviation equipment in 1981. On 25 December 1991, the day that the Soviet Union was dissolved, Sergei Krikalev became the last Soviet man in space, having launched from the USSR and returning as a Russian citizen.[40] For him, the transition was smooth. His cosmonaut colleagues from Kazakhstan and Ukraine had to choose between their nationality and citizenship.

In 1992, the Soviet space program became the Russian space program. Those involved in human spaceflight of the Soviet Union began combing over the details of renegotiating Soviet identity. For cosmonauts who had a non-Russian nationality, the choice was a simple one between the location of their births and that of their careers. For the public of the successor countries, the choice was much more complicated. They had to pick through the threads that had formed the tapestry of the space program and determine for themselves where their affiliation lay, either to the heroes of the Cold War or the experiences of their childhood. Alternately, personal affection could be held for the design of the small pins and postage stamps or monuments and museums that the Soviets had distributed throughout all republics.

While the public and individuals were deciding what role the legacy of the Red Stuff played in their lives, the cosmonaut corps took definitive steps to define their organization, mixing both secular and religious rituals, including the open institutionalization of the preflight ritual blessing of Russian Orthodox priests.[41] Priestly blessings before a flight were a public break with the official atheism of the USSR even as enforcement of the policy had dwindled since Khrushchev.[42] It was nowhere near the first flirtation with religion among the cosmonauts. There had been over thirty

Figure 17. One of the secular rituals that cosmonauts introduced to their launch preparation is autographing a large photograph of the *Soyuz* launch vehicle immediately prior to launch. This photograph is on public display at the Baikonur Museum on the premises of the launch complex.

years of unspoken and oblique references to Russian Orthodoxy in the human spaceflight program. The most obvious had been the timing and rituals followed for Yuri Gagarin's funeral that obliquely mimicked those of the Orthodox ritual. The obvious and clumsy substitution of a party corner for the traditional icon corner in his mother's house was done in such a way that the thin veneer of atheism was even more transparent. The first prelaunch blessing by an Orthodox priest was the final step in making the Soviet space program Russian. It was clear that the centralized control of the Red Stuff was losing its impact. Growing public skepticism of the government received fuel from the broad reach of centrally broadcast television, which, instead of strengthening control, allowed individuals their own domain of interpretation.

Television

Soviet television, during the mid-1980s, had the broadest reach to the Soviet population. By the 1980s, there was a television in most Soviet households. While television was relying on the excitement over spaceflight to hold the public's interest, there was a fringe movement among Soviet artists to deconstruct the Red Stuff.

In contrast to American television, which began to play a role in domestic politics through its broadcast of the civil rights movement, the 1960 presidential campaign, and the Vietnam War, Soviet television was not a significant media until the 1980s. During the period between Brezhnev's death in November 1982 and the collapse of the USSR, there were official attempts to address public dissatisfaction with the government through the reorganization of Soviet television. The limited scope of national broadcast coverage and programming during the early 1960s made television an insufficient force to popularize spaceflight.[43] As television service grew, Soviet state television became the eager recipient of feeds from the space program. However, through the 1960s and 1970s it acted like video news service more than a traditional Western reporting enterprise. Soviet secrecy placed strict limitations on material that Soviet television could broadcast and when it could be shown. As mentioned in the previous chapter, only certain Soviet elites viewed the live broadcast of the *Apollo 11* Moon landing in July 1969. Even though Soviet cinematographers filmed every space launch beginning with *Sputnik*, the first live national broadcast of a Soviet space launch did not take place until 1975 for the launch of *Soyuz*

19, the Soviet spacecraft that would dock with the American *Apollo* as part of the *Apollo-Soyuz* test project. At a time when Soviet confidence should have been high for a well-practiced launch for a much-rehearsed space mission, even the coolest Soviet diplomat, Anatoli Dobrynin, was deeply apprehensive:

> Nothing like the live *Soyuz* launching had ever been shown before by our television to the Soviet public—let alone the rest of the world. I must admit that I was in a state of nervous excitement as the broadcast was beamed into the State Department conference hall on July 17 before a broad American audience. There was the President, the Secretary of State, other cabinet members, members of Congress, media, and all the rest. "What if something goes wrong?" I thought. But nothing went wrong, and everybody in the hall relaxed.[44]

It would be over ten years before changes took place in the Soviet television that reflected a more relaxed and self-confident attitude in broadcasting the news to the Soviet population. They held launch film until the nightly broadcast of the news, not interrupting daily programming for the event. Nowhere was this as immediately apparent as the transformation of the nightly central news in the late 1980s.

Before the transformation in 1986, the old *Vremia* news opening was remarkably dry and punctual. A few seconds before 9 p.m. Moscow time a black-and-white image of a clock appeared on television screens. As the second hand passed the twelve and the clock struck nine, one of a cast of rotating news announcers appeared behind a desk with the image of the Moscow Kremlin projected onto a screen behind him. While the dry opening granted the *Vremia* presentation official status, it also drew contempt as representative of all that was wrong with the Soviet system in the 1980s.[45] Under Brezhnev, the program had the international reputation for alternating between outright propaganda, tractor production celebrations, and harvest reports. After Brezhnev's death, there were attempts to improve the popularity and credibility of the news.

Political scientist Ellen Mickiewicz has scrutinized the content of Soviet news broadcasts during the mid-1980s and determined the importance of science and space reporting to the upbeat and visual aspect of the news.[46] Mickiewicz had pointed out that even before significant reforms took place in Soviet television, during 1984 and 1985, space reporting took a leading role in broadcast news:

Space stories on the news had high status. Space stories were frequently among lead stories on the nightly news—in our coding period, a space story was the lead story one night. Given the rigid and inflexible order of the Soviet news, with official pronouncements first, then economic stories, then international stories, and only then science and arts stories, this signals extraordinary attention to the space program.[47]

With only a sprinkling of available space stories in the old Brezhnev news format, when opportunities for criticism opened in the late 1980s, the newspaper *Literaturnaia gazeta* carried letters from readers criticizing the program.[48] That had been the first public airing of discontent with television. Mikhail Gorbachev responded to these complaints, first by courting the Soviet television audience through a direct appeal with appearances on television news programs.[49] In 1989, Gorbachev made a radical change by announcing the appointment of a new editor of the evening news, Eduard Sagalaev, who had begun his career in more upbeat, modern, and internationally focused programs that had proliferated on Soviet television since Gorbachev's arrival to power.[50] With Sagalaev's arrival, gone was the drab black-and-white ticking clock announcement of the news. Instead, a lively color montage of film clips from Soviet history with a voiceover of Yuri Gagarin declaring "Poekhali" (Let's go!) proclaimed the start of the nightly news.[51] A sound clip from an event that was nearly three decades old signaled a new, outward focus for the nightly news, *Vremia*. Public familiarity with that clip was overwhelming. The association was with something exciting, new, and confident, even though the flight took place in 1961. Spaceflight provided a fertile opportunity to enliven Soviet television programming.

The Red Stuff lost ground with the Soviet audience in the Perestroika and Glasnost years after 1985. New programming had flooded the expanded channels, and satellite and cable broadcasts allowed the affluent in European Russia to watch Western programs. Shows on history and talk shows dominated.[52] Outside of news programming, there were only two opportunities to see the current state of the Red Stuff on television. The first was a program called *Chelovek, zemlia, vselennaia* [Man, Earth, Universe]. The daily, 15-minute program reviewed the current activities of the space program. The show usually included reports on scientific satellites and civilian launch vehicle production. The second was called *The Bulletin:*

Journal of Scientific Commentary, after the print journal of the same name (*Vestnik: Nauchno-publitsisticheskii zhurnal*). This documentary program produced stories on new technologies in various branches of science and engineering.

Conclusion

From the last decade of Brezhnev's life through the collapse of the USSR, there was a noticeable but slow loosening of state control of the Red Stuff, with only an occasional effort to reassert a central identity. The lack of official support did not lead to stagnation of the Red Stuff. To the contrary, it showed the beginning of independent reshaping of the idea. Some of the changes resulted from intentional new directions on the part of artists, designers, writers, and directors, and some resulted from the long-brewing alienations from the Soviet state. By 1991, there was no longer a single way for the public to appreciate the program. Artists and writers had adapted the space program to fit into the strictures of socialist realism, yet still cast an acrid pall over its heroes. The cosmonaut corps had taken control over their own culture, rituals, and taboo topics. Even in the cases when the Soviet state attempted to resurrect the image of the national hero, the efforts were short-lived and halfhearted. Instead of a single image, there were many.

Just as the Soviet Union was disintegrating, piece by piece, the public face of the space program was decentralizing. Political, social, organizational, and cultural forces constructed their versions of the Soviet cosmonaut. Of all, the political side suffered from diminished credibility and audience; like the Brezhnev Doctrine, creditability evaporated, but it was never wholly refuted. The end of the Soviet Red Stuff was anticlimactic. Throughout the former Soviet Union, myths continued to flourish about the taboo topics of the program. The long-unfinished monuments first designed and conceptualized in the 1960s did not become real until the 1980s, just in time for their official status to be undercut.

In the short thirty years that passed between Yuri Gagarin's first orbit in space and the dissolution of the Soviet Union, Soviet cosmonauts did not get much farther from Earth than Gagarin had, but they flew much longer duration missions along a similar orbit trajectory. By the time the USSR dissolved, the Red Stuff no longer solely illustrated the heroism of the Cold Warriors. Rituals of the late Soviet period took on their own identity. Cosmonauts became diplomats when they hosted cosmonauts from allied countries for orbiting photo opportunities. Unofficial participants in the Red Stuff served as ironic foils with whom to complain about the Soviet system in stagnation. Everyone knew cosmonauts were occupants of a cobbled-together space station that suffered from a decrepitude that matched the Brezhnev infrastructure.

7

Remembrance of Hopes Past

Nostalgia and Editing Public History

The Red Stuff did not end with the collapse of the USSR, but the country's collapse did offer a watershed moment from which to examine the Red Stuff. The USSR no longer existed. It was as though all the component parts of the USSR were set free in the wind to be collected for free examination by whoever paused to examine one. For intelligent people seeking meaning from the Bolshevik Revolution, reexamination was a visceral response to the collapse. For a generation, the USSR developed its own, intricate civil society that at once prepared people for a change and steadfastly denied that one was coming. In the words of anthropologist Alexei Yurchak, the collapse of the USSR was paradoxical: "The spectacular collapse of the Soviet Union was completely unexpected by most Soviet people and yet, as soon as people realized that something unexpected was taking place, most of them almost immediately realized that they had actually been prepared for that unexpected change."[1] These words apply to the reexamination of the Red Stuff. It was both everlasting and carried the seeds of its own destruction. A generation of waning official direction and its consequent growth of independent cultures set Russia up to reinterpret the meaning of Yuri Gagarin's 1961 flight around the Earth and all that followed.

On Christmas Day 1991, Soviet leader Mikhail Gorbachev resigned in a speech, which was the formal acknowledgment that the USSR could no longer continue to exist without the use of force. This marked the end of the Communist Party rule that had bound the republics to the central Russian Republic. In his speech, Gorbachev conceded the inevitability of an autopsy of the USSR. Many historical and experiential autopsies began during the Brezhnev era. Some of them guided new appraisals of the Red Stuff.

Prescient Western historians made a literary assessment of the USSR, declaring it a tragedy. The first was Martin Malia before the collapse of the USSR and then there was Orlando Figes, several years later. University of California, Berkeley, historian Malia wrote of the USSR as tragedy before its dissolution in the prescient essay "To the Stalin Mausoleum," which he signed as "Z."[2] Malia's 1990 article opened with a quote from Alexis de Tocqueville, "The most dangerous moment for a bad government is when it begins to reform."[3] He took this as a warning that previous understandings of the USSR were no longer valid once Gorbachev's reforms had been set in motion. Malia later expanded his argument in *The Soviet Tragedy: A History of Socialism in Russia, 1917–1991*, in which he contrasted the ideological Soviet Union with the reality of the Soviet lives.[4] Orlando Figes wrote his account of the Soviet Revolution, *A People's Tragedy: The Russian Revolution, 1891–1924*, in 1996, which further embellished the tragic nature of Russian socialism in the early twentieth century.[5]

Other writers have taken different literary directions in their assessments of the USSR. In his book *Lenin's Tomb*, David Remnick referred to the waning years of the Soviet Union as a farce rivaling the plays of the Paris theater of the absurd.[6] Remnick used playwright and Czech president Václav Havel as his literary guide to Russian history. These literary assessments of the USSR are very apt, well-documented, and well-argued. However, they focus on the role that a single ideology, Bolshevism as developed in Russia under the guidance of Lenin and Stalin, played in Russia from the late nineteenth through the early twenty-first century.

To the Soviet people, the assessment had more nuance. No one experienced it the same, as tragedy or farce. Some individuals and populations thrived while others suffered. The economic, social, political, and military histories of the USSR share tragic elements, but have outcomes that were more ambiguous. If one can disqualify the Red Stuff as being neither uniformly tragic nor absurd, then it is necessary to find other devices to measure the experience of the Red Stuff.

Domestic reexamination of the Red Stuff occurred in the same way that foreign observations did. The initial efforts set the population in opposing camps, one that celebrated the Soviet experience and one that denigrated it. The case of the Red Stuff was much more complex and nuanced, however. Yuri Gagarin became an icon of the Soviet state in 1961. Moreover, he has remained a favorable icon in the former Soviet Union and Russia for over fifty years after his ninety-minute orbit around the Earth. Despite the failure of the Soviet Union to send men to the Moon, despite the failure of the government to effectively explain Gagarin's death, and even though the USSR did not achieve communism by 1980, the image of Yuri Gagarin remained. The official efforts to infuse the Red Stuff with the optimism of the early 1960s were not entirely successful. The establishments that created the Red Stuff failed to respond at times when the successes were rare and failures occurred. This unresponsiveness left private individuals to draw their own conclusions. While the early culture was well-suited to address failure, the question remains as to how well the culture endured over time resulting from its initial successes. Moreover, even though the symbolism of Gagarin's ascent into orbit had seemed vigorous, even as dissent began to grow in the USSR in the 1980s, the question remained as to how indefatigable the image could be generations after the fall of the Soviet Union.

The measures of the relative post-Soviet endurance of the Red Stuff are wide and varied. The Russian population relied on not only the remembered experience, but also the arts and culture and the official ideology of human spaceflight. The measures also include careful comparisons to other events in Soviet history. Moreover, recent years have witnessed the birth of polling within the former USSR. Recent evidence on public attitudes about the space program includes post-Soviet surveys and writings on the public memory and nostalgia for the former Soviet Union. These surveys measure how people have remembered the Red Stuff.[7]

During the twentieth century, Russia underwent frequent political, military, economic, and cultural transformations, beginning at the start of and ending at the conclusion of the twentieth century. Civil war and cultural and economic upheaval preceded the unmatched impact of World War II. Stalin's death after the war left the country in decades of uncertainty. The nation that began the century with the optimism and violence of impending revolution ended it with the jaded cynicism over the failed Bolshevik experiment. As the Soviet state stagnated and withered away, a disappointed population cannily reappraised the icons of the Soviet experience. Each examination, save one, generated a harsh judgment against the motivations of the state and shunned the fond memories. Even the seemingly sacrosanct experience of the Soviet Union during the Great Patriotic War suffered from these reexaminations. In contrast, during this period of reexamination, those early years of human spaceflight seemed to have survived as a point of pride among the population.

In contrast to the Great Patriotic War, the space program avoided the tendentious reexamination that plagued the war memory and remained a favorite and positive topic of interest. The image of the cosmonaut from the 1960s survived the deterioration and dissolution of the Soviet Union in many ways, but it lost its significance in others. Until the recent recentralization of the evening news and television programming, the sound of Yuri Gagarin's voice kicked off the Russian national nightly television news program. While other monuments to Soviet-era heroes were the target of late and post-Soviet disillusionment and rage, monuments to cosmonauts and space museums remained in place and suffered no direct attacks even in the newly independent states.

Crowds removed statues of Lenin throughout the former Soviet Union and Eastern Europe, while monuments to the cosmonauts remained in place.[8] The statue of Feliks Dzerzhinsky, the head of the Soviet secret police (Cheka/GPU/NKVD) in the 1920s, came down from Dzerzhinsky Square in response to the August 1991 abortive coup against Gorbachev. At the same time, the monuments and museums to the Red Stuff survived, and cultural references to the cosmonauts continued to flourish in post-Soviet culture.[9] The material remnants of the popular culture, stamps and znachki, retained their status as collectible items among enthusiasts. In addition, because of the introduction of private markets, collectors exchanged their merchandise in public and multinational marketplaces, selling off personal memorabilia from the earlier space age of Russia. Artists, too, continued to reflect on cosmonauts from the 1960s. Modern, post-Soviet artistic references to cosmonauts isolated the cosmonauts in a generally favorable light, while condemning the bureaucratic infrastructure that surrounded them.

During the three decades of Soviet history after Yuri Gagarin orbited the Earth, the human space programs of the USSR became routine and limited to operations in a series of space stations in low-Earth orbit. The Soviet Union offered only sporadic public challenges to the United States' civilian human space program, primarily in the form of personal firsts, as with Svetlana Savitskaya. The human spaceflight program continued at the same altitudes of Gagarin's flight through the rest of the century, never achieving the science-fiction aspirations of travel to the Moon or another planet.[10] After the Soviet Union dissolved, human spaceflight remained a popular concept but was a low funding priority and eventually declined to sub-contractor status to the US-led International Space Station program.[11] The Soviet and Russian governments did not step back from the Red Stuff, but they ceased to rely on it heavily as a measure of their standing in the world and at home. Throughout this period, the public maintained a remarkable level of widespread support for the annual celebrations of the cosmonauts' accomplishments of the 1960s. Public opinion, literature, film, and television of the late Soviet period reflected this continued popular recognition. The public and popular culture of that period indicate that despite skepticism over everything closely associated with the Soviet state, interest in and the popularity of the early program remained intact in Russia and the other former Soviet republics. For many, the iconography of spaceflight of the 1960s withstood the iconoclastic 1990s.

From the ashes of this seeming chaotic reexamination of the USSR rose specialized fields in the arts that have come to represent specific judgments. Even the lowest periods of Soviet history did not succumb to uniform denouncement. Neither did the most optimist periods avoid criticism. As is true with all national experiences, all had positive and negative elements. In the case of the Red Stuff there were many threads, both positive and negative, to untangle. The process becomes clear when examining what happened in public opinion, the arts, and literature. To understand the distinctiveness of the Red Stuff's perseverance in post-Soviet culture, it is first necessary to look at the example of the decline of an old and ostensibly impenetrable cult, that of the Great Patriotic War. Unlike spaceflight, every Soviet citizen either had direct experience of World War II or had a relative who did, no matter what his or her age. The positive, culturally binding, public memory of the war crumbled along with the Soviet Union. Examining the process of the decline of the cult of World War II offers an analogy to the cultural shifts that the Red Stuff experienced during the same time.

Rethinking the Great Patriotic War

To understand the significance of the endurance of cosmonaut popularity, one must first examine the history of the unraveling of the public memory of a formerly cherished event, the Great Patriotic War. The one event that seemed least likely to be de-sanctified was the Soviet experience in World War II. During the war, the Germans had invaded the most densely occupied regions of the country. With losses calculated to be 25–27 million deaths over the course of four years of fighting, most of which took place on Soviet lands, one might expect that a unified national appreciation of the meaning of the Great Patriotic War would be unshakable to Soviet and, later, Russian citizens. Nonetheless, questions over the legacy of the war began even before the Gorbachev era in the 1980s. Dissident questions grew more probing as dissatisfaction over the Soviet Union under Brezhnev grew and international pressures were applied to reevaluate the legacy of World War II. Even when official support for the legacy of the war revived at the beginning of this century, the events that touched off reevaluation of the war remained unexamined. This movement for reevaluation had a decidedly different impact on the Red Stuff.

Historians Nina Tumarkin and Anna Krylova have discovered that despite the near-universal experience that World War II was for the Soviet population in the long term, the memory of it suffered from its association with the Stalinist state.[12] Since the 1980s, external and then internal criticism of the conduct of the war mounted. The decline of the cult of the Great Patriotic War serves as a blueprint for the deconstruction of public culture in the former Soviet Union. The cult of World War II was a Khrushchev invention. Construction began on the first national war memorial, the Tomb of the Unknown Soldier, under his watch. Brezhnev turned commemoration of the war into full-blown national ritual. Even while the cult of the war grew, events during the 1970s and early 1980s added to both domestic and international pressures to confront the myths and legends that had grown around the war. One example of myth took form in 1943, when Stalin had denounced Germany's call for the Red Cross to investigate the truth behind the discovery of the graves of Polish officers in the Katyn Forest.[13] Many people in the Soviet Union accepted Stalin's accusations that the Germans had killed the Polish officers. When Soviet reinvestigations began in the 1980s, however, forensic evidence proved that Stalin had lied. German actions at the time of the discovery preserved the forensic evidence for fifty years for a final judgment.[14] Soviet forces had killed the Polish officers to short circuit independent Polish nationalism after the Soviet occupation of eastern Poland in 1939.

Questions over the murders at Katyn contributed one of the many uncertainties about the war legacy that lingered for nearly a half-century. Among the least internationally known questions were ones about both the Soviet leadership's competence in the war and the economic and personal price the country had paid, and in many cases, continued to pay. Khrushchev's "Secret Speech" in 1956 was the first time that a party official publicly articulated the doubts about Stalin's war leadership.[15] By the time that the Soviet Union ended, questioning about the war continued at an increasingly aggressive pace. The collapse of the one-party system left the state without its primary support system. In anticipation of public acrimony over war memory, Russian president Boris Yeltsin canceled the Victory Day celebrations in 1992 for the first time since Brezhnev had instituted the holiday in 1967.[16] Yeltsin's decision marked a milestone in Russian public culture.[17] For the first time the government had canceled the commemoration of a seemingly non-political and state-supported anniversary.

By the 1990s, the positive memory of the war and the celebrations of its veterans were no longer a given in Russia. Much of the internal dissent over the war's legacy was about disentangling the legacy from Russian culture. People were trying to sort out the difference between the Soviet Union as a whole and personal experiences. Over the course of the discussion about war legacy and the simultaneous dissolution of the Soviet Union, a long-postponed national monument to the memory of World War II underwent design and redesign to fit rapidly changing and widely varying national sensibilities. Early designs focused on the CPSU contribution to the war effort.[18] The monument to the memory of World War II in Moscow was in truth a monument to the uncomfortable resolution of the issues that post-Soviet Russia confronted over the war.[19] The Moscow monument acknowledged the national wartime experience of veterans and civilians while marginalizing the wartime role of the CPSU and ignoring the history that led up to the war. Similarly, war memorials through the Soviet republics placed an emphasis on the local contributions to the war, making the local attachment to the memory stronger.

Boris Yeltsin's efforts to isolate the Kremlin from the emerging controversy did not mark a permanent official position on the war. When he took office as acting president at the end of 1999, Vladimir Putin immediately sought to revive official Russian nationalism that had lost favor during the unraveling of the USSR. Putin was keenly aware of the importance of public symbolism to national culture. In his first year in office, he restored the Stalin-era Soviet national anthem as the Russian anthem in defiance of public and political opposition.[20] He also directed an initiative to transform the historiography of the Soviet Union through the curriculum in schools that began with the conclusion of World War II and ended with his term in office. The new curriculum made a clear arc between victory and the resurgence of Russian power under Putin.[21] In his direct statements regarding World War II, Putin dispensed with any doubts as to the causes and domestic consequences of the war, emphasizing the role that the war had in strengthening the unity of the Soviet Union.[22] For him, any tragedy associated with the war was due to Nazi Germany; the Soviet Union had only a heroic role. In his speech that commemorated the sixtieth victory anniversary, Putin made clear its heroic role: "For us, the Second World War was the most tragic event of the last century. But at the same time, it was also the most heroic event of that era."[23]

While Putin dispensed with any discussion of the Soviet preparation for and conduct of the war, his actions subsumed discussions of the war into broader discussions of post-Soviet Russian politics. The controversy of Katyn Forest began to shrink in significance relative to the issues of Russia's political freedom and its role in the twenty-first century. When Putin took power, the fiftieth anniversary of Victory Day (1995) had been the last major celebration of the war in which living and ambulatory veterans were likely to participate. Their role in the war was no longer the center of the debate on the legacy of the Soviet Union.

As this debate over the legacy of the war ran its course between the fiftieth and sixtieth anniversaries, the public image of human spaceflight remained relatively unchanged. The significant difference between the two was that the monuments, museums, and memorials to spaceflight appeared sooner after the events in question compared to those of World War II. The timeliness of these constructions avoided the perils of prolonged public reevaluation. The official celebrations of World War II began in the Soviet Union when the Tomb of the Unknown Soldier opened in the Aleksandrovskii Garden in December 1966, on the twenty-fifth anniversary of victory outside Moscow.[24] A full generation had lived before its dedication on Victory Day in 9 May 1967. Monuments and memorials to the Red Stuff emerged within a decade of the first Soviet ventures into space. The timing spared spaceflight the burden of generations of reinterpretation. Construction took place with little public national doubt about recent events. The Monument to the Conquerors of Space appeared in October 1964 before the *Vostok/Voskhod* missions had finished.[25] In fact, the monuments and museums might have preempted doubts and standardized national memories. Two generations hence, the image of the early years of the Soviet space program held ground in public opinion, nostalgia, literature, television, and film at a time when the debate over Soviet legacy was at its hottest. Even the most critical works saved some affection for the early cosmonauts, no matter how savage their critique of the environment in which the cosmonauts worked.

Public Opinion, Nostalgia, and the Red Stuff

The combination of the economic collapse of the USSR and Gorbachev's legacy of openness permitted Western journalists to ask probing questions about personal experiences in the USSR and emboldened former Soviet citizens to answer those questions frankly. Historians and social scientists began to conduct formal and widespread interviews among the former Soviet population. Eventually, Russian and other journalists took on the opportunity to publish informal surveys of the population for entertainment purposes. This opportunity to document the opinions of Russians on the street did not last for long, but it was an improvement on the previous and anecdotal and anonymous reporting that had dominated public opinion reporting in the USSR. The Soviet space program and the Red Stuff featured prominently in the memories of all of those who responded to questions and surveys.

In 1997, the year before the new Russian economy overheated, *New York Times* Moscow Bureau Chief Michael Specter wrote an extended article for the newspaper's Sunday magazine.[26] He interviewed four people, two each at the front and the back of the economic wave that was overtaking the country and especially Moscow. Two interviewees were doing exceptionally well in post-Soviet Moscow. In contrast, two others were suffering from the loss of the Soviet support system. One of his interviewees feeling the downside of the economy was a sixty-four-year-old pensioner, Mira Pavlovna Ivanova, who had been a computer programmer at a military installation during her career. Cradle-to-grave protection for those employed in military installations ceased to exist when the Soviet Union collapsed. When Specter met her, Ivanova was attempting to supplement her fifty-dollar-a-month government pension by redeeming empty bottles that she collected from trash cans and the streets.[27] During her conversations that touched on her unanticipated losses in life, including sharing a small apartment with her estranged daughter and her family, Mira Ivanova reflected on a happier time:

> She mentions that Saturday will be the anniversary of the day Yuri Gagarin became the first man in space. Cosmonauts Day is a major holiday in Russia, a rare chance to celebrate both the glory of the Soviet Union and a time when its future seemed to shimmer. "How I remember that day," she says, dreamily. "It was 36 years ago, and my baby was 3 weeks old. Everything seemed possible." She stops talking before she starts to cry.[28]

In theory, one could measure the endurance of the Red Stuff through public opinion surveys over the years. Only two formal surveys have attempted to capture the public memory of the Soviet population from the later 1950s and early 1960s in the years since the collapse of the Soviet Union. Each study provides insight into the popular thought about spaceflight among Russians who lived through the era. Sociologists Howard Schuman and Amy D. Corning have published their research on collective knowledge of public events in the Soviet Union.[29] In 2006, Indiana University Press published Donald J. Raleigh's collections of interviews with members of the "*Sputnik* generation"—the first postwar generation that was old enough to remember the first artificial Earth satellite that orbited three and a half years before Yuri Gagarin.[30] Twenty-nine members of Saratov's School No. 42, Class of 1967, answered questions about their coming of age during the Space Race.[31] In each case, the researchers found connections between official propaganda and public memory. Raleigh noted the degree to which people linked spaceflight and the Soviet state in their memories: "The *Sputnik* generation came of age at the zenith of Soviet socialism, only to see the system crumble some three decades later."[32] The interviewees' process of explaining the growing disintegration of the USSR provides insight into the independent preservation of the image of the cosmonauts of the 1960s.

Raleigh did not directly question his respondents on Gagarin and subsequent spaceflights.[33] Schuman and Corning conclude that state efforts to popularize space events were successful but did not match the effect that having personally witnessed the event did. In discussing the differences between those who personally witnessed the flight of *Sputnik* 2 (3 November 1957), which contained the dog Laika, and those who learned about the event later in school and books, the sociologists note,

> If we exclude the cohorts from 1941 to 1956 that show the best knowledge, most other Russians alive at the time of Laika's flight demonstrate significantly more knowledge as a group than those born after the event—an overall period effect attributable simply to being alive in 1957. Of course, even those born after *Sputnik* learned about Laika, but their considerable exposure through schools and mass media to the triumphs of the Soviet space program could not match the live experience of the event itself.[34]

Those who witnessed events firsthand had an advantage in memory over those who did not. Propaganda had the effect of increasing awareness of an event, but it did not provide detailed knowledge of the events that Schuman and Corning surveyed:

> Individuals alive when an event occurs are more knowledgeable than those born even soon after the event has ended. In our data, we find

evidence that this process applies to knowledge of Laika, the Cuban Missile Crisis, [the book] *One Day in the Life of Ivan Denisovich,* the Yezhovshchina, the Twentieth Congress of the CPSU, and the Doctors' Plot.[35]

In Raleigh's book, when members of the "*Sputnik* generation" spoke about their memories of spaceflight, they instinctively linked memories of the accomplishments of spaceflight directly to the declarations of the Twenty-Second Party Congress of the CPSU,[36] also known as the "Congress of the Builders of Communism." The 1961 Congress declared that the Soviet Union would achieve communism by 1980. The party largely based its declarations on the previous four years' accomplishments in space as indications of the country's explosive development that would continue indefinitely.[37] And they buttressed their belief by citing the American reaction to the Soviet space program. American opinion carried weight when it implicitly or explicitly recognized Soviet activities. Its criticism carried little or no weight.

The Soviet Union's successes in space exploration compelled the capitalist world to take a different view of the achievements of socialist society, of the advance of science and industry in the USSR. Khrushchev quoted US diplomat Chester Bowles at the Congress as having said that until the first Soviet *Sputnik,* "Almost no one had questioned America's industrial, military, and scientific superiority. Then suddenly there was *Sputnik,* ringing the earth, and millions began to ask whether communism was not the winning side after all (Animation. Applause)."[38] The rhetoric of the Congress adopted the principle that spaceflight was a demonstration that all other political and economic claims of the Soviets were irrefutable.

When interviewed about her memories of the time over forty years later, Natalie Aleksandrovna Belovolova, one of Raleigh's respondents, seemed quite comfortable in linking Gagarin's flight to the promise of communism, as had been done for her as a schoolchild. The pride of spaceflight associated closely with the pride of building communism in her mind:

> I went to school in 1957 and don't remember much about the early period of my life. When I began to realize that these were the Khrushchev years, it was already the 1960s. The "Moral Code of the Builder of Communism" was posted at school, and we had to memorize it. You know, there was absolutely nothing wrong with this, since it's almost like the Ten Commandments. Gagarin's achievement was our

greatest pride. For the most part, we were raised to be proud of our country.[39]

Belovolova's classmate Arkadii Olegovich Darchenko conceded that although he harbored seeds of skepticism about Khrushchev's promises, he did not perceive that there was an alternative to believing in the blueprint for progress that the party congress outlined. At the time, the space program and the promises of a bright future from the CPSU were the only sources of hope for the future:

> Back then, we were too young and most likely too trusting. I was confident that we should genuinely achieve communism by 1980. Why not? Indeed, people believed that we were on the correct path, and that soon we'd achieve it. But when we were a bit older, we realized that we're not heading in that direction . . . Even during the Khrushchev years, we understood. Yes, we knew that nothing would come of this, because there was this feeling that we should be getting more prosperous, yet things began to get worse.[40]

The post–World War II War generation harbored hopes and aspirations that times would be better than the ones in the past. Along with promises, the state had provided crisp demonstrations of improvements. At the time, it was possible to accept the idea that the country was overtaking the United States. The immortalization of the Red Stuff in stamps, pins, buildings, museums, and public celebrations served as a constant reminder, as did museums, monuments, and meager space-themed consumer goods. These images remained part of the Russian culture even after the image of the Soviet Union had begun to wither. Nevertheless, within forty years, the well-forged link between spaceflight and the utopian Twenty-Second Party Congress began to unravel, especially among the post-Soviet generation—those who had no firsthand memory of the early years of the Red Stuff.

If living through an event is the most significant predictor of the strength of memory, that assertion begs the question about the retention of affinity for cosmonaut culture among children and young adults, most of whom have no direct memory of the Soviet Union. If celebrations of World War II could cease, what assurances were there that the Red Stuff would endure? While there have been no systematic surveys that would reveal answers to this question, local newspapers conducted informal surveys of young people on anniversaries of Gagarin's flight to find out what they knew about the cosmonaut and the Red Stuff. The surveys were informal and unscientific but did contain kernels of information on the perspectives of the post-Soviet generation, which did not grow up reading *Pioner* or *Smena* magazines,[41] never aspired to Komsomol or Communist Party membership, did not take school trips to the Kosmos Pavilion in its heyday, and for whom spaceflight had always been a multinational endeavor.

The results of three informal surveys appeared in 2001 and 2002. *Surgutskaia tribuna* asked teenagers and young adults what they knew about Yuri Gagarin.[42] That same year *Komsomolskaia Pravda* asked elementary school students: Who said "Poekhali!"?[43] In 2002, the newspaper in Iakutiia asked similar questions of secondary school students.[44] Not surprisingly, the youngest respondents had the most difficulty in deciding who Yuri Gagarin was and who the first man in space was.[45] However, the most interesting answers came from the adolescents from Surgut in Siberia. Even though they all seemed to be aware that Gagarin had flown in space, few of them seemed sure about the nature of his flight, and all were doubtful about the future of Russian human spaceflight.[46]

The high school students in Iakutiia frequently conflated many myths and legends about Gagarin and the early spaceflight program with the facts. Two of them thought that Gagarin had flown to the Moon.[47] Several mentioned the dogs that had flown before him. Everyone seemed to be under the impression that Gagarin died in an experimental aircraft and not a routine training flight.[48] One girl, Nina, knew that these events had occurred well before her time, under Stalin, she suspected. Another child, Dzhulus, knew that Gagarin's flight was more recent, answering, "After the flight to the Moon in *Vostok*, he met with Brezhnev and Gorbachev."[49] Finally, and most powerfully, in 2001, twelve-year-old Vasia Maslakov was uncertain who said "Poekhali!" He suspected that German Formula One champion sports car driver Michael Schumacher was the one who said it.[50]

A less quantitative but more personal measure of the Soviet experience emerged from the US-Russian émigré community. For a short period, many were able to travel between their birthplaces and adopted countries and contribute to the interpretation of their youth. Mira Ivanova's wistful nostalgia about the first human spaceflight reflected her sensibility about the Soviet space program of the 1960s. To many of that generation who left for America, it appears the early 1960s had been a unique and fleeting period in Soviet history when everything was within reach. During stressful times, it is not surprising that many people were nostalgic for this period.

However, a similar nostalgia also existed for those who had not suffered disappointment in the new post-Soviet era. Writer and literature professor Svetlana Boym, who immigrated to the United States as a young woman in 1981, has written about the historical stages of nostalgia.[51] Her reminiscences about that period in the 1960s echoed Mira Ivanova's:

> When I returned to Leningrad-St. Petersburg, I found myself wandering around the miniature rockets rusting in the children's playgrounds. Crash-landed here three decades ago, they reminded me of the dreams of my early childhood. I remembered that the first things we learned to draw in kindergarten in the 1960s were rockets. We always drew them in mid-launch, in a glorious upward movement with a bright flame shooting from the tail. The playground rockets resembled those old drawings, only they didn't fly very far. If you wanted to play the game, you had to be prepared to glide down, to fall, not fly. The playground rocket began to appear in courtyards during the euphoric era of Soviet space exploration, when the future seemed unusually bright and the march of progress triumphant.[52]

In discussing these childhood memories, Boym has created a metaphor for what happened in the Soviet Union since the heyday of human spaceflight. The memory of Soviet spaceflight was optimistic, like the children's drawings of rockets and spacecraft ascending toward their objectives. If the expectant minds of children were not enough explanation of the persistence of this hope for spaceflight, Boym points out the link between spaceflight and the government's promise of the new and imminent era of communism. The state and private individuals had repeated as much for decades after German Titov first articulated it in 1961.[53] Moreover, even though the promise of communism exceeded reality, the post-Soviet population retained an attachment to that utopian aspiration. That attachment was far more profound for those who failed materially, such as Mira Ivanovna. By the late 1990s, those who could replace the promises of spaceflight and communism with wealth had done so. The remainder of the population still had the remnants of past hopes:

> Soon after the first man flew into space, Nikita Khrushchev promised that the children of my generation would live in the era of communism and travel to the moon. We dreamed of going into space before going abroad, of traveling upward, not westward. Somehow, we failed in our mission. The dream of cosmic communism did not

survive, but the little rockets did. For some reason, most likely for lack of an alternative, neighborhood kids still played on these futuristic ruins from another era that seemed remarkably old-fashioned. On the playgrounds of the nouveau riche, the attractions have been updated in the spirit of the time. Brand-new wooden huts with handsome towers in a Russian folkloric style have supplanted the futuristic rockets of the past.[54]

In her writings on the historical and literary concept of nostalgia, Boym accepts that it is an imprecise recollection of the past shaped by present circumstances: "As a public epidemic, nostalgia was based on a sense of loss not limited to personal history. Such a sense of loss does not necessarily suggest that what is lost is properly remembered and that one still knows where to look for it."[55] The spaceflights of the 1960s were finite and occurred with such infrequency that they could not have supplanted the shortcomings of Soviet life at the time. However, in retrospect, memories of the bold promise for the future overshadowed memories of the drudgery of everyday life. Boym's accounting of nostalgia ties the personal experience to a place, not necessarily to events, however.[56] Nevertheless, her interpretive framework adapts well to the study of nostalgia about events in the history of the USSR.

The nostalgia that Mira Ivanova felt was a longing for a sense of community.[57] In the 1960s, she was part of the anonymous collective of Soviet people who shared acclaim for the accomplishments in the space program. She had been one of the legions of specialists and engineers who made the technology work. In addition, unlike other historical events of shared acclaim, the space program, and specifically the period of rapid and repeated space accomplishments that appeared to overtake the United States, remains unique from all other collective Soviet experiences. Neither Ivanova nor Boym expressed a longing for World War II or the Brezhnev era. Decades later, after the last icons had fallen, the space program retained a hopeful nostalgia that was immune to the iconoclasm of the late and post-Soviet Union. During the ensuing years, spaceflight, and the Red Stuff in particular, would play a prominent role in the negotiations over the legacy of the Soviet Union. As researchers found, personal experiences were not the sole nurturers of these memories. The post-Soviet era was rich in literature and film reflections on the Red Stuff, all of which had a potential impact on individual interpretations of the legacy.

Literature

Although scholars have studied socialist realism, modernism, and post-modernism pretty thoroughly in Russian and Soviet literature, only recently have they paid attention to the three styles' convergent dissatisfaction with the Soviet state at the end of the USSR. Socialist realist authors found unique ways to comment on the Soviet Union while still using approved stylistic methods. Modernist authors echoed the unaligned politics of their literary ancestors of the 1920s. Russian postmodernist literature emerged as a response to the decline of the USSR much in the way that science fiction had been the literary response to industrialization. These three genres offer insight into the popular perception of the 1960s space program. The modernist literature of the 1920s and utopian science fiction began with speculations on the effects that space travel and contacts with other civilizations would have on Soviet society.[58] Later, socialist realism served as the model of Soviet heroic fiction of the 1930s.[59] This style underwent resurgence during the 1960s as the traditional heroic biographical treatment addressed the lives of cosmonauts. It is fitting that this examination of the cultural and public history of the Soviet golden age of spaceflight conclude with an examination of the final literature of the Soviet Union that reflected on the accomplishments of the 1960s.

Victor Pelevin is a Moscow-born fiction author raised during the golden age of the Red Stuff. His mother was an English teacher and his father taught military technology at the prestigious Bauman Technical Institute. Pelevin studied creative writing at the Maxim Gorky Literary Institute, which expelled him in 1991—the year that the USSR dissolved. Pelevin published his first novella, *Omon Ra,* in 1992, the first full year after the Soviet Union. The author and book won the Russian Booker Prize in the second year of the award. The hero of *Omon Ra,* like Edigei in Aitmatov's 1980 book *The Day Lasts Longer than a Hundred Years* discussed in the previous chapter, has two names. Double names are an expression of the duality of existence. The protagonist's original given name was Omon Krivomazov.[60] OMON was the acronym for the Soviet special police forces, which is known in the US as SWAT.[61] His older brother had a name of similar bureaucratic origins: Ovir, for the Office of Visas and Registrations. His father chose to name his sons to express his desire that the older boy become a diplomat and that Omon become a police officer.[62] Omon takes the last name Ra, for the Egyptian sun god, only after he decides that he wants to become a cosmonaut.

Виктор
ПЕЛЕВИН
ОМОН РА

Figure 18. This postmodern and post-Soviet literary work took a new perspective on the Soviet human spaceflight program, echoing satirical novels of nineteenth-century Russia.

Omon Krivomazov has a classically dystopian childhood of his age—deceased or otherwise absent mother, a brother felled by infectious disease, and an abusive and drunken father all in a world surrounded by aviation and spaceflight motifs. It is at the Exhibitions of Economic Achievements' Kosmos Pavilion where he meets another boy, Mitek, who is destined to become his best friend: "This happened one winter evening, when I was walking around VDNKh. I went into an empty, dark, and snow-covered alley; suddenly from the left sounded a hum that resembled the ring of an enormous telephone. I turned around and saw him."[63] Charmed by the idea of spaceflight display at the exhibition and the aviation culture that pervaded the USSR, they both embark on careers as cosmonauts, first by enrolling at the Maresev Military Academy.

It is at the flight school that the layers of state deception begin to reveal themselves to the two young men. The name of the Maresev Academy honors an actual World War II hero. Aleksei Maresev was a pilot shot down behind enemy lines.[64] He crawled back home ultimately to have both legs amputated. Of course, in the Soviet heroic tradition, he volunteered for frontline service again. He is an example for all pilots that followed him. However, Pelevin's postmodern Soviet Union can no longer afford an Air Force, so the only way available to replicate the heroism of Maresev is to create copies of Maresev-type heroes by amputating the legs of pilot candidates.[65] When his friend Slava suffers a similar fate, no one at the academy comments on the loss of his legs. The reason behind these amputations is the corruption of the system. The "heroes" are for show because there was no real flight school. They do not fly airplanes; therefore, there are no crashes. The pilots do not have to crawl from wrecks. There is no front and combat, only heroism. As one literary historian has written, Omon's experiences at the Maresev Academy marks the "unmaking of the Soviet man" when everything was in decline.[66]

As Omon makes his way through his training, he continues to learn of the pervasive heroic deceptions throughout the system. Among the parade of ersatz heroes who come to the flight school to inspire the cadets is Major Popad'ia. Popad'ia's previous job was to pose as the game for high official hunting parties. Once in the past, a wild boar had killed a party official who had been hunting. To prevent the further loss of party officials, who would never consider abandoning their blood sport, Popod'ia and his son, Marat, dressed as wild animals with bulletproof vests and simulated death when shot. The Popad'ia family career ends when Henry Kissinger joins the hunt and enthusiastically stabs his quarry with a knife instead of using a gun.[67] Marat dies as result of Kissinger's attack, and the major retires to become yet another professional hero.

Omon continues to learn of the depth of official deception as he moves closer to his goal of being a cosmonaut. Once he passes through the selection process for the program, along with his childhood friend Mitek, he learns that the Soviet Union lacks the technology to send a man to the Moon and return him to Earth. His commander calmly and rationally explains that:

> The goal of the space experiment for which we are preparing, Omon, is to show that we have not technologically fallen behind the West and can send a mission to the Moon. We are not capable of sending a crew to the Moon and bring it back. But we do have the capability of sending an automated team that would consist of eleven cosmonaut candidates each.[68]

To maintain the ruse that the country is still competing against the United States in the Moon race, the planners have orchestrated an elaborate hoax. The plan is to send a group of cosmonauts to the Moon, each manually operating a stage of the rocket and spacecraft instead of relying on the faulty Soviet guidance and computer technology. Once each stage of the mission is complete, the cosmonaut who has completed that stage kills himself.

As a final test, before the cosmonaut selection for each stage of the mission takes place, the cosmonauts must reveal their true selves to the selection committee as a test of ideological purity. The men accomplish renewal through drinking a reincarnation drink to assure that all the cosmonaut candidates have impeccable social and political credentials in their past lives. The reincarnation drink re-creates "past lives" as a hallucinogenic background investigation. Omon passes the final test before selection. His friend Mitek, however, turns out to have been a Nazi pilot in one of his previous lives, so his commanders execute him, and no one speaks of him again.[69] This scene, of course, parodies the party membership interviews that were as prominent as the rite of passage in the cosmonaut biographies during the 1960s.

Undaunted by the prospect of suicide on the surface of the Moon or the loss of his friend, Omon remains enticed by the thrill of spaceflight and continues with the mission. He takes instructions well and turns on and off switches as told by his "ground controllers" through the launch and landing of the spacecraft on the Moon. As he finishes his mission, he listens to the radio tribute to the eleven cosmonauts "shift workers" who have given their lives for the Moon mission. When his final heroic moment comes, his gun misfires, and the ultimate deception becomes apparent.[70] Omon Ra had not been on the Moon at all but is in a tunnel adjacent to the Moscow Metro. His most profound dream of spaceflight, just like aviation, has been a farce. He flees the mock mission and escapes via the Metro in another allusion to Russian literature, this time to Dostoevsky's *Notes from the Underground.*

The authors Aitmatov and Pelevin were from two different generations, came from different regions of the Soviet Union, and wrote in different and often ideologically opposing styles. Yet, they came to similar conclusions. The cosmonauts in Aitmatov's novel suffer the same affronts as his main hero, Edigei. The repressive and paranoid state denies them all access to enlightenment. While the government represses Edigei's efforts to bury his friend, the cosmonauts defy orders and continue their journey with the knowledge that they would never be able to come home again. Pelevin's Omon Ra suffers from similar governmental offenses. Where Aitmatov's state is paranoid, Pelevin's is absurdly byzantine to conceal its deficiencies. Pelevin's would-be cosmonaut, too, escapes the clutches of the government. Instead of escaping to the *lesnogrudtsy* (the plural of inhabitants of the planet Lesnogrud), Omon flees in a traditional Russian manner—to the underground. In each case, the cosmonauts must break the bond with the state to find their true destiny.

Film

The collapse of the USSR brought down with it the state-run film industry in the USSR. There was little institutional structure immediately put into place to replace the production and film distribution system. This severed the previous state-sanctioned relationship between filmmakers and official ideology that Lenin had valued so much. During the early 1920s, funding that was originally directed toward feeding factory workers found its way toward financing Soviet cinema. The funding also contributed toward retrieval of Imperial Russian film artists who had fled the country during World War I and the Bolshevik Revolution.[71] Unlike the early 1920s when film production relied on heavy investment from the Ministry of Enlightenment to re-form its prerevolutionary self, an alternative, nongovernmental source of funds rescued the post-Soviet film industry. In the 1990s, money from the early Russian oligarchs flooded into the Russian film industry, doubling the number of films produced. In the absence of normative rules and regulations, which would take a generation to replace, these new-money films took on the challenges of social change of the era, including oppression, crime, Russian national identity, and the national favorite—literary adaptations.[72] And included among these fresh takes on Soviet history, the Red Stuff was present.

There are three twenty-first-century Russian feature films, Aleksei Fedorchenko's *Pervye na lune* [First on the Moon] (2005),[73] Aleksei Uchitel's *Kosmos kak predchuvstvie* [Space as Premonition] (2005),[74] and Aleksei German's *Bumazhnyi soldat* [Paper Soldier] (2008),[75] that address the legacy of the golden years of Soviet spaceflight in their unique manner. Each film places spaceflight into the context of a specific period. Uchitel's and German's films take place in the early 1960s, and Fedorchenko's begins in the 1930s. Each dissects the origins of the culture of spaceflight using either historical footage or staged footage to resemble the historical record.

All three directors of these films have the first name Aleksei. Each one has paved a unique course into Soviet and Russian cinema. Aleksei Stanislav Fedorchenko was born 29 September 1966, in Sol-Iletsk, in the Orenburg region of Siberia. He initially studied engineering at college and worked on space defense projects in a factory in Sverdlovsk (Ekaterinburg). *First on the Moon* was his first film. Aleksei Efimovich Uchitel was born 31 August 1951, in Leningrad. Uchitel's debut as a feature film director was in 1995 with the *Giselle's Mania* film, telling the story of the famous Russian ballerina Olga Spesivtseva. *Space as Premonition* (2005) was Uchitel's third theatrical film, the screenplay written by Alexander Mindadze. The film received the Grand Prix of the 27th Moscow International Film Festival. The youngest of the three Alekseis has the most extended pedigree in Soviet cinema arts. Aleksei Alekseevich German, also known as Aleksei German Jr. was born 4 September 1976, in Moscow. His father, Aleksei Yurievich German, was a successful film director during the postwar Soviet period and continues to make movies today. In turn, his father was the writer Yuri German, famous for his wartime novels. Like Fedorchenko, German Jr. graduated from VGIK (All-Russian State University of Cinematography named after S. A. Gerasimov). *Paper Soldier* was German's sixth feature film as director and the fourth for which he wrote the screenplay. All three directors are of the post-*Sputnik,* post–Space Race generation, growing up and coming of age during the decline of the Soviet era. Each grew up within pockets of privilege within Soviet society. Moreover, each witnessed the crumbling and collapse of the USSR.

Fedorchenko's *First on the Moon* is closer in tone to Viktor Pelevin's novella *Omon Ra* in its take on the space program. Produced as a mockumentary, this film fabricates the existence of a secret Cheka film archive of a Stalinist program to send men to the Moon during the 1930s. *First on the Moon* portrays the cosmonauts as tragic victims of the state. The events of this film take place during the late 1930s and present footage of the selection and training of cosmonauts for a secret flight to the Moon. The purported documentary reports on the uncovered mission and the search for the survivor, the cosmonaut Ivan Sergeevich Kharlamov, his journey from the crash site in Chile and ultimate return to the Soviet Union. Although this is a parody, partly of the Stalinist Falcons' flights of the 1930s and the 1960s space program, the treatment of the cosmonauts is genuinely affectionate. They, too, are hapless and blameless in their efforts.

The cosmonaut candidate and the leading role in the film is a character named Kharlamov. He started his career as an aviation pioneer along with Chkalov and Baidukov, the first heroes of the Soviet Union. At some point in his career, he joins a team of Soviets and Germans who were cooperating on rocket development. From that team, he joins the secret recruitment for the first Soviet cosmonaut. The recruits make up a diverse group. Among the finalists for the mission are Kharlamov; Nadia, a young woman, also known as the Komsomol princess; a Central Asian, Kharif Ivanovich Fattakhov; and a dwarf who has no name.[76] All endured the final testing. The chief designer makes the final selection of Kharlamov for a launch in 1938. After the launch, ground control loses all contact with the spacecraft, which throws the entire program into turmoil. The chief designer dies by suicide. Mysterious men sedate and kidnap the remaining cosmonaut candidates from their barracks. They destroy all evidence of the mission, saving only a scale model of the spacecraft and films that remain in the NKVD archives until researchers discover them.

What could have been a complete cover-up of the program's existence was, however, imperfect. In March 1938, Chilean peasants reported seeing a fireball in the sky. This episode refers to a real event, a meteor landing in the country. In Fedorchenko's film, it is not a meteor, but Kharlamov's spacecraft that lands in Chile. He had survived his mission. However, without official status, Kharlamov has no resources with which to return home and has to become personally resourceful to do so. The NKVD interviews trace his steps along the return home. He travels via Mongolia through battles between the USSR and Japan, after crossing the Pacific Ocean via boat and through China. Speaking only gibberish upon his return to the USSR,

Kharlamov ends up spending time in a psychiatric hospital. It is then that the NKVD takes notice of his return. Always not close enough on his trail, thirty NKVD agents gather information on his life as they follow him.

The bulk of the film contains interviews with those who knew or saw him. The materials that the NKVD agents salvaged include film footage of the draconian psychiatric evaluations that included electroshock and insulin therapy that bear a remarkable resemblance to the flight-testing footage. Agents also interviewed his wife after he left the psychiatric hospital, and later interviewed Fattakhov and the dwarf. The interviews that the NKVD conducted with Fattakhov inquiring about Kharlamov's background were particularly instructive. Fattakhov has become a builder of giant mechanical insects for children's museums. His profession makes the analogy to Franz Kafka's *Metamorphosis* visible, mainly because the insects are always on their backs, save one final scene with Fattakhov at a museum.

The NKVD's trail of Kharlamov goes cold after they track down the dwarf, who returned to his original profession—a circus performer. At one point, Kharlamov joins him there, where, for a while, he plays the part of a circus version of Aleksandr Nevskii repelling the Teutonic invaders.[77] The NKVD gives up the pursuit of Kharlamov at this point and eventually opts to destroy all evidence of the lunar program and cease their chase of Kharlamov. They neglect one thing: the NKVD could not destroy one remaining source of evidence. In the closing scenes, the movie takes the viewer to the natural history museum in Chile, near where the peasants had seen the fireball in 1938. This museum retains the footage from Kharlamov's lunar mission and the hardware from his flight. The movie closes with film footage of a lunar landscape, a silent cosmonaut sitting inside his spacecraft. *First on the Moon* weakened the links between the hero cosmonaut and the Soviet state. Kharlamov was loyal to his country to the end. His devotion is evident in his struggles to return home as a non-existent person. Even his final known role was that of the legendary and publicly manipulated image of Aleksandr Nevskii, whom Stalin used as a surrogate for preparing for war.[78] His reward for all this had been pursuit and abuse by the system that created him.

Aleksei Uchitel's film *Space as Premonition* takes a nostalgic approach in which the early Soviet space program provides the background for a story about the illusion of nostalgic optimism. The film takes place between the time of the launch of *Sputnik* and Yuri Gagarin's flight (1957–1961). The protagonist of the story is a hapless young man, Konek, whose naivete

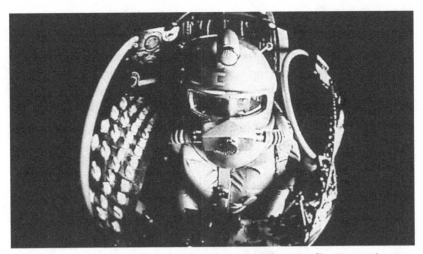

Figure 19. Stalin-era cosmonaut character Kharlamov from the film *First on the Moon* sits impassively inside his spacecraft while traveling between the Moon and Earth. This was the long-lost evidence of his top-secret mission.

has benefited him. Unaware of the injustices around him, he can wander through life unaffected by it. The main character is a cook whose real name is Viktor, but he goes by the nickname "Konek" (Horsie). The plot rides on Konek's relationship with a former sailor and dockworker, German. German, who is also known as "Lefty," is a former sailor who is trying to defect to the West. His persona allures Konek, a man who is haplessly living with his mother and indecisive about committing to his girlfriend, Lara. In contrast, German is worldly and sophisticated. The men establish bonds; both are war orphans and relish fights with sailors. To Konek's mind, German is exotic and mysterious, possessing superior skills and knowledge about the world as well as material possessions, including an East German radio that picks up BBC. Despite his seeming sophistication, German cannot articulate the English words properly to declare his intention to defect. Ironically, his new hapless friend demonstrates the ability to mimic the voices on BBC radio almost perfectly, although he has no ambitions for contact with the West and understands little of what he is saying.

Over the course of their relationship, Konek begins to dress and act like German. He goes so far as to practice swimming in the harbor. Konek assumes that this activity is to improve his athletic performance; he is clueless that his friend intends to defect by swimming out to a foreign ship. Sailors eventually beat him up for his attitude and flagrantly walking around town with the forbidden radio. As their relationship develops, German confesses that his difficult secret assignment is to seek out the ten cosmonauts who are training for the first spaceflight in Kustanay in the Kazakh Republic. For a while, German insists to Konek that soon men will travel to the Moon every day, but ultimately, he confesses again that he had been in prison in Kustanay. He left the Navy after a conviction for making wisecracks. German's last appearance in the film shows him swimming toward a shipping vessel marked *Lake Michigan* as the ship moves away. In response to German's escape, Konek marries Rima, Lara's sister, and they take a train to Moscow.

In *Space as Premonition*, space is a metaphor for inexplicable hope. In the movie, Lara asks Konek if he could see the *Sputnik* after German seduces her, as a plea for reassurance. A second use of the metaphor occurs during Konek and Rima's trip to Moscow. While on the train, Konek crosses paths with an equally unassuming young pilot, whom the hero and audience believe to be Yuri Gagarin. When speaking to Gagarin in the train, Konek asks him if he is going to fly rockets. Gagarin responds by asking if he is referring to the predictions of Tsiolkovsky. Konek replies, "No, German." Gagarin replies that he has not heard of that scientist. Konek replies, "He is not a scientist, but he has already flown in space." When the pilot arrives at his stop, Konek asks the young pilot his name and notices that his shoelace is untied. Later Konek recognizes Gagarin by this untied shoelace. At this time, Gagarin has made his flight and is walking down the red carpet to greet Khrushchev.

It is through this meeting that the director projected the hapless life of his hero onto the space program. The experienced and knowledgeable character, German, is determined to escape the Soviet Union, even if it costs him his life. The more hapless of the two, Konek, identified most closely with Gagarin. One film reviewer described the time between *Sputnik* and Gagarin's mission as "the two moments of Soviet triumph in space that, the contemporary audience knows, led nowhere and that provide the bookends of the film (the flights of *Sputnik* and Gagarin)."[79] These two moments of triumph represented a vital period that encouraged the nation in its naivete but provided no objective improvement in its circumstances.

Aleksei German's film *Paper Soldier* takes a more intimate and introspective look at the Soviet space establishment in 1961. The director has taken the title from a line from Georgian singer/songwriter Bulat Okudzhava's 1959 song:[80]

THE PAPER SOLDIER

Once there lived a soldier-boy,
quite brave, one can't be braver,
but he was merely a toy
for he was made of paper.
He wished to alter everything,
and be the whole world's helper,
but he was puppet on a string,
a soldier made of paper.
He'd bravely go through fire and smoke,
He'd die for you. No vapour.
But he was just a laughing-stock,
a soldier made of paper.
You would mistrust him and deny
your secrets and your favour.
Why should you do it, really, why?
'cause he was made of paper.
He dreads the fire? Not at all!
One day he cut a caper
and died for nothing; after all,
he was a piece of paper.
    —1959[81]

Okudzhava had provided a musical, social, and functional critique of
the Soviet system in the 1970s and became a composer of the anthem of
late and post-Soviet space culture in the cases of *White Sun of the Desert*
(1970) and *Paper Soldier* (2008). His largely dissident following expanded
as he gained popularity among the Russian intelligentsia—mainly in the
USSR at first, but soon among Russian-speakers in other countries as well.
German made a statement by selecting his poem for his film.

The movie is set in the remote Baikonur launch facility near Tiuratam,
Kazakhstan in early 1961, and based loosely on fact. The central character
is Daniel Pokrovsky (Merab Ninidze), a medical officer who works for the
first Soviet cosmonaut troop, the twenty men recruited to fly in space in
late 1959. While at the launch facility and preparing for Gagarin's launch,
the married Pokrovsky finds himself in a complicated and yet tender relationship with a young girl, Vera. Pokrovsky also finds himself haunted by
an elderly local merchant, Olbrykhskii, who is hawking a garish portrait of

Stalin framed by makeup lights, reminding the doctor of the dark origins of the space program and his own work.

As a physician to these young cosmonaut candidates, Pokrovsky tries not to be just a doctor for the cadets, but also their friend. He finds the fact that these young men could have to sacrifice their lives for the country troubling. This is not his private concern, but one that he shares with the cosmonaut trainees. One of the corps, Valentin, confides to Pokrovsky that they know that others call them "Laiki." This is the plural of the ill-fated dog that that orbited the Earth in November 1957 with no hope of ever returning alive. At home in Moscow, his wife Nina feels the same way. Facing the prospect that he is participating in a project that could put human lives at risk, she repeatedly asks her husband to quit his job. Pokrovsky chooses instead to leave his wife and return to Kazakhstan to Vera. Before he has a chance to do so, one of the cosmonaut cadets dies. This episode is an allusion to the actual death of Valentin Bondarenko, a twenty-four-year-old cosmonaut candidate. Bondarenko died less than three weeks before Yuri Gagarin's flight. The pilot was coming near the end of a fifteen-day isolation experiment in a pure oxygen, low-pressure altitude chamber when he inadvertently ignited an alcohol-soaked cotton swab. The real Bondarenko's death caused years of recrimination between the cosmonaut corps and the physicians at the institute that designed the experiment. In *Paper Soldier,* Pokrovsky bears the full emotional burden of the cosmonaut's death and has an emotional breakdown in the face of his worst fears.

Pokrovsky recovers quickly enough within weeks to return to Kazakhstan to prepare the launch of the first man into space. His wife follows him. When she arrives, she comments on the contrast between the optimistic, official space-enthused propaganda of the period and the reality of life at the Tiuratam-Baikonur complex. In this film, as in reality, Baikonur is a railroad stop that is distant from the launch facility. In the end, Pokrovsky remains in conflict and haunted, but "Yuri" climbs on board the launch vehicle. During the final scene with "Yuri's" launch in the distance, Olbrykhskii appears, declaring "everything for sale."

Both Uchitel and Fedorchenko drew on the traditions of realistic science fiction in their films. In each case, spaceflight is technically accurate and not metaphysical. These films refer to the traditions that Zhuravlev and Klushantsev had pioneered in the 1930s and 1960s, respectively. The difference between these post-Soviet filmmakers and their predecessors is that they have portrayed Soviet cosmonauts stripped of the either implicit (*Kosmicheskii reis*) or explicit (*Planeta bur'*) ideological discipline. Uchitel's Gagarin lived in a morally reprehensible system that only the hapless can ignore. Fedorchenko's Gagarin surrogate, Kharlamov, returned to a country whose state apparatus is determined to remove all evidence of his existence. For these directors, the Red Stuff does not represent the power of autonomous actors. The system has trapped the cosmonauts and predetermined their lives.

German makes a more explicit and overt critique of the Soviet system. Pokrovsky is the tragic figure that is incapable of extricating himself from a situation that he knows to be morally wrong. German uses the character as a surrogate for the legions of engineers, scientists, and technicians who willingly participated in building credibility for the USSR while knowing what the cost could be. In *Paper Soldier,* the cosmonauts are hapless pawns in an elaborate game in which they have no deciding role. Pokrovsky is like Gagarin walking the red carpet with an untied shoelace. Sheer luck saved them from embarrassment or an even worse fate.

Diminished Official Interest in the Red Stuff

The official post-Soviet Russian attitude about the Red Stuff existed within a shroud of ambivalence in the late twentieth and early twenty-first centuries. There had never been a grassroots effort to challenge its elevated role in Soviet history. However, the flood of other, long-suppressed concerns in the country seemed to diminish the attention paid to it. The space program was never on the cutting block when the parliament prepared budgets, but neither did it receive its full, appropriated annual amount. Even in international engagement, pressing concerns dwarfed the importance of the Red Stuff. Boris Yeltsin signed an agreement in 1992 with the Americans to collaborate on the construction of a new space station. That agreement placed a limit on the life expectancy of the existing Soviet station, Mir. The once popular and government-supported publication of hagiographic biographies of the cosmonauts and their deeds stopped as state-funded publications evaporated. The space-themed books that replaced them were the "tell-all" and "tell-some" tales of the secrets of the space program. Underemployed and retired engineers, technicians, and cosmonauts sought to set the record straight and earn hard currency in a time when government pensions fell to laughably low levels. Russia continued to send humans to the Mir space station and welcomed Americans arriving on board the shuttle to their nearly continuously inhabited space station. The Red Stuff was no longer an official statement. One of several places where the decline in attention to the Red Stuff was most obvious was the tourist monument to it—the Kosmos Pavilion of the Exhibition of Economic Achievements.

The location where the Kosmos Pavilion of the VDNKh stood had been the display area of the pivotal technology of the era since the late 1930s. Before World War II, the original pavilion housed exhibits on the mechanization of agriculture under the watchful eye of a statue of Stalin. During the early 1960s, spacecraft replaced tractors inside the pavilion, and an R-7 (*Sputnik*) rocket replaced the statue of Stalin outside.[82] For nearly a quarter of a century, the Kosmos Pavilion was a world-renowned icon of the space program, even while it housed a personal collection of displays and artifacts.

In 1991, everything began to change.[83] With Boris Yeltsin's signature, the exhibition lost its national allegiance, federal subsidy, and the authority of the Academy of Sciences and changed its name yet again. In 1992, Boris Yeltsin re-created the VDNKh into the Vserossisskii vystavochnyi tsentr (All-Russian Exhibition Center, VVTs) leaving the existing management to fend for itself for funding. Their first response was to rent out the largest open spaces for commercial interests. In the early 1990s, a car dealership brushed aside the Academy of Sciences' displays of scientific satellites and began selling cars inside the pavilion.[84] Later, Asian electronics merchants installed and sold radios and televisions to consumer goods–starved Muscovites.[85] Finally, in 1998, RKK Energiia, the legacy organization of Sergei Korolev's OKB-1, reclaimed the human spaceflight hardware that remained housed among the electronics kiosks that had replaced the car dealership. RKK Energiia retained the collection for its formally private, corporate museum.[86] This one grew out of Korolev's collection of flown hardware. Besides, even though the museum existed behind the steel gates of the formerly secret enterprise, Energiia began to encourage visitors and school groups, like the ones that used to attend excursions to the Kosmos Pavilion twenty years earlier.[87]

The sign, "Kosmos," remained above the entrance of the pavilion, as did the *Vostok* launch vehicle that had replaced the statue of Stalin. For over two decades, the pavilion, as had been the case for so many problematic issues of the former USSR, seemed to face destruction through neglect and not by deliberate action. It seemed almost inevitable and appropriate that for a time the only displays in the building were kiosks of imported consumer electronics. In the 1990s, imported household goods were the pivotal technology of the era. In the last decade, new changes have come

Figure 20. This is what remains of the *Buran* hangar. Despite the complete destruction of the flight vehicle and the loss of eight lives from the roof collapsing, the outer shell of the building remains attached to the vehicle assembly building in which the *Soyuz* and other spacecraft are prepared for launch.

to the pavilion. In 2014 a new mayor took renewed interest in the pavilion. On 14 May 2014, he announced that the structural test model of the Soviet space shuttle *Buran,* which had been an attraction and restaurant in Gorky Park, would move to the *VDNK*h. The move was complete by 21 July as part of a planned rejuvenation of the park, including a rededication of the Kosmos Pavilion. The move of the *Buran* began plans to revitalize the old Kosmos Pavilion.

While these plans were in action, few people in Russia noted the searing irony that *Buran,* a failed and abandoned program, was being using to point to a bright future in Russian aerospace. The program had never flown with a crew. Its sole orbital mission in November 1988 became the focal point of a very public dispute between Soviet space sciences and the Soviet Ministry of Defense Industries. The financial collapse of the USSR led the Russian government to abandon the program, with occasional faint attempts to market components to foreign customers. The final insult to the program occurred on 12 May 2002, when the hangar in which the flown article was stored collapsed, killing eight workers. And, as almost comic relief to the undignified end of the program, the remnants of the collapsed hangar remained in full view of international space tourists who flocked to Baikonur to witness launches.[88]

On 12 April 2018, Vladimir Putin toured the newly refurbished pavilion, now under the name the Cosmonautics and Aviation Center. This new center featured a museum and exhibition area, an educational and research section for the implementation of projects proposed by children and young people, and recreation and family entertainment areas.[89] The exhibition area featured the *Vostok* capsule from which Yuri Gagarin ejected in 1961 and other hardware from the private corporate aerospace museums throughout the Moscow region. What was striking about this redo of the Kosmos Pavilion was the absence of the Red Stuff. It was as though the official Russian strategy had done a 180 degree turn from the strategies of previous generations. Instead of using the cosmonauts as a shield against Soviet hardware secrets, the new center had shunned cosmonauts in favor of showing hardware. More striking was the fact that at the opening events, where cosmonauts were present, they appeared in business suits, indistinguishable from their corporate and political counterparts. The photo mural around the dome of the pavilion showing Yuri Gagarin cradling the dove in his hand remained, even when the hall was empty. Managers arranged the spacesuits, the most human-scale displays in the pavilion, as tradeshow exhibits not representing the Red Stuff.

Conclusion

The Red Stuff did not undergo the brutal disassembly that broke up the cult of the Bolshevik Revolution, Stalin, and World War II. Reexamination of what had been sacrosanct memories of the Soviet experiment were almost inevitable given its seemingly sudden and unceremonious collapse. Whereas criticism and sarcasm about the USSR had become a national pastime during the late Soviet period, critique became a national mandate after 1992. And unlike the previous era, even the staidest members of professional organizations were now free to join in the intellectual dismantling of the USSR without the fear of being labeled a dissident. Dissent became a respectable part of Russian culture.

Late- and post-Soviet artists and writers felt free to criticize the Soviet government of that time, but the critique of the culture was subtle. The appraisals severed the link between party ideology and spaceflight without directly attacking the history of the program head on. For the generations that grew up among the excitement and material culture of the program, cosmonauts have remained untouched, cherished heroes of a bygone and hopeful era. In literature, film, and even television, spaceflight remained an optimistic enterprise, not tainted by the disappointment and losses that savaged the shared memory of World War II. It offered the promise of hope and enlightenment. However, each artistic version distanced the dream of spaceflight so predominant in the 1960s from its Soviet legacy.

Pelevin distanced spaceflight from the Soviet state by peeling layer after layer of subterfuge and corruption. He revealed a state that was incapable of having accomplished anything without ruse and lies. This fact flew in the face of the youthful optimism of Omon and his friend Mitek. The only way that Omon could survive was to run away. Each author clearly stated that the USSR was incapable of reaping the benefits that spaceflight promised. Similar distancing took place in the films of the time. Fedorchenko's *First on the Moon* made the distance literal and political as well as metaphorical. Kharlamov had to trek back to the USSR from Chile and then evade the NKVD to survive. In *Space as a Premonition,* German inserted an inside joke for the audience to make space between the official portrayal of Gagarin and his hapless reality. Finally, Dr. Pokrovsky, the featured character in *Paper Solider,* has a breakdown when facing the chasm between the risks of human spaceflight and his heroic image in public.

Even after the end of the Soviet Union, the memory of spaceflight retained an air of wistfulness that other Soviet experiences did not. Memories of the Red Stuff had set it as a phenomenon that had been born of Soviet efforts to justify its existence, but the memory of the repressive and reckless state did not burden it. It is somewhat paradoxical that the Soviet space program did not meet its full potential of competing directly with the United States, which allowed its decoupling from the state. The state failed to reach the promise of the space program and was thus subject to critique. The human spaceflight program seemed to shed blame because it had not fulfilled the optimism of the time through no fault of its own.

Nostalgia for the hopes and promise of spaceflight is distinct from nostalgia for the communist system. Mira Ivanova, who had expressed a wistful longing for the day that man first flew in space, was clear on that point: "She admits to nostalgia, but not for Communism. 'I would return to the old Russia,' she says, 'but only to return to my youth.'"[90] Boym diagnoses Ivanova's nostalgia as a particular type of longing—a restorative nostalgia:

The past for the restorative nostalgia is a value for the present; the past is not a duration but a perfect snapshot. Moreover, the past is not supposed to reveal any signs of decay; it has to be freshly painted in its 'original image' and remain eternally young.[91]

While Ivanova and her Saratov classmates looked at that time as a period devoid of destitution (either personal or societal), or at least of lessened strife than their parents had recalled, the writers and filmmakers that I have discussed have juxtaposed the cosmonauts to the decay of the Soviet system. They show agreement on the source of the decay—the corrupt and dishonest state that tried to conceal itself through technological achievements.

Human spaceflight was an inextricable component of the public culture of the post-Stalinist Union of Soviet Socialist Republics during the 1960s. It had also been the source of utopian fantasy before the beginning of the Soviet Union. Even after utopian science fiction ended abruptly because of Stalin, the fantasy continued through the popularization of science in the 1920s and 1930s. Once Gagarin orbited Earth, the fantasy revived. This time, the chief authors of this utopian fiction associated with the government. They created a renewed utopian fantasy of communism around spaceflight instead of the other way around. Besides, when the fantasy of communism crumbled, all that remained was the memory that the Soviet Union had launched men and a woman into space and created material and public culture where a civic society had not existed before. After the end of the Soviet Union, the dream of spaceflight remained, but instead of a future ambition, those who lived through the experience had only the past. Svetlana Boym argued that this ill-fated link between spaceflight and communism had created a change in Russia:

> Before cyberspace, outer space was the ultimate frontier. More than merely a displaced battlefield of the cold war, the exploration of the cosmos promised a future victory over the temporal and spatial limitations of human existence, putting an end to longing. Now that the cosmic dream has become ancient history, new utopias are neither political nor artistic, but rather technological and economic.[92]

If one assumes that all post-Soviet reflections on the space program are nostalgic, then the Soviet Red Stuff remained a utopia that existed in the past among the rubble that was the Soviet Union. Nevertheless, not everyone had found himself or herself longing for a childhood place where the

Red Stuff once dwelled. Some sought to explain why those places were no longer desirable. The major preoccupation among writers and other artists was the question of what went wrong. Victor Pelevin and post-Soviet filmmakers have taken the stories and culture of the space programs of the 1960s to reconstruct the past not as a comfort to themselves, but as a challenge to the past and a challenge to the well-crafted official reality that dominated in the 1960s. *Omon Ra* and *First on the Moon* portray the past as a staged farce much in the way that Gogol and other satirical writers savaged Imperial Russian reality in the nineteenth century.

The Red Stuff in the films *Space as a Premonition* and *Paper Soldiers* is not comforting, but a source of personal torment. The main characters in both cases faced painful introspection, even though Konek turns away from it. Moreover, Boym's theories about longing for a place in the past only apply to those with firsthand memories of that time. Mira Ivanova longed for the promise of her youth, while the post-Soviet generation conflated the heroes and accomplishments of fifty years ago, with no unique attraction to the 1960s.

What Boym does share with others in this post-Soviet retrospective of the 1960s spaceflights is that in each case she argues that people take care to choose precisely which components to contribute to the personal memories of the time. Boym and Ivanova chose a seemingly carefree and promising youth that the Soviet government marketed to the Soviet population during the 1960s to divert attention from the failure of the communist promise. In contrast, others have chosen to juxtapose the realities of Soviet life with what the official cosmonaut images omitted from the promises and placed into a starker reality.

8

Epilogue

April, 12, 2011, marked the fiftieth anniversary of Yuri Gagarin's orbit of the Earth. Throughout the human spaceflight–obsessed world, two sharply contrasting models of celebration marked the anniversary. In Russia official tributes to Gagarin's flight combined images of him and his monuments. The celebrations featured a panoply of Russian space icons, including the grandfatherly image of Konstantin Tsiolkovsky and the stern father image of Sergei Korolev, flashing on screens in front of audiences of the surviving cosmonauts, technicians, and engineers whose labor made the flight of the first human in orbit possible and contemporary politicians. Cosmonauts from the previous half-century and invited foreign space dignitaries took turns to claim their role in the formal national celebrations. In contrast, in fashionable bars, clubs, and science centers around the world, including Moscow, young people, born well after Gagarin's flight, and some even after the collapse of the USSR, gathered to celebrate their own version of the Red Stuff. These celebrations were a postmodernist and somewhat ironic use of Gagarin's image and the anniversary of his flight as an excuse for a world party. "Yuri's Night" has become a global pop culture flash event. It began at the turn of this century to "bring the world together to dream about where we're going."[1] These more youthful and independent celebrants of Gagarin's legacy use a line drawing of Yuri Gagarin from the Center for Cosmonaut Training located outside Moscow as their logo. The origins of these divergent celebrations are the same Red Stuff, but they represent how far an idea can drift from its starting point in half a century. Both were created in the reality of superpower competition, and they diverged as a result of being nurtured in different realities and by a broad array of actors. A modern participant in either event could be excused for not recognizing their common origin.

The image of Yuri in the Gagarin Monument in Moscow was one created under Khrushchev. The Gagarin in the monument is a socialist realist,

Stalinist New Man in post-Stalinist circumstances. Khrushchev had borrowed Stalin's authority and dressed Gagarin up in a post-Stalinist style to assure that although the foundation, regnant myths, and taboo topics remained the same, this was a narrative of the Soviet ideal. The official Gagarin Red Stuff had a single origin and mission. In the words of the 1961 song "Gagarin March," which echo the first three lines of a 1925 poem written by Russian Futurist avant-garde poet Vladimir Mayakovsky:

We're leaving for space to work,
Seated in orbiting ships
And everything starts from the first flight,
Gagarin's first turn around the Earth.[2]

Yuri Gagarin was the origin story of a promising Soviet future—the Red Stuff. Gagarin was a sweet Russian boy with a broad smile, recognized as excellent by all—Soviet citizens and members of the Western world. This picture of Gagarin was tolerated and nurtured under Khrushchev and grew into its own iconography when the Cold War imagery lost its appeal. Sometime during the last half-century, Russian human spaceflight had transformed from being a futuristic idol and measure of national status to a yardstick against which to measure the past and an icon unencumbered by historical context. Those changes in meaning occurred during Brezhnev's rule as the public curation of the cosmonaut developed independently of a still repressive state that saw the space program as having declining importance to national prestige.

There is no doubt that the organizers of the human spaceflight program in the Soviet Union kept Stalin's aviation exploits in mind as they constructed new public memories around the cosmonauts. Nikolai Kamanin, the first head of the cosmonaut corps, had been one of the first aviation heroes of the 1930s. The public culture of spaceflight was very different from aviation on a fundamental level. Stalin promoted aviation to encourage acceptance of industrialization and create an air-minded population prepared for the ever-present danger of foreign invaders. His goal had been to adapt the population to a quickly approaching militarized future. In contrast, Khrushchev used spaceflight as a decoy to distract the Soviet population and the rest of the world from the consequences of World War II. The economic, military, and political woes that challenged the USSR in the 1960s originated during the war and did not heal quickly. Khrushchev offered them a fanciful future as a distraction. In contrast to Stalin's aviation program in the 1930s, the Soviet space program in the 1960s did not have an underlying goal of creating an army of trained cosmonauts and support personnel that would support a new move toward modernization. There was never a plan to create armies of spacemen marching toward the future. Planners squeezed the cosmonaut program out of an already established ICBM program to make a public justification of the costly consequences of the government's secret activities. In many ways, the primary goal of the space program was public visibility. The visible side hinted at unrevealed strengths. Khrushchev was not trying to mobilize the Soviet population; he was trying to placate it. To do so, he used images that he drew from the past, to give the impression of looking forward to a brighter future.

The public and material culture of spaceflight used Stalinist socialist realism and neo-modernist images to drill into the consciousness of the Soviet population. It was based on this culture that the Twenty-Second Party Congress was able to declare that orbiting men around the Earth was evidence that the USSR would achieve communism by 1980. Despite the apparent disappointment in early promises, reminders of Yuri Gagarin still sparked excitement and interest among the Soviet population even after the country was in decline.

The metaphoric ground on which Khrushchev tilled was first cultivated earlier in the century. The Soviet population already had in place a set of cultural guidelines within which they could interpret the significance of the flight when TASS announced Yuri Gagarin's successful mission in 1961. The idea of cosmonauts—men and women flying in space—was familiar. Science-fiction literature and film had portrayed them in earlier generations. Before that, human space travels were associated with utopianism and early Bolshevism. Before the revolution and the implementation of the New Economic Policy (NEP), science-fiction stories and novels included space travelers who were at the forefront of the socialist revolution. After Lenin and then Stalin clamped down on the independent interpretations of utopianism in science fiction, spaceflight and its fiction became the dream of scientists, who augmented politically orthodox space science fiction with technical accuracy about the effects of spaceflight. During the 1920s and 1930s, scientists had promoted a concept of national science education that permanently broadened the idea of scientific participation through public education. Their activities included the publication of science fiction. Theorists from all occupations, such as Konstantin Tsiolkovsky, who had previously experimented in isolation, joined forces to turn science into a national pastime. The popularity of this vision of spaceflight broadened when they adapted it to the movies. The relatively new medium of film showed promise as an instrument of propaganda for science as much as politics. Even when directors eschewed the heavy-handed political messages that dominated much of the culture of the time, the films were effective in conveying the excitement and technical aspects of spaceflight to the entertainment-starved population.

At the dawn of the space age, Khrushchev discovered and tested the limits of the propaganda power of space events. Realizing that reactions to the launch of *Sputnik* reverberated far beyond the borders of the USSR, he came upon the idea that effective demonstrations of the Soviet Union's rocket capabilities could convey the impression that it was far more technologically advanced and robust than it was. By using the handful of existing rockets, Khrushchev was able to produce a far more significant impact than stockpiling them would have had. With these components in place, Soviet planners only had to build on the existing culture to create a cosmonaut one. In these ways and more, the early Soviet human spaceflight program reprised successful scenes from the early Soviet past. In the early 1960s, at least on the surface, rockets had replaced airplanes and space had replaced the Arctic, but instead of exhorting the population to look forward, the government was encouraging the population to take comfort in the past as an assurance that the future would soon be better.

Cosmonaut biographies proliferated after Gagarin's mission in newspaper stories, magazine articles, and, later, books. These biographical and autobiographical stories told highly ritualized tales that resembled the heroic biographies of the 1930s. In the absence of Stalin and rapid industrialization, the cosmonauts thanked the party and frequently referred to their childhood experiences during World War II. The biographies carried over in tightly managed identities that space managers, especially Commandant of the Cosmonauts Kamanin, orchestrated to convey a message of hope for the near future. In a circular argument, the party, in turn, pointed to the existence of the cosmonauts as an indication that the immediate future would be bright. With this, the government tied the Khrushchev-promised 1980s utopia to the cosmonauts.

The uses of the heroic image of the cosmonaut multiplied after Gagarin's flight. Politicians relied less on cosmonauts to propagate their philosophy, but cosmonauts remained representatives of their country, and are therefore likely to be adherents to its political philosophy. What has happened over time is that the cosmonaut became less visible as a symbol of political power and more visible as a profession and a symbol of past glory. The cosmonaut heroes became heroes in perpetuity. Many features remained constant among the personal biographies of the old cosmonaut heroes, such as Leonov, who devoted most of their careers to popularizing the idea of spaceflight. One feature not written into the books is the fact that cosmonaut monuments survived the collapse of the USSR.

Human spaceflight was not the only new thing in the Soviet Union in the early 1960s. Stalin's death and Khrushchev's de-Stalinization unleashed other forces. Among them, architects and designers found new independence with which they could reassert the modernism that had been popular among them in the 1920s and 1930s. They created a new aesthetic for the post-Stalinist era. Besides that, pent-up consumer demand and the rising independence of Soviet youth combined to create a palpable dissatisfaction with the current economic situation. That prompted the government to pay lip service to satisfy the demand for consumer goods. Individual factories and enterprises joined to produce consumer goods that might satisfy the market or at least deflect its hunger. The combination of human spaceflight and these changes in the material culture of the Soviet Union offered many opportunities.

Space museums and monuments to spaceflight frequently combined modernist and neo-classical architecture. For example, socialist realist sculpture lines the approach to the modernist Monument to the Conquerors of Space. The content, however, was similar no matter what the design. Public spaceflight exhibitions during this period followed a very narrow focus—to encourage national interest and celebration, but at the same time not to be so instructive as to reveal tightly held secrets. The models revealed little of the actual spacecraft but the displays frequently paired them with models of ambitious plans. As had been the case with cosmonauts themselves, there was no inducement to participation beyond the superficial celebrations.

Nevertheless, they attracted steady streams of Soviet visitors who yearned for any positive portrayal of their national life. Exhibitions of the 1960s promised the continuation of Soviet space achievements without revealing many technical details. The result was a sense of rationed abundance during the Khrushchev period that created an atmosphere of high expectation. These exhibitions were but one avenue through which the Soviet government could convey its message of optimism for the future. Moreover, even given their shortcomings when compared to their chief space rival, the United States, Soviet citizens rarely had the opportunity to make comparisons between the two sets of displays. Thus, space exhibits complemented and reinforced the official message about human spaceflight. It was a key component for replacing the cult of personality in the Soviet Union. Hardware, exhibits, and allusions to anonymous engineers stood where statues and references to Stalin once stood. Spaceflight replaced industrialization and mechanization of agriculture in the public culture of the Soviet Union. Much in the same way that exhibits did not call for mass participation, the exhibits that replaced the Stalinist exhibits were not the authentic hardware that had once stood in the shadow of Stalin. The actual hardware remained tucked away among the engineers who built it. They sought public support to maintain their enterprises, but no one was trying to transform the country using this hardware.

Stamps and znachki, too, offered the opportunity for comparison of the old and new style of portraying public events. Both embraced spaceflight as subject matter. Each responded to the popular demand for collectibles that had grown out of dissatisfaction with material life in the USSR. The Ministry of Post and Telegraph joined with the rest of the country and relaxed scrutiny on private collecting. However, they changed little in their approach to their product. Spaceflight encouraged manufacturers to expand their production of znachki, which had only recently established them as souvenirs. There were other differences between the two forms of collectibles. While the Ministry released over one hundred space stamps, there were thousands of individual space znachki from nearly two dozen independent manufacturers. The Ministry of Post, with no competitors in the market, had no reason to attempt innovative designs. The pin manufacturers were competing in this new market and had incentives to make their products exciting and attractive.

The existence and unlimited numbers of pins were an indication that the collecting and possession of material goods not only became acceptable in the 1960s Soviet Union but also gained encouragement through official channels. The design of the pins hearkened back to a more favorable time when constructivism and modernism reigned in Soviet art and architecture. Moreover, in the absence of systematic exhibits and a focused public relations program to promote the space program, znachki took on the role of telling the tale of Soviet spaceflight. Children and students learned its lessons through Pioneer and youth organizations that encouraged collecting through articles and columns that announced new issues. Znachki also differed from stamps because they represented a significant departure from previous public culture movements. Their decentralized manufacture had no authority dictating the content and message on all pins. Finally, the pins were significant for their endurance. Large collections remained intact and, much like modern American baseball cards, they had taken on a following of their own beyond the subject that they illustrated.

Events that occurred beginning with the ouster of Khrushchev in October 1964 and concluding with the death of Gagarin in March 1968 shattered the public culture of the Soviet human spaceflight program. For three and a half years, accepted wisdom about the trajectory of the Red Stuff fell by the wayside. The space program that lost political and managerial support had sustained the appearance of a wild success. The removal of Khrushchev and Korolev's death deprived the country of the ability to camouflage technical failures as routine tests and spin repetitive flights into great technological achievements. Finally, the most recognized symbol of the pinnacle of its accomplishments, Yuri Gagarin, died an inexplicable death. The previously effusive Soviet press offered little explanation about these challenges. The resulting divergence between the public understanding and official mythologies grew. Engineers continued to salvage what remained of the space program through space station missions. At the Politburo, Defense Minister Ustinov maintained a commitment to space activities, with an eye to mimic American accomplishments, as he did with other military technologies. Work continued with the crewed lunar mission until the mid-1970s, albeit without the public knowledge, enthusiasm, confidence, and successes of the beginning of the decade. On the public side, Yuri Gagarin became an icon of the glorious years of Soviet power with monuments and museums in every capital and regional library.

Yuri Gagarin's death startled the nation and the world. When his trainer aircraft went down along with his instructor, the Soviet Union lost national heroes from two generations: World War II and the space age. The Gagarin death had more considerable reverberations than that of Col. Seregin. What was more shocking than Gagarin's death was the absence of an explanation for it. The cause of the crash was not an immediate, identifiable, and isolated cause. Not able to point to a single, correctable mistake, the state could not readily fix the situation. The specter of a failure within the system would continue to haunt the space program through the rest of the century.

Gagarin's death closed the first age of spaceflight for the Soviet Union. The death and the domestic reaction to it offered significant insight into the society as a whole and the relationship between the space program and the developing post–World War II Soviet civil society. There are three features of Gagarin's death and burial and the events that preceded it that are important to the history of the role that spaceflight has played in Soviet culture. First, it signified the resounding end of an era of the easily accomplished space firsts. Second, the way in which Gagarin's death was officially commemorated reveals much about the extent to which the postwar Soviet state was flexible in reaching out to the sensibilities of the Soviet—mostly Russian—population to assure the preservation of Gagarin's memory, and by association, its legitimacy. His burial and posthumous commemoration of his life and that of his family corresponded to deeply held Orthodox beliefs. Third, the reasons for Gagarin's crash frustrated the minds of the Soviet population as well as those of the aerospace community. To this day, there is neither popular nor scholarly agreement on the reasons for his crash, even after a long and painstaking examination of the crash site and all available avionics evidence. Neither an official explanation nor the full and ambiguous report of Gagarin's death was ever published. In response to the absence of an official cause of death, Soviet popular culture generated rumors to account for it. Unlike previous mysteries in the USSR, the precise cause of Gagarin's death remained a mystery despite thorough and deliberate inquiry. The broader, more general conclusions about his death brought forth an indictment against the organization of one of the most esteemed parts of the Soviet state. Gagarin, who had been a hero of the Soviet Union, became its most famous victim.

The loss of Gagarin coincided with the exhaustion of national faith in the post–World War II optimism that fueled public aspirations that the Soviet Union could match or overtake the United States in spaceflight. After the reorganization, the public space program turned to a lower-key, less ambitious program than the one that Korolev and Khrushchev had begun. Once again, Soviet space missions methodically repeated themselves with reliable and well-tested hardware. This time, however, the program flourished without the flamboyance of Khrushchev's overt challenge to the United States, while still maintaining a Cold War pace to keep up with the Americans.

Even in the face of political and technological shortcomings, spaceflight did not lose its widespread appeal. The Red Stuff did not undergo the brutal disassembly that broke up the cult of World War II starting in the late 1980s. Late twentieth- and early twenty-first-century criticisms were subtle and effective in undermining faith in Soviet authority. Instead of attacking the program head on, critiques sought to sever the link between party ideology and spaceflight. For the generations that grew up among the excitement and material culture of the program, cosmonauts remained cherished heroes of a bygone and hopeful era. In literature, film, and even television, spaceflight remained an optimistic enterprise. The disappointed expectations that savaged the shared memory of World War II did not taint it in the same way. The space program offered the promise of hope and enlightenment. However, each artistic version distances the 1960s utopian dream of spaceflight. Socialist realist author Aitmatov does this by having the state that refuses the proper burial of Edigei's friend join with the United States and physically bar any contact with the intelligent civilization from the distant planet that had contacted their international space station. The state cuts its ties with the past. Its policies prohibit the nation from benefiting from the extraterrestrial contact. By launching the protective satellite seal around the Earth, the government once and for all severs itself from its scientific utopian ideals. It isolates itself in the present.

Russian postmodernist writer Viktor Pelevin, too, distances spaceflight from the Soviet state by peeling layer after layer of subterfuge and corruption. Pelevin reveals a system that is so corrupt that it is no longer capable of accomplishing anything that is not a ruse. *Omon Ra* is a warning to youth. Discovery of the corruption and decrepitude of the state flies in the face of the youthful optimism of Omon and his friend Mitek. The only way that Pelevin's hero, Omon, survives is to run away and escape the state through its own erected facade. In the case of both writers, the USSR is incapable of producing or accepting the benefits that spaceflight promised.

Post-Soviet films, too, utilize a powerful and popular medium to analyze the brief heroic period of spaceflight. Filmmakers come to many of the same conclusions as had fiction writers. In film, the cosmonaut is either the hapless hero who thrives despite his naivete or a victim who somehow miraculously survives the ordeal that the state lays out for him. The hero of *Pervye Na Lune* and the Gagarin character in *Paper Soldier* are nearly identical characters with little agency of their own. However, the scale of the costs of film production have left Russian filmmakers in this century dependent on the existing film infrastructure at the very least, and the enduring space establishment in the case of any realistic representation of spaceflight. Knowledge of this dependency calls into question the sincerity and incisiveness of any critique of the Red Stuff in recent space films.

The public culture of spaceflight translated well to television. It was the lure to entice the viewers to accept effective electronic propaganda. However, electronic broadcasts did not have a national audience until the 1970s, after the most significant celebrations of cosmonauts. Consequently, television did not have the opportunity to make use of those heady days of the *Vostok* cosmonauts until a generation after the fact. It is ironic that it was a generation-old event that still symbolized optimism for those born after the 1960s. In 1986, Gagarin's voice became a signal for a changing culture that challenged the shared values of his time.

Even after the end of the Soviet Union, people remembered the *Vostok* cosmonauts with an air of wistfulness that other Soviet experiences did not engender. For these people, the phenomenon that had been born of Soviet efforts to justify its existence was no longer tethered to the memory of the repressive and reckless state. It is somewhat paradoxical because it was the Soviet space program not meeting its full potential of overtaking the United States that allowed its decoupling from the state. The state failed the promise of the Red Stuff and was thus subject to critique. The Red Stuff shed association with the government for the same reason.

Human spaceflight became an inextricable component of the public culture of the post–World War II, post-Stalinist culture of the USSR during the 1960s. The years 1961–1965 provided the USSR with a series of international victories unencumbered by either the warfare or ideology that had dominated life in Russia since the beginning of the twentieth century. Spaceflight had also been the source of utopian fantasy since before the beginning of the Soviet Union. Even after utopian science fiction ended abruptly, the fantasy continued through the popularization of science in the 1920s and 1930s. The actual feat of launching a man into orbit revived the fantasy. This time, the chief authors of this utopian fiction associated with the party and government. When the fantasy of communism crumbled, all that was left was the fact that the Soviet Union had launched men and a woman into space and created material and public culture where a civil society had not existed before. In attempting to explain the end of the Soviet Union, the dream of spaceflight remained, but instead of a future ambition, those who lived through the experience had only the past. The Red Stuff retained meaning for those who lived through the era. They remembered the atmosphere of the era and recognized the remnants of the material culture. The response to the Red Stuff that has been written and enacted in the last twenty-five years has been as much a response to the Bolshevik experiment as it has been to its most sacred memory.

For all its efforts to preserve and burnish the legacy of the USSR and Imperial Russia, the Putin government has not enthusiastically and whole-heartedly embraced the Red Stuff. The choice to emphasize the hardware over the human idealization at the revitalized Kosmos Pavilion reflected a deliberate decision to promote economic development over celebration of a Russian ideal of a hero. Without official sanction, the Red Stuff remains a cultural touchpoint of Russia. Recent biopics feature the first flight of Yuri Gagarin and the first spacewalk of Aleksei Leonov, which serve as a reminder that the hagiography of early cosmonauts persists.[3] Today, when almost none of the participants remain alive, the Red Stuff endures in the memories of the Russian population, however uneven those experiences were. It was no longer essential to the official Russian identity, but one of many Soviet experiences that defined the twentieth century, and it persists as part of the Russian identity.

Chronology

30 September 1857: Birth of Konstantin Eduardovich Tsiolkovsky

September 1924: *Aelita* opens in Moscow

13 February 1934: Stalin awards the first Hero of the Soviet Union medals to the rescuers of the Cheliuskin

1936: Vasilli Zhuravlev's *Kosmicheskii Reis* [Spaceflight/Cosmic Voyage] opens

2 November 1938: Valentina Grizodubova becomes first woman to receive Hero of the Soviet Union Award

3 August 1955: Soviet academician Sedov announces intention to launch satellite during the International Geophysical Year (IGY)

25 February 1956: Khrushchev gives "Secret Speech" at the Twentieth Party Congress of the CPSU denouncing Stalin's "cult of personality"

1 July 1957: International Geophysical Year (IGY) begins (1 July 1957 to 31 December 1958)

21 August 1957: Launch of the first Intercontinental Ballistic Missile (ICBM)

4 October 1957: Launch of *Sputnik,* first artificial satellite of the Earth

3 November 1957: Launch of *Sputnik 2,* carrying dog, Laika. She lives only a few hours into orbit

6 December 1957: Failed launch of American *Vanguard* satellite

31 January 1958: Launch of first successful American satellite, *Explorer 1*

17 April–19 October 1958: Brussels World's Fair, "Atomium"

24 June 1958: First space models on display at VDNKh in Moscow

29 June 1959: USSR Exhibition opens in New York

24 July 1959: American National Exhibition opens in Moscow

7 October 1959: USST publishes first photographs from the far side of the Moon from *Luna 3*

12 April 1961: First human orbits the Earth, Yuri Gagarin on board *Vostok*

15–21 April 1961: Failed Cuban exile invasion of Cuba at Bay of Pigs

25 May 1961: US president Kennedy gives speech before Congress outlining the *Apollo* Program

13 June 1961: Yuri Gagarin lays cornerstone for Konstantin E. Tsiolkovsky State Museum of the History of Cosmonautics in Kaluga

6–7 August 1961: German Titov becomes first to spend a day in space on board *Vostok 2*

13 August 1961: Construction begins on Berlin Wall

1962: Pavel Klushantsev's *Planeta bur'* [Planet of the Storms] opens

20 February 1962: John Glenn becomes first American to orbit the Earth on board *Friendship 7*

21 April–21 October 1962: Seattle World's Fair

14–28 October 1962: Cuban Missile Crisis

August 1962: First simultaneous flight of two spaceships, *Vostok 3* and *Vostok 4*

16–19 June 1963: First flight of a woman in space, Valentina Tereshkova on board *Vostok 6*

1964: First broadcast of national nightly news, *Vremia*

22 April 1964–17 October 1965: New York World's Fair featuring NASA-Department of Defense Space Park

4 October 1964: Monument to the Conquerors of Space opens

12 October 1964: Launch of first multi-passenger crew on board *Voskhod* spacecraft

14 October 1964: Nikita Khrushchev resigns as First Secretary of the Communist Party of the USSR and Chairman of the Council of Ministers of the USSR

18 March 1965: Aleksei Leonov performs first spacewalk from *Voskhod 2* spacecraft

29 April 1965: First public display of model of *Vostok* capsule at the Mechanization of Agriculture Pavilion at VDNKh

1966: Mechanization of Agriculture Pavilion at VDNKh renamed the Kosmos Pavilion

14 January 1966: Chief designer of the Soviet space programs, Sergei Korolev, dies during surgery

27 April–29 October 1967: Montreal World's Fair, Expo '67, opens featuring the USSR Pavilion that was later reconstructed at VDNKh in Moscow

24 April 1967: Crash landing and death of Vladimir Komarov in *Soyuz 1*

3 October 1967: Konstantin E. Tsiolkovsky State Museum of the History of Cosmonautics opens in Kaluga

27 March 1968: Death of Yuri Gagarin in training flight outside Moscow

21 February 1969: First failed test-launch of Soviet Lunar rocket *N-1*

3 July 1969: Second failed test-launch of Soviet Lunar rocket *N-1*

21 July 1969: Armstrong and Aldrin take first steps on the Moon

19 April 1971: Launch of the world's first space station, Salyut 1

7 June 1971: First human occupation of Salyut space station with crew of *Soyuz 11*

24 June 1971: Third failed test-launch of Soviet Lunar rocket *N-1*

29 June 1971: Death of the crew of *Soyuz 11* due to decompression upon reentry

24 May 1972: US president Richard Nixon and USSR premier A. N. Kosygin sign Agreement Concerning Cooperation in the Exploration and Use of Outer Space for Peaceful Purposes, paving the way for the *Apollo-Soyuz* test project

22 November 1972: Fourth and final failed test-launch of Soviet Lunar rocket *N-1*

15 May 1974: Soviet program to send humans to the Moon officially canceled

July 1975: Completion of the joint *Apollo-Soyuz* test project

1980: A serialized version of the novel *The Day Lasts Longer than a Hundred Years* published in *Novyi mir*

4 July 1980: Gagarin Monument in Moscow completed

10 April 1981: Memorial Museum of Cosmonautics opens at the base of the Monument to the Conquerors of Space

1985: *The Man Who Flew into Space from His Apartment* first displayed in Moscow

1986: Primary Soviet news program, *Vremia,* undergoes overhaul with new opening credits

15 November 1988: Sole orbital flight of the uncrewed Soviet shuttle *Buran*

25 December 1991: Mikhail Gorbachev concedes the dissolution of the USSR

1991: Russia and the Kazakh Republic sign a 99-year lease agreement for the Russian use of the Baikonur Cosmodrome

June 1992: US president George H. W. Bush and Russian president Boris Yeltsin agree to cooperate on space exploration by signing the Agreement between the United States of America and the Russian Federation Concerning Cooperation in the Exploration and Use of Outer Space for Peaceful Purposes

1993: *Omon Ra* published

September 1993: US vice president Al Gore and Russian prime minister Viktor Chernomyrdin announce plans for a new space station

9 August 1999: Vladimir Putin becomes Prime Minister of Russia

23 March 2001: Russia deorbits its last space station, Mir, into the Pacific Ocean

12 May 2002: *Buran* hangar at Baikonur collapses, killing eight workers

2005: Aleksei Uchitel's *Kosmos Kak Predchuvstvie* [Space as Premonition] opens

2005: Aleksei Fedorchenko's *Pervye Na Lune* [First on the Moon] opens

2007: Aleksei German's film *Paper Soldier* released

2008: Yuri Kara's biopic on Korolev released

21 July 2011: NASA retires the space shuttle program, leaving the Russian *Soyuz* as the only access to the International Space Station for almost a decade

2013: Pavel Parkhomenko's biopic on Yuri Gagarin released

2017: Dimity Kiselyov's biopic of the first spacewalk, *Spacewalk,* released

12 October 2017: Klim Shipenko's film *Salyut 7* released

Spring 2018: Putin opens the rededicated Cosmonautics and Aviation Center inside the Kosmos Pavilion in Moscow

17 April 2020: Maiden flight of Space X's Crew Dragon Demo-2, restoring access to the ISS from US soil

Library of Congress System
Russian Transliteration Table

RUSSIAN		ENGLISH	
Capital	Small	Capital	Small
А	а	A	a
Б	б	B	b
В	в	V	v
Г	г	G	g
Д	д	D	d
Е	е	E	e
Ё	ё	Ё	ё
Ж	ж	Zh	zh
З	з	Z	z
И	и	I	i
Й	й	Ĭ	ĭ
I	i	Ī	ī
К	к	K	k
Л	л	L	l
М	м	M	m
Н	н	N	n
О	о	O	o
П	п	P	p

(continued)

RUSSIAN		ENGLISH	
Capital	Small	Capital	Small
Р	р	R	r
С	с	S	s
Т	т	T	t
У	у	U	u
Ф	ф	F	f
Х	х	Kh	kh
Ц	ц	$\widehat{TS}$	$\widehat{ts}$
Ч	ч	Ch	ch
Ш	ш	Sh	sh
Щ	щ	Shch	shch
Ъ	ъ	"	"
Ы	ы	Y	y
Ь	ь	'	'
Ѣ	ѣ	$\widehat{IE}$	$\widehat{ie}$
Э	э	Ė	ė
Ю	ю	$\widehat{IU}$	$\widehat{iu}$
Я	я	$\widehat{IA}$	$\widehat{ia}$
Ѳ	ѳ	Ḟ	ḟ
Ѵ	ѵ	Ẏ	ẏ

Notes

Chapter 1. Introduction

1 Walter A. McDougall, . . . *the Heavens and the Earth: A Political History of the Space Age* (New York: Basic Books, 1985); James E. Oberg, *Red Star in Orbit* (New York: Random House, 1981). Tom Wolfe, *The Right Stuff* (New York: Bantam Books, 1980).

2 Vasili P. Mishin and N. I. Panitskiy, *Osnovy aviatsionnoy i rakety* [The Basis of Aviation and Rockets] (Moscow: MAI, 1998).

3 It would be too cumbersome to list more than a few of the most critical memoir and diary publications here. Please see the bibliography for a more comprehensive list. Boris E. Chertok, *Rakety i liudi* [Rockets and People] (Moscow: Mashinostroenie, 1995); Boris E. Chertok, *Rakety i liudi: fili podlipki tiuratam* [Rockets and People: Creating a Rocket Industry] (Moscow: Mashinostroenie, 1996); Boris E. Chertok, *Rakety i liudi: goriachie dni kholodnoi voiny* [Rockets and People: The Hot Days of the Cold War] (Moscow: Mashinostroenie, 1997); and Boris E. Chertok, *Rakety i liudi: lunnaia gonka* [Rockets and People: The Moon Race] (Moscow: Mashinostroenie, 1999).

4 Nikolai Petrovich Kamanin, *Skrytyi kosmos. Kniga pervaia: 1960–1963 gg* [Hidden Space. Book One: 1960–1963] (Moscow: Infortekst—IF, 1995); Nikolai Petrovich Kamanin, *Skrytyi kosmos. Kniga vtoraia: 1964–1966 gg* [Hidden Space. Book Two: 1964–1966] (Moscow: Infortekst, 1997); Nikolai Petrovich Kamanin, *Skrytyi kosmos. Kniga tret'ia: 1967–1968 gg* [Hidden Space. Book Three: 1967–1968] (Moscow: Novosti Kosmonavtiki, 1999); and Nikolai Petrovich Kamanin, *Skrytyi kosmos. Kniga chetvertaia* [Hidden Space. Book Four: 1969–1978] (Moscow: Infotekst, 1997).

5 The Rocket and Space Corporation, Energiia, was the organizational descendant of Sergei Korolev's legendary Special Design Bureau that had launched the space age. In 1996 it began the publication of its historical documents with the first of what was to become three volumes: *Raketno-Kosmicheskaia Korporatsiia "Energiia" Imeni S.P. Koroleva*, ed. Iurii Pavlovich Semenov (Moscow: Raketno-kosmicheskaia korporatsiia "Energiia" imeni S.P. Koroleva, 1996); and then in

2001, published the second: *Raketno-Kosmicheskaia Korporatsiia Energiia Imeni S.P. Koroleva Na Rubezhe Dvukh Vekov, 1996–2001,* ed. Iurii Pavlovich Semenov (Korolev: Raketno-kosmicheskaia korporatsiia Energiia im. S.P. Koroleva, 2001).

6 Asif A. Siddiqi, *Challenge to Apollo: The Soviet Union and the Space Race, 1945–1974* (Washington, DC: NASA, 2000).

Chapter 2. The Birth of the Cosmonaut

1 Throughout this book, the name German is directly transliterated from the Russian using Library of Congress standards, except in the case where the name is quoted or published as either "Gherman" or "Herman." I have endeavored to use Library of Congress transliterations in all cases except those in which a more common usage prevails. For example: Yuri is used throughout, except in the case of transliterated titles.

2 Iurii Sal'nikov, Director, *Iurii Gagarin,* Film (Moscow: Ekran, 1969).

3 Valentin Alekseevich Gagarin, *Moi brat Iurii: povest* [My Brother Yuri: A Story] (Moscow: Mosk. Rabochii, 1979), 328.

4 Aleksandr Fursenko and Timothy Naftali, *Khrushchev's Cold War: The Inside of an American Adversary* (New York: W. W. Norton & Company, 2006), 348.

5 A. N. Nesmeianov, "Utro novoi ery" [The Morning of a New Era], *Priroda,* 4 1962: 4.

6 *Pravda* (Moscow), 7 August 1961, Extra, 2.

7 One of the distinguishing features that differentiated the US and USSR space programs was that the Soviets lacked an independent civilian space agency, like NASA, through which they could isolate their public from their secret activities in space. The organizations that worked on the space program were also working on strategic weapons. This arrangement dated back to Stalin's organization of ballistic missile research in 1946 and 1950, which Khrushchev did not change. *Raketno-kosmicheskaia korporatsiia "energiia" imeni S.P. Koroleva* [S. P. Korolev Energiia Rocket and Space Corporation], ed. Iurii Pavlovich Semenov (Moscow: Raketno-kosmicheskaia korporatsiia "Energiia" imeni S.P. Koroleva, 1996), 41–42.

8 Nikolai Ivanovich Kilbal'chich (1853–1881) was the member of the revolutionary organization Narodnaia volia [People's Will] that built the bomb that killed Tsar Aleksandr II in 1881. In his final days in the Peter-Paul Fortress, Kilbal'chich wrote several treatises on possible methods of landing a man on the Moon.

9 Khrushchev titled this speech before the Twentieth Party Congress formally "On the Cult of Personality and its Consequences." Although the speech took place on 25 February 1956, before a closed session, the official publication of the text of the speech did not appear until the following summer. It should be noted that this was not the only "secret" speech that Khrushchev delivered, only the most famous one. Roy Medvedev, *Khrushchev,* trans. Brian Pearce (Garden City, N.Y.: Anchor Press/Doubleday, 1984), 84–87.

10 Fursenko and Naftali, *Khrushchev's Cold War,* and William Taubman, *Khrushchev: The Man and His Era* (New York: W. W. Norton, 2003).

11 Neither was this unique to the Russian and Soviet culture. Historians and anthropologists have written about the utility of ritual in bonding communities throughout modern history. Daniel J. Boorstin, *The Image: A Guide to Pseudo-Events in America* (New York: Vintage Books, 1992). Eric Hobsbawm, "Introduction: Inventing Tradition," in *The Invention of Tradition,* ed. Eric Hobsbawm and Terence Ranger (Cambridge: Cambridge University Press, 1984), 1–14.

12 For a discussion of the military and political benefits of the polar flights, see Kendall E. Bailes, "Technology and Legitimacy: Soviet Aviation and Stalinism in the 1930s," *Technology and Culture* 17, no. 1 (January 1976): 55–81. This article was later reprinted as Chapter 14 in Kendall E. Bailes, *Technology and Society Under Lenin and Stalin: Origins of the Soviet Technical Intelligentsia, 1917–1941,* Studies of the Russian Institute, Columbia University (Princeton, N.J.: Princeton University Press, 1978).

13 John McCannon, *Red Arctic: Polar Exploration and the Myth of the North in the Soviet Union, 1932–1939* (New York: Oxford University Press, 1998), 64.

14 Ibid., 65.

15 Asif A. Siddiqi, *Sputnik and the Soviet Space Challenge* (Gainesville: University Press of Florida, 2002), 163.

16 One of the better-known examples of Khrushchev's attempts to create a perception of parity with or even superiority to the United States was the "bomber gap," when a squadron of ten Bison (Miasishev, M-4, *Molot*) bombers circled back around and repeated overflights of Red Square in order to create the illusion of numerous bombers. This event created the impression of the "bomber crisis" among some but not all Western observers. "Talbott Gives Answer to the New Soviet Bomber," *New York Times,* 8 June 1955, 14. Stewart Alsop, "Red Air Might Dampens a 'Fair,'" *The Washington Post and Times Herald,* 10 July 1955, E5. Hanson W. Baldwin, "U.S. Air Power Still Superior to Russia's," *New York Times,* 29 May 1955, 107. The USSR continued this strategy to create an analogous illusion about their rocket technology. Walter A. McDougall, *The Heavens and the Earth: A Political History of the Space Age* (New York: Basic Books, 1985): 250–63.

17 Vladislav Zubok and Hope M. Harrison, "The Nuclear Education of Nikita Khrushchev," in *Cold War Statesmen Confront the Bomb: Nuclear Diplomacy since 1945,* ed. John Lewis Gaddis (Oxford: Oxford University Press, 1999), 152.

18 Ibid., 150.

19 Chertok, *Rakety i Liudi: Fili Podlipki Tiuratam,* 198.

20 "Soviet Space Medicine, Visual Tour," Oral history interview, 9551, Institute for Bio-Medical Problems, vol. transcript (Moscow, USSR, 30 November 1989), Smithsonian Videohistory Program, 24–25.

21 IMBP scientists made complete disclosure on the fate of Laika forty years after she died. See Alan Philips, "Scientist Sorry for Space Dog's Death," *Daily Telegraph,* 27 June 1998, International: 15 and "Doomed Soviet Space Dog Gets Tribute 40

Years Later," *Buffalo News* (Buffalo, New York), 4 November 1997, City Edition, News: 3A.

22 The story of Laika provides one example of how official Soviet reports were falsi-
fied. The original story of Laika stated that she was humanely euthanized after
her mission was complete. The official Soviet documents say that she died four
days after launch. Siddiqi, *Challenge to Apollo,* 252–59 (this is the two-volume
NASA version, not the single volume previously cited). Recent first-hand reports
centering around the 55th and 60th anniversaries and other commemorations of
her flight admit that the cabin temperature of *Sputnik* 2 was close to 40 degrees
centigrade after entering orbit. It is doubtful that she could have survived hours
at that temperature. Vladimir Ischenkov, "Russia Opens Monument to Space Dog
Laika," *USA Today,* 4/11/2008, http://usatoday30.usatoday.com/news/world/2008
-04-11-177105809_x.htm.

23 Slava Gerovitch, "'New Soviet Man' Inside Machine: Human Engineering, Spacecraft Design, and the Construction of Communism," in *The Self as Project: Politics and the Human Sciences,* ed. Greg Eghigian, Andreas Killan, and Christine
Leuenberger, OSIRIS, vol. 22 (Chicago: University of Chicago Press, 2007),
135–57. Gerovitch explains the divergence in cultures between the United States
and the Soviet programs. In the US case, the astronauts gained an advantage in
the design process, making the *Gemini* spacecraft a "pilot's" craft. On the other
hand, Soviet cosmonauts never gained a commanding position in the engineering
of their spacecraft.

24 For example, at an unspecified historical conference in Moscow in 1999, Derevianko reported on Gagarin's pre-flight physiological testing: Iu. M. Derevianko,
"Resultaty kliniko-fiziologicheskogo issledovania kosmonavta No. 1, Iu A. Gagarina v TsNIAGe" [Results of the Clinical and Physiological Investigations of Cosmonaut No. 1, Yuri A. Gagarin at TsNIAG], Moscow, Russia, 1999.

25 In the case of Yuri Gagarin, photographs of him and his family did not appear
until after the heat of the technical and political coverage of his flight had died
down.

26 Andrew L. Jenks, *The Cosmonaut Who Couldn't Stop Smiling: The Life and Legend
of Yuri Gagarin* (DeKalb: Northern Illinois University Press, 2012). Jenks provides an in-depth study of the interaction between man and public legend Gagarin and how he became a generation's archetype of behavior and image.

27 Gagarin's flight was only ninety minutes long, thus, leaving little opportunity for
official notification before his flight was over. Later, during more extended duration missions, it was the tradition to announce space missions once the cosmonauts had achieved orbit around the Earth.

28 I have consulted the following cosmonaut biographies. Biographies of Yuri
Gagarin: Aleksei Adzhubei and Azizian Ateik Kegamovich, *Utro kosmicheskoi
ery* [The Dawn of Spaceflight] (Moscow: Gos Polit, 1961); Sergei Mikhailovich
Belotserkovskii, *Pervoprokhodtsy vselennoi: zemlia—kosmos—zemlia* [The First
Travelers into Space: Earth—Space—Earth] (Moscow: Mashinostroenie, 1997);
A. Dikhtiar', *Prezhde chem prozvuchalo: "Poekhali"* [Before Anyone Said, "Let's

Go"] (Moscow: Politizdat, 1987); Iurii Alekseevich Gagarin, *Doroga v kosmos* [The Road to Space] (Moscow: Detskaia Literatura, 1978); Gagarin, *Moi brat Iurii;* Anna Timofeeva Gagarina, *Slovo o syna* [A Word About My Son] (Moscow, USSR: Molodiia gvardiia, 1983); M. I. Gerasimova and A. G. Ivanov, *Zvezdnyi put* [The Starry Road] (Moscow: Izd-vo polit. lit-ry, 1986); Iaroslav Kirilovich Golovanov, *Our Gagarin* (Moscow: Progress Publishers, 1978); Sal'nikov, *Iurii Gagarin;* A. Komarova, A. Kobzarev, and K. Petrov, "Pervym v kosmos poletel Gagarin" [Gagarin Was the First to Fly in Space], *Novosti Pskova* (Pskov); P. L. Nechaiuk, *Den Gagarina* [The Day of Gagarin] (Moscow: "Sovremennik," 1986); Mitchell R. Sharpe, *Yuri Gagarin; First Man in Space* ([Huntsville, Ala.]: Strode Publishers, [1969]); Viktor Stepanov, *Iurii Gagarin, Zhizn' zamechatel'nykh liudei* [Yuri Gagarin, the Life of Remarkable People] (Moscow: Molodaia Gvardiia, 1987); Mariia Zaliubovskaia, *Syn zemli i zvezd: liricheskaia povest' o Gagarin* [Son of the Earth and the Stars: A Lyrical Story about Gagarin], Second edition (Kiev: Izdatel'stvo TsK LKSMU, 1984), 304. Biographies of German Titov: Adzhubei and Kegamovich, *Utro kosmicheskoi ery;* Wilfred G. Burchett and Anthony Purdy, *Gherman Titov's Flight into Space* (London: Hamilton, 1962); M. I. Kuznetskii, *Titov: vtoroi kosmonavt planety* [Titov: The Second Man in Space] (Krasnoznamensk: Vladi, 2005); German Titov, *Golubaia moia planeta* [My Blue Planet] (Moscow: Voenizdat, 1973); German Titov, *Na zvezdnykh i zemnykh orbitakh* [In Starry and Earth Orbits] (Moscow: Detskaia literatura, 1987); *Gherman Titov, Gherman Titov: First Man to Spend a Day in Space,* Documents of Current History, vol. 21 (New York: Crosscurrents Press, 1962); Gherman Titov and Martin Caidin, *I Am Eagle!* (Indianapolis: Bobbs-Merrill Co., 1962); Herman Titov, *700,000 Kilometres through Space: Notes by Soviet Cosmonaut No. 2,* trans. R. Daglish, ed. Nikolai Kamanin (Moscow: Foreign Languages Publishing House); Aleksandr Ivanovich Volkov, *Vetv' sibirskogo kedra* [A Branch of the Siberian Cedar] (1962). Biographies of Valentina Tereshkova: A. Kokhov, *Nasha "chaika"* [Our "Seagull"] (Moscow: Izogiz, 1963); Vasilii Dmitrievich Sokolov and R. P. Smirnova, *Snova k zvezdam!* [Once Again to the Stars!] (1964); Mitchell R. Sharpe, *"It is I, Sea Gull,"* *Valentina Tereshkova, First Woman in Space* (New York: Crowell, 1975). Biographies of Andrian Nikolaev: A. Iu. Ishlinskii, ed., "A. G. Nikolaev" (Moscow: Nauka, 1987), 482–85; Andriian Grigor'evich Nikolaev, *Vstretimsia na orbite* [We Will Meet in Orbit] (Moscow: Voennoe Isdatel'stvo ministerstva oborony sssr, 1966); Iurii S. Ustinov, *Kosmonavt no. 3 Andriian Nikolaev* (Moscow: "Geroi otechestva," 2004). Biographies of Aleksei Leonov: Aleksei Avdeev, *Na zemli on takoi* [On Earth He Is] (Moscow: Detskaia literature, 1983); Aleksei A. Leonov, Andrei Sokolov, and I. A. Golovanov, *Chelovek i vselennaia* [Man and the Universe] (Moscow: Izobrazitel'noe iskusstvo, 1976); David Scott and Alexei Leonov, *The Other Side of the Moon: Our Story of the Cold War Space Race* (New York: Thomas Dunn Books, 2004). Joint biographies: Sokolov and Smirnova, *Snova k zvezdam!* [Again to the Stars!]; Pavel Romanovich Popovich, *Beskonechnye dorogi vselennoi* [The Never-Ending Road to the Universe] (Moscow: Sovetskii Pisatel, 1985).

29 The military affiliations of *Mercury* 7 astronauts before joining NASA were: Car-

penter, Navy; Cooper, Air Force; Glenn, Marine Corps; Grissom, Air Force; Schirra, Navy; Shepard, Navy; Slayton, Air Force.

30 Tereshkova and all the female cosmonaut-candidates in the 1960s wore the uniform of a Junior Lieutenant in the Air Force. Valentina Leonidovna Ponomareva, *Zhenskoe litso kosmosa* [The Women's Face of Space], ed. L. V. Golovanov (Moscow: Gelios, 2002), 140.

31 The birth years of the Soviet cosmonauts: Yuri Gagarin, 1934; German Titov, 1935; Andriian Nikolayev, 1929; Pavel Popovich, 1930; Valerii Bykovskii, 1934; Valentina Tereshkova, 1937. The birth years of the American *Mercury* 7 crew were: M. Scott Carpenter, 1925; L. Gordon Cooper, 1927; John H. Glenn, 1921; Virgil "Gus" Grissom, 1926; Walter M. Schirra, 1923; Alan B. Shepard, 1923; Donald "Deke" Slayton, 1924. Among the Americans, only Cooper and Grissom were too young to enlist in the war.

32 Titov, *Gherman Titov,* 54. German Titov's account of his recruitment for the space program makes it understandable that there were no pre-announced recruitments. It was only during the preliminary interview that the interviewers mentioned rockets and satellites. The recruits had no idea that they were interviewing for spaceflights until that time.

33 One of the chroniclers of Gagarin's life describes the "pitiless" rejection of candidates in October 1959. Even after the initial medical screening, Gagarin took part in intellectual and political vetting: Viktor Mitroshenkov, *Zemliia pod nebom: khronika zhizni Iuriia Gagarina* [The Earth Under the Skies: A Chronicle of the Life of Yuri Gagarin], Izdania vtoroe, dopolnennoe (Moscow: Sovetskaia rossiia, 1987), 115–21.

34 I. A. Marinin, S. Kh. Shamsutdinov, and A. V. Glushko, *Sovetskie i rossiiskie kosmonavty: 1960–2000. Spravochik* [Soviet and Russian Cosmonauts: 1960–2000. A Handbook] (Moscow: OOO Informatsionno-izdatel'skii dom "Novosti kosmonavtiki," 2001), 17–195.

35 "Popularnist' imen v Moskve i oblasti v xx veke" [Popular Names in Moscow and its Surroundings], http://names.mercator.ru (23 January 2015).

36 Aleksei is the sole exception to this statement, but the increase in that name occurred in 1975, a decade after Aleksei Leonov's spacewalk and the year in which he was the Soviet commander of the *Apollo-Soyuz* test project. In 1975 Aleksei was the third most popular boy's first name in Moscow and eighth overall in the twentieth century. It should also be noted that although German is not a Russian name, it is readily identified by any Russian schoolchild as the name of the antagonist in Alexander Pushkin's tale "Pikovaia Dama" [The Queen of Spades]. There is no doubt that German Titov was named after this character, as his father was a literature teacher and subsequent references from Titov indicate that he grew up in a household that was immersed in Russian literary culture.

37 Apprehension about the flight surgeon is a universal emotion among professional pilots. Pavel Barashev and Yuri Dokuchayev, *Gherman Titov: First Man to Spend a Day in Space,* Documents of Current History, vol. 21 (New York: Crosscurrents Press, 1962), 54.

38 Mitroshenkov, *Zemliia pod nebom* [The Earth above the Clouds], 116–17.

39 *Pravda,* 13 April 1961, 2.

40 *Izvestiia,* 13 April 1961, 3.

41 Naftali and Fursenko, *Khrushchev's Cold War,* 348.

42 "Obrashchenie tsentral'nogo komiteta KPSS, prezidiuma verkhovnogo soveta SSSR i pravitel'stva Sovetskogo Soiuza" [Address of the Central Committee of the CPSU and the Leadership of the Soviet Union], *Komsomolskaia Pravda* (Moscow), 13 April 1961, No. 88 (11028), 1. Soobshcheniie TASS, "Prizhok vo vselennuiu" [A Hop into Space] *Komsomolskaia Pravda* (Moscow), 13 April 1961, No. 88 (11028), 1. "Velikoe sobytie v istorii chelovechestva!" [A Great Event in the History of Mankind!] Soobshchenie TASS, *Pravda* (Moscow), 13 April 1961, No. 103 (18503), 1.

43 *Pravda,* 26 April 1961, 4.

44 *Pravda,* 20 April 1961, 1.

45 Gagarin, *Moi Brat Iurii,* 331.

46 This exchange and excerpts from Gagarin's travels throughout the world after his flight, are contained in the documentary film by Iurii Sal'nikov: Sal'nikov, *Iurii Gagarin.*

47 N. Denisov and I. Novikov, "Praga privetstvuet sovetskogo kosmonavta Iuriia Gagarina" [Prague Greats Soviet Cosmonaut Yuri Gagarin], *Pravda* (Moscow, USSR), 29 April 1961, 1.

48 Mitroshenkov, *Zemliia pod nebom,* 199.

49 Ibid., 208.

50 Ibid., 211.

51 Ibid., 216.

52 Ibid., 216.

53 Ibid., 217.

54 Ibid., 218.

55 Ibid., 309.

56 Interestingly enough Gagarin and the Soviet press never made an issue of the fact that any American *Mercury* astronauts had combat flight experience. Despite their rhetoric of peace, no comparison arose between the motivations of American and Soviet space explorers.

57 The military uniform did not necessarily have militarist implications and throughout the world translated as an indication of professionalism. One dissertation has examined the widespread cultural implications of commercial pilots wearing uniforms. For example, all civilian airline pilots wear uniforms that resemble military ones: Suzanne Lee Kolm, "Women's Labor Aloft: A Cultural History of Airline Flight Attendants in the United States, 1930–1978" (Ph. D. diss., History, Brown University, 1995), 100–101.

58 Sal'nikov, *Iurii Gagarin.*

59 Gagarin, *Moi brat Iurii,* 337.

60 *Pravda,* 14 April 1961, 5.

61 Kamanin, N. P. (2001), "Entry for January 11, 1969," *Hidden Cosmos,* vol. 2, p. 332.

62 Vasilli Zhuravlev, *Kosmicheskii reis* [Spaceflight]. S. Komarov, K. Moskalenko, V. Gaponenko, V. Kovrigin, N. Feoktistov. Gosudarstvennoe upravlenie kinematografii i fotografii (GUKF), 1936, 70 min.

63 "O zapuske v Sovetskom Soiuza chetvertogo korablia-*Sputnika*" [About the Soviet Launch of the Fourth Orbiting Spacecraft] *Pravda* (Moscow), 10 March 1961, 1.

64 One of the earliest uses of the term was published in the *New York Times* in a story covering the thoughts of French aircraft designer and spaceflight theorist Robert Albert Charles Esnault-Pelterie: "Astronautics," *New York Times,* 1924.

65 The earliest Soviet publication calling the future space travelers "cosmonauts" was Ario Abramovich Shternfel'd, *Soviet Space Science,* 2d rev. and extended ed. (New York: Basic Books, 1959).

66 Morton Benson, "Russianisms in the American Press," *American Speech* 37, no. 1 (February 1962): 44–45.

67 Joseph Campbell, *The Hero with a Thousand Faces* (Princeton, N.J.: Princeton University, 1949).

68 Ibid., 113.

69 Ibid., 329.

70 Ibid., 136.

71 Ibid., 133.

72 Ibid., 133.

73 Ibid., 149–71.

74 The most common publishers for biographies of cosmonauts were publishing houses that specialized in youth literature. The largest nation-wide publishers were: Molodaia gvardiia [Young Guard] and Detskaia literatura [Children's Literature] directed at adolescent and grade school readers, respectively. The print runs for such biographies were usually in the 100,000 copies per publication range with biographies of Gagarin having the largest number of copies with as many as 300,000. For example, Viktor Stepanov's biography of Gagarin, which was part of the "notable lives" series that Maxim Gor'kii established, had the largest press run of 300,000; Stepanov, *Iurii Gagarin.*

75 One ambiguity in Gagarin's biography stems from the fact that his maternal grandfather had been a factory worker in St. Petersburg before the revolution. Even today, the details of her family's decision to return to Gzhastk are unclear, but one assumes the post-revolution famine in the city to be a likely motivator as it had been for many others.

76 Titov's father's background became a point of argument among space watchers during the early space program. Stepan Titov's employment as a teacher was once thought to have been a disqualifying mark against his son becoming the first man in space, hinting at a classist hierarchy in the USSR. The younger man's name itself became a bone of contention. As any Russian schoolchild would know, Titov was named for the anti-hero in a Pushkin short story and later a Tchaikovsky opera. Titov and Caidin, *I Am Eagle!* 25.

77 Avdeev, *Na zemli on takoi,* 48.

78 Ibid., 48.

79   Campbell, *The Hero with a Thousand Faces,* 136.

80   Avdeev, *Na zemli on takoi,* 81–83.

81   Ibid., 83.

82   Ishlinskii, ed., "A. G. Nikolaev," 482.

83   Ibid., 483.

84   Popovich, *Beskonechnye dorogi vselennoi,* 92.

85   Gagarina, *Slovo o syna,* 39. This book bears no indication that Anna Gagarina had a ghostwriter for the book. However, it is likely that she had some assistance. Gagarina spent her childhood in St. Petersburg before the revolution, where her father was a factory worker. Therefore, she was more likely to have grown up literate than the average milkmaid in the Russian countryside, but there is no indication that she authored anything other than this book about her son.

86   Gagarin, *Moi brat Iurii,* 323.

87   Georgii Bergovoi (b. 1921) was a decorated World War II pilot.

88   Gagarin, *Moi brat Iurii,* 98.

89   Khrushchev sought to court the Third World through personal attention and peaceful rhetoric as a means to gain strategic advantage and minimize the strategic and military dimensions of the Cold war conflict with the United States. Taubman, *Khrushchev,* 348.

90   Gagarin, *Moi brat Iurii,* 98.

91   Paul R. Josephson, "Atomic-Powered Communism: Nuclear Culture in the Postwar USSR," *Slavic Review* 55, no. 2 (Summer 1996): 302.

92   The circumstances behind Korolev's anonymity remain unclear. It could be that he was as much a victim of bad timing as of Soviet secrecy. The identity of his contemporary Igor Kurchatov, designer of the Soviet atomic bomb, was published in the Great Soviet Encyclopedia (BSE) in October/November 1954, see Harry Schwartz, "Soviet Identifies Atomic Physicist," *New York Times,* 14 November 1954, 28. It might be that Korolev was unfortunate that the timing of his contributions to spaceflight did not allow a timely disclosure. Kurchatov took on a public role in Soviet foreign policy almost immediately after the "Secret Speech" when he and aircraft designer Andrei Tupolev accompanied Khrushchev and Bulganin on a trip to Britain; see Christopher Smart, *The Imagery of Soviet Foreign Policy and the Collapse of the Russian Empire* (Westport, Conn.: Praeger, 1995).

93   Ezra Samoilovich Vilenskii and K. Taradankin, *Vzlet* [Flight] (Moscow: Izdatel'stvo Detskoi Literatury, 1939), 53. Grigorii Karlovich Grigor'ev, *Sledy v nebe* [I Follow in the Sky] (Moscow: Izdatel'stvo DOSAAF, 1960), 28–29.

94   The most common English translation of the Russian phrase *glavnyi konstruktor* is the literal "chief constructor." The phrase "chief designer" is a more accurate translation of the role that Korolev played. He directed the working groups of technicians and engineers, who designed, built, and managed the hardware components of rocket engines, launch vehicles, and spacecraft.

95   Titov, *Gherman Titov,* 76.

96   Titov and Caidin, *I Am Eagle!* 113.

97   Leonid I. Brezhnev, "Akademik Sergei Pavlovich Korolev," *Pravda* [Academician Sergei Korolev] (Moscow), 16 January 1966, No.16 (17333), 4.

98   Stepanov, *Iurii Gagarin,* 188–211.

99   Ibid., 188–211.

100 Katerina Clark, *The Soviet Novel: History as Ritual* (Chicago: University of Chicago Press, 1981), 99.

101 McCannon, *Red Arctic.*

102 Clark, *The Soviet Novel,* 99.

103 Ibid., 125–26.

104 John McCannon, "Tabula Rasa in the North: The Soviet Arctic and Mythic Landscapes in Stalinist Popular Culture," in *The Lands of Stalinist: The Art and Ideology of Soviet Space,* ed. Evgeny Dobrenko and Eric Naiman (Seattle: University of Washington Press, 2003), 248.

105 Nikolai Petrovich Kamanin, *Moia biografiia tol'ko nachinaetsia* [My Biography Has Only Begun] (1935).

106 The medal of the Hero of the Soviet Union, the highest honor in the USSR, was created in April 1934 primarily as a reward for the seven pilots who took part in the Cheliuskin rescue: Anatolii Liapidevskii, Mikhail Vodop'ianov, Vasilli Molokov, Ivan Doronin, Nikolai Kamanin, Sigismund Levanevskii, and Mavriki Slepnev. McCannon, "Tabula Rasa in the North," 256.

107 Jenks, *Cosmonaut Who Couldn't Stop Smiling.*

108 Titov's flight coincided with the publication of individual party programs for discussion. For example, see the programs for construction of electric stations, I. Novikov, "Obsuzhdenie proekta programmy kommunistichekoi partii sovetskogo soiuza" [Discussion of the Project Program of the Communist Party of the Soviet Union], *Pravda* (Moscow, USSR), 7 August 1961, 4. This overlap further associated Titov's mission and the upcoming party congress.

109 Titov, *700,000 Kilometres through Space,* 127. It is worth noting that cosmonaut number three was in fact Nikolayev.

110 Ibid., 17.

111 Jenks, *Cosmonaut Who Couldn't Stop Smiling,* 235.

112 Hope Harrison, *Driving the Soviets up the Wall: Soviet-East German Relations, 1953–1961* (Princeton, N.J.: Princeton University Press, 2005), 139–201.

113 Titov arrived in Berlin on 1 September 1961 for a two-day visit. "German Reds Bar Prelate's Return," *New York Times,* 2 September 1961, 5.

114 Heather L. Gumbert, "Soviet Cosmonauts Visit the Berlin Wall: The Spatial Contradictions of the Cold War," in *The Cultural Impact of the Cold War Cosmonaut,* chair Diane P. Koenker, American Association for the Advancement of Slavic Studies National Convention, Washington, DC, 17 November 2006.

115 Decorating the elka (fir tree) is a recurring theme in the portrayal of Gagarin. His brother's biography opens with their search for the ideal tree. Gagarin, *Moi brat Iurii,* 13. Sal'nikov's biographical film features scenes from Gagarin's adult home life which include helping his daughter decorate the tree. It was eventually rein-

troduced in the 1930s. Jennifer McDowell, "Soviet Civil Ceremonies," *Journal for the Scientific Study of Religion* 13, no. 3 (September 1974), 267.

116　Gagarin, *Moi brat Iurii,* 341.

117　Comments at World's Fair, *Seattle Daily Times* (Seattle, Washington), 7 May 1962, 2.

118　Vladimir Mayakovsky, "The Flying Proletariat. II. Daily Life in the Future (1925)" in Rosy Carrick, ed., *Volodya: Selected Works* (London: Enitharmon Press, 2015): 143–163.

119　Historian Victoria Smolkin has written about this episode as the beginning of the decline of scientific atheism in the USSR because the statements were crude and ineffective. Victoria Smolkin, *A Sacred Space Is Never Empty: A History of Soviet Atheism* (Princeton, N. J.: Princeton University Press, 2018): 87–94.

120　Sonja Luehrmann, "Recycling Cultural Construction: Desecularisation in Post-Soviet Mari El," *Religion, State & Society* 33, no. 1 (March 2005): 35–36.

121　Kamanin, *Skrytyi kosmos. Kniga pervaia: 1960–1963,* 109. The Central Committee made an unusual decision to allow Titov's wife, Tamara Vasil'evna, to travel with the May 1962 delegation to the United States. Television news coverage of Titov's visit to Washington, DC: Inc. Columbia Broadcasting System, Meeting of the Astronauts [Motion Picture] ([n.p.]: CBS News, 1962).

122　*The Flight of Vostok II. Pravda, August 7, 1961, Extra Edition,* Jet Propulsion Laboratory. National Aeronautics and Space Administration no. NASW-6, by J. L. Zygielbaum, trans. (Pasadena, 2 October 1961).

123　Kamanin, *Skrytyi kosmos. Kniga pervaia: 1960–1963,* 97.

124　Mitroshenkov, *Zemliia pod nebom,* 317.

125　Nechiporenko, "Na rodine atstekov," pp. 173–74. Christopher Andrew and Vasili Mitrokhin, *The World Was Going Our Way: The KGB and the Battle for the Third World* (New York: Basic Books, 2005), 20.

Chapter 3. The Women

1　Ponomareva, *Zhenskoe litso kosmosa,* 146–48.

2　Neither *Vostok* nor *Voskhod* had maneuvering engines, which would allow the craft to change orbit after launch. The small engines on the exterior were only adequate for orientation. When the Soviets finally unveiled the *Soyuz* in 1967, that was the first public concession that *Vostok* had not been maneuverable. S. Borzenko and N. Denisov, "'Voskhod'—'Soiuz,'" *Pravda* (Moscow), 24 April 1967, No. 114 (17796), 3. In contrast, the American *Gemini* was capable of making orbital maneuvers.

3　Natasha Kolchevska, "Angels in the Home and at Work: Russian Women in the Khrushchev Years," *Women's Studies Quarterly* 33, no. 3/4, Gender and Culture in the 1950s (Fall Winter, 2005), 114–37.

4　Boris Groys, *The Total Art of Stalinism: Avant-garde, Aesthetic Dictatorship, and Beyond,* trans., Charles Rougle (Princeton, N.J.: Princeton University Press, 1992). Elena Prokhorova, "The Post-Utopian Body Politic: Masculinity and the Crisis of

National Identity in Brezhnev-Era TV Miniseries," in Helena Goscilo and Andrea Lanous, eds., *Gender and National Identity in Twentieth-Century Russian Culture* (Dekalb: Northern Illinois University Press, 2006), 131–50.

5 Lynn Attwood, "Rationality versus Romanticism: representations of Woman in the Stalinist Press," in Linda Edmondson, ed. *Gender in Russian History and Culture* (Houndmills, Basingstoke, England: Palgrave, 2001), 158–76.

6 Barbara Alpern Engel, *Women in Russia, 1700–2000* (Cambridge: Cambridge University Press, 2004), 239–49.

7 Attwood, "Rationality versus Romanticism," 161.

8 Very pointedly, Attwood points to the coverage of the flight of the Rodina in 1938 as an example of characterizing labor as female and caring. Ibid., 161.

9 Ibid., 171–72.

10 Ibid., 171–72.

11 Ibid., 172–73.

12 After high school, Tereshkova went to Yaroslavl, where she received a quick succession of promotions first at a tire factory and then at a nearby textile factory. Sharpe, *"It is I, Sea Gull,"* 20–26.

13 We know this because Valentina Ponomareva received her invitation directly from Keldysh. Valentina Leonidovna Ponomareva, *Zhenskoe litso kosmosa,* [The Female Face of Space] ed. L. V. Golovanov (Moscow: Gelios, 2002), 38.

14 Ibid., 39–41.

15 Ibid., 45.

16 *Spaceflight* 54, no. 6, 216–17.

17 Ponomareva, *Zhenskoe litso kosmosa,* 125–33.

18 Marinin, Shamsutdinov, and Glushko, *Sovetskie i rossiiskie kosmonavty xx vek* [Soviet and Russian Cosmonauts of the 20th Century], 13, 79, 114, 144–45, 165–66, and 169–170.

19 Ponomareva, *Zhenskoe litso kosmosa,* 45.

20 Siddiqi, *Challenge to Apollo,* 364–66. At the time of its publication, Siddiqi did not have access to Ponomareva's memoirs that clearly state Gagarin's aversion to flying with a mother. While Gagarin said this publicly, he did not commit his opinion to official documents that have survived in the archives.

21 Siddiqi, *Challenge to Apollo,* 365.

22 Isaak P. Abramov and A. Ingemar Skoog, *Russian Spacesuits* (Chichester, UK: Springer Praxis Publishing, 2003), 56–58. The authors discreetly do not mention the necessity of providing a solution for female urine collection in their description of the modifications for the SK-2. Yet, it was a well-known engineering challenge and a source of jokes among all who have flown in space.

23 Nikolai Kamanin, *Skrytyi kosmos: tom 1* (Moscow: Isdatel'stvo "RTSoft," 2013), 231. Kamanin was concerned that he could not have the six to eight days that he had planned to test the four finalists in the new SK-2 suits when he heard on 29 April 1963 that it would be a few more days.

24 Abramov and Skoog, *Russian Spacesuits,* 58.

25 The first multiple-member crew flight did take place in October 1964. Cosmo-

nauts Vladimir Komarov, Konstantin Feoktiskov, and Boris Yegorov packed themselves into the slightly redesigned *Vostok* spacecraft that was rechristened *Voskhod* on 12 October 1964.

26 Ponomareva, *Zhenskoe litso kosmosa,* 158.

27 Roshanna P. Sylvester, "'Let's Find out Where the Cosmonaut School Is': Soviet Girls and Cosmic Visions in the Aftermath of Tereshkova," in *Soviet Space Culture: Cosmic Enthusiasm in Socialist Societies,* ed. Julia Richers Eva Mauer, Monica Ruthers, and Carmen Scheide (Basingstoke, England: Palgrave Macmillan, 2011); Roshanna P. Sylvester, "She Orbits Over the Sex Barrier: Soviet Girls and the Tereshkova Moment," in *Into the Cosmos: Space Exploration and Soviet Culture,* ed. James T. Andrews and Asif A. Siddiqi (Pittsburgh: University of Pittsburgh Press, 2011).

28 Kamanin cited the examining doctor's report that Tereshkova's space sickness was due to her emotional distress during flight. Kamanin, *Skrytyi kosmos. Kniga pervaia: 1960–1963,* 304; Chertok, *Rakety i Liudi: Goriachie dni kholodnoi voiny,* 237–38.

29 Tereshkova attributed the cracked helmet to the difficulty she had in removing the film cassette from the movie camera, which struck the helmet. Kamanin, *Skrytyi kosmos. Kniga pervaia: 1960–1963,* 300.

30 Nina Popova, "Geroicheskaia doch' strany kommunizma" [The Heroic Daughter of the Country of Communism], *Pravda* (Moscow), 17 June 1963, No. 168 (16389), 4. TASS, "'Vostok-6' vyshel na orbitu" [*Vostok* 6 Has Arrived in Orbit] *Pravda* (Moscow), 17 June 1963, No. 168 (16389), 1. N. Denisov, "Na orbite— 'Chaika'" [Seagull Is in Orbit], *Pravda* (Moscow), 17 June 1963, No. 168 (16389), 2. "Kosmonavt shest' Valentina Vladimirovna Tereshkova" [The Sixth Cosmonaut, Valentina Vladimirovna Tereshkova], *Izvestiia* (Moscow), 17 June 1963, No. 143 (14306), 1.

31 *Izvestiia* (Moscow), 17 June 1963, No. 143 (14306), 1.

32 Komsomolskaia, *Pravda,* 14 April 1961, 1.

33 The next Soviet woman in space was Svetlana Savitskaia, the daughter of the Marshall of the Soviet Air Forces, Evgenii Savitskii. Savistskaia flew in space on board the *Soyuz T-7* to the Salyut 7 space station on 19 August 1982, over 19 years after Tereshkova's flight. Her flight did, however, preempt Sally Ride's first mission as the first American woman in space. Savitskaya preempted the American Kathleen Sullivan's spacewalk less than two years later.

34 In her profile, Valentina Grizodubova declared her devotion to the cause of Soviet aviation supremacy at age nineteen. Lazar Brontman, *The Heroic Flight of the Rodina* (Moscow: Foreign Languages Publishing House, 1938), 91.

35 Ibid., 105.

36 Reina Pennington, *Wings, Women, and War: Soviet Airwomen in World War II Combat (Modern War Studies)* (Lawrence: University Press of Kansas, 2002), 12–15.

37 Reina Pennington, "From Chaos to the Eve of the Great Patriotic War, 1921–41," in *Russian Aviation and Air Power in the Twentieth Century,* ed. Robin Higham,

John T. Greenwood, and Von Hardesty, Studies in Air Power (London: Frank Cass, 1998), 126–27.

38 Pennington, *Wings, Women, and War,* 91–93.

39 Osoaviakhim was the prewar precursor of the Soviet paramilitary civil defense organization DOSAAF. Its official title in English was Union of Societies of Assistance to Defense and Aviation-Chemical Construction of the USSR. The government formed the organization in January 1927 when it merged three component organizations.

40 Kazimiera Janina Cottam, *Women in War and Resistance: Selected Biographies of Soviet Women Soldiers* (Newburyport, MA: Focus Publishing/R. Pullins Co., 1998).

41 Reina Pennington, *Amazons to Fighter Pilots: A Biographical Dictionary of Military Women* (Santa Barbara, CA: Greenwood, 2003), 352.

42 Perhaps the most profound demonstration of this public memory was the famous Soviet film of 1966 that recounts the poignant post-war life of a woman pilot, Larisa Shepitko, "Wings (Krylia)" (Moscow: MosFilm, 1966).

43 Lillian Kosloski and Maura J. Mackowski, "The Wrong Stuff," *Final Frontier,* May/June 1990, 22.

44 Randy Lovelace, M.D., director of the clinic, knew that the Soviets were planning to launch a woman from a trip that he took to Moscow in 1959. The Central Intelligence Agency had predicted the launch of a woman, too. Margaret A. Weitekamp, *Right Stuff, Wrong Sex: America's First Women in Space Program* (Baltimore: Johns Hopkins University Press, 2004), 162.

45 Ibid., 163–65.

46 Jacqueline Cochrane, Lyndon Baines Johnson Papers, Vice Presidential Papers, 23 March 1962, Jacqueline Cochrane to Miss Jerrie Cobb, Austin, Texas, National Archives and Records Administration. Johnson Library.

47 Jacqueline Cochrane, Lyndon Baines Johnson Papers, Vice Presidential Papers, 14 July 1962, Jacqueline Cochrane to Mr. James Webb, Austin, Texas, National Archives and Records Administration. Johnson Library.

48 Clare Boothe Luce, "But Some People Simply Never Get the Message," *Life,* 28 June 1963, 31.

49 Susan Bridger, "The Cold War and the Cosmos: Valentina Tereshkova and the First Woman's Space Flight," in *Women in the Khrushchev Era,* eds. Melanie Ilic, Susan E. Reid, and Lynne Attwood (New York: Palgrave Macmillan, 2004), 230.

50 The fact was that almost anyone who could fit into the spacecraft could become a cosmonaut. The piloting skills were minimal, even under manual control. This limitation inside the spacecraft was primarily the result of the complex hierarchy within the Soviet engineering community that placed engineering above piloting skills. Tereshkova's flight brought to light the tension in the Soviet rhetoric between statements about cosmonaut proficiency in technology and the reality that the cosmonauts had almost no control over their spacecraft, especially when contrasted to the different situation with the American astronauts. Gerovitch, "'New Soviet Man' Inside Machine," 135–57.

51 Robert L. Griswold, "'Russian Blonde in Space': Soviet Women in the American Imagination, 1950–1965," *Journal of Social History* 45, no. 4 (summer 2012), 881–907.

52 Ibid., 893.

53 Valentina Leonidovna Ponomareva, "Roman s kosmonavtikoi" [Romance with Astronautics] (Moscow: 1995) (unpublished memoir), Ponomareva cites the article as being in the March 1985 *Rabotnitsa,* but I have checked 1983–1986 and cannot find it.

54 During this time, the Soviets were also planning a mission to send a man to the Moon. However, the first test launch of the rocket that would carry a cosmonaut to the Moon failed just before the launch of the American *Apollo 11.* There were three subsequent launch failures until the program ended. For a complete account of the Soviet lunar program and its impact on the space program, see Chapter 4.

55 "Po dorogam kosmosa" [On the Road to Space], *Ogonek* (Moscow), 2 April 1961, 2.

56 The couple officially divorced in 1982 but had lived separately before that time.

57 Ponomareva, *Zhenskoe litso kosmosa,* 268–71. The stories of the few women who worked in polar science and exploration offers another intriguing analogy to the women in pioneering fields. E. Kalemeneva and J. Lajus, "Soviet Female Experts in the Polar Regions," Chapter 18 in Melanie Ilic, ed., The *Palgrave Handbook of Women and Gender in Twentieth-Century Russia and the Soviet Union* (London: Palgrave Macmillan, 2018), 267–83.

58 Weitekamp, *Right Stuff, Wrong Sex,* 162.

59 Probably the greatest weakness in the scientific results from Tereshkova's mission is the fact that the mission results were a single data-point based on her lone flight as a woman. There were no other women in space with whom to compare her, nor were there any ground controls. Tereshkova also refused to complete some of the experiments in flight because she thought them too difficult or awkward. Kamanin, *Skrytyi kosmos. Kniga pervaia: 1960–1963,* 300.

Chapter 4. New Cultures of the Cosmonaut

1 Jeffrey Brooks, *Thank You, Comrade Stalin! Soviet Public Culture from Revolution to Cold War* (Princeton, N.J.: Princeton University Press, 2000), xvii–xviii.

2 Susan E. Reid and David Crowley, eds., *Style and Socialism: Modernity and Material Culture in Post-war Europe* (Oxford: Berg, 2000).

3 Iurii Gerchuk, "The Aesthetics of Everyday Life in the Khrushchev Thaw in the USSR (1954–64)," in Reid and Crowley, *Style and Socialism,* 81–100.

4 Nina Tumarkin, *Lenin Lives! The Lenin Cult in Soviet Russia* (Cambridge: Harvard University Press, 1983); Jeffrey Brooks, *Thank You, Comrade Stalin! Soviet Public Culture from Revolution to Cold War* (Princeton, N.J.: Princeton University Press, 2000); and Karen Petrone, *Life Has Become More Joyous, Comrades: Celebrations in the Time of Stalin* (Bloomington: Indiana University Press, 2000).

5 The Lenin All-Union Pioneer Organization (Vsesoiuznaia pionerskaia organizat-

siia imeni V. I. Lenina), known as the Pioneers, was a mass, scouting organization for children ages ten through fifteen in the Soviet Union. Komsomol, the Communist Youth Union (Kommunisticheskii soiuz molodëzhi), was the youth wing of the party for students ages fourteen through twenty-eight, after which one could petition to full party membership.

6   Grant, "Socialist Construction of Philately in the Early Soviet Era," 476–93.

7   Both economic and social studies of the Soviet postwar population have indicated that the need to rebuild the economy and continue economic growth that would maintain the USSR's role in world affairs built expectations among the populace that the victor against fascism would reward its population with some portion of its new status and growing wealth. Susan J. Linz, "World War II and Soviet Economic Growth, 1940–1953," in *The Impact of World War II on the Soviet Union,* edited by Susan J. Linz (Totowa, N.J.: Rowman and Allenheld, 1985); and Elena Zubkova, *Russia after the War: Hopes, Illusions, and Disappointments, 1945–1957,* translated by Hugh Ragsdale, in *The New Russian History Series,* edited by Donald J. Raleigh (Armonk, N.Y.: M. E. Sharpe, 1998).

8   Stamps and znachki were not the only small objects that were for sale and collected in the former Soviet Union. There also existed *palekh*-style miniatures (*palekh* are the traditional Russian lacquered miniatures that were initially religious icons, but under the Soviet rule they became secular and marketed for export) that portrayed Gagarin and other space themes. See in Boris Groys, *Ilya Kabakov: The Man Who Flew into Space from His Apartment,* installation review (London: Afterall Books, 2006), plates 20–23: 20. K. V. Kukulieva, B. N. Kukuliev, and O. V. An, a portrait of Yuri Gagarin, lacquer painting; 21. K. V. Kukulieva, B. N. Kukuliev and O. V. An, a portrait of Konstantin Tsiolkovsky; 22. K. V. Kukulieva, B. N. Kukuliev and O. V. An, a portrait of Yuri Gagarin and Sergei Korolyov, lacquer painting; 23. K. V. Kukulieva, B. N. Kukuliev, and O. V. An, a portrait of Yuri Gagarin, lacquer painting, from a book of postcards Syn Rossii (Son of Russia) (Moscow: Izobraziteľnoe iskusstvo, 1987).

9   Stamp collecting had its origins in British and American industrial capitalism, and these origins might have tainted the hobby for midlevel Soviet officials who determined stamp issues. Steven M. Gelber, "Free Market Metaphor: The Historical Dynamics of Stamp Collecting," *Comparative Studies in Society and History* 34, no. 4 (October 1992): 742–69. Until the official allowance of independent collecting societies in 1961, Soviet sanction of domestic philately was grudging at best. The philately journals appeared sporadically over the years, and the Philately Society reported directly to the NKVD; see Grant, "Socialist Construction of Philately in the Early Soviet Era," 476–93.

10   Brooks, *Thank You, Comrade Stalin!* xvii–xviii, states the philosophy behind Khrushchev's belief: "After Stalin's death, Khrushchev increased public expectations about the quantity and quality of the state's gift to society, but his promises to match and surpass Western Living standards went unfulfilled. . . . Along with the emphasis on consumerism, Khrushchev opened society to a degree by limiting repression as easing censorship. Intellectuals took advantage of Khrushchev's

'thaw' to champion 'sincerity', the antithesis of the formative ethos, but left much of the performative culture intact. In the end, neither Khrushchev nor his successors were willing to discard the ritualistic certainties from which they derived legitimacy. Their performative culture lingered on in a semi-moribund state until Brezhnev inadvertently turned it into self-parody." S. E. Reid, "Cold War in the Kitchen: Gender and the De-Stalinization of Consumer Taste in the Soviet Union under Khrushchev" *Slavic Review* 61, no. 2 (Summer 2002): 211–252, explains the practical significance of Khrushchev's activities.

11 Victor Buchli, "Khrushchev, Modernism, and the Fight against 'Petit-Bourgeois' Consciousness in the Soviet Home," *Journal of Design* 10, no. 2 (1997), 161–76.

12 Levant, "Soviet Union in Ruins," 97.

13 The "kitchen debate" was an impromptu debate between Nixon and Khrushchev at the opening of the American National Exhibition in Moscow, on 24 July 1959, among the American household appliances display that had been a focus of the US exhibit at the 1958 Brussels World's Fair.

14 Reid, "Cold War in the Kitchen," 223–24.

15 Marquis Childs, "Moscow Exhibits Stress *Sputniks,*" *Washington Post and Times Herald,* 24 June 1958, A16.

16 Jonathan Grant, "Socialist Construction of Philately in the Early Soviet Era," *Comparative Studies in Society and History* 37, no. 3 (July 1995): 476–93.

17 Ibid., 476.

18 Ibid., 481.

19 Ibid., 484.

20 Ibid., 492–93.

21 The Soviet Ministerstvo Sviazi is the Ministry of Communications; however, the traditional translation is Ministry of Post and Telegraph, alluding to the pre-twentieth-century origins of the name.

22 Anthony Swift, "The Soviet World of Tomorrow at the New York World's Fair, 1939," *Russian Review* 57 (July 1998): 376.

23 The Ministerstvo sviazi issued stamps in five denominations in 1961 (after revaluation of the ruble in 1961): one, three, four, six, and ten kopeks. The one-kopek stamp paid for the delivery of domestic postcards. The four-kopek stamp delivered domestic envelopes. The six- and ten-kopek stamps delivered international postcards and letters, respectively.

24 "Marki dlia kollektsii," *Pioner* (Moscow), August 1961, n.p. (inside back cover).

25 *Vostok Stamp with Khrushchëv Quote, 6 k,* in *Moscow, USSR,* postage stamp, Smithsonian Institution National Air and Space Museum, USSR Ministerstvo Sviaz, April 1961.

26 *Vostok 3* launched on 11 August 1962, with Andriian Nikolayev on board. While that craft was still in orbit, Pavel Popovich launched onboard *Vostok 4* on 12 August 1962. Both craft remained in orbit until 15 August 1962.

27 The first model of a *Vostok* went on display inside what was soon to become the Kosmos Pavilion at the Exhibition of Economic Achievements in Moscow: "Vostok Model Is Shown to Public in Moscow," *New York Times,* 30 April 1965, 10.

28 *Voskhod 1* launched on 12 October 1964 and landed on 13 October. The mission's technological significance was that the craft carried a crew of three, making it the first multi-passenger human space mission. The mission's added political significance came from the fact that the crew was in orbit at the time when the Politburo removed Nikita Khrushchev from power.

29 *Voskhod 2* was the second mission of the modified *Vostok* hardware. Pavel Belaev commanded the mission, and pilot-cosmonaut Aleksei Leonov performed the first spacewalk on this daylong mission beginning 18 March 1965.

30 "Marki Dlia Kollektsii," *Pioner* (Moscow), August 1961, n.p. (inside back cover).

31 Samuel A. Tower, "Looking into New Soviet Issues," *New York Times*, 22 June 1975, 143.

32 The Pioneer organization magazines, *Pioner* and *Smena*, continued to run "Marki dlia kollektsii" (Stamps for collectors) and "Dlia kollektsioner" (For the collector) articles on an alternating basis in the monthly magazines throughout the 1960s. The latter followed the stamp issues. The former announced the release of znachki, coins, and special-issue medals.

33 *Pioner* and *Semena* were the official magazines of the Vladimir Lenin All-Union Pioneer Organization and the Little Octobrists organization, respectively. The former scouting-like organization was for young people ages 9–15, and the latter was for ages 7–9.

34 S. I. Ozhegov, *Slovar' russkogo iazyka* (Moscow: Izdatel'stvo "Sovetskaia entsiklopediia," 1968), 228.

35 V. N. Il'inskii, *Znachki i ikh kollektsionirovanie (posobie dlia fileristov)*, 1976, *Izdanie vtoroe pererabotannoe i dopolnenie* [Znachki and Collecting (A Guide to Collectors)] (Moscow: Izdatel'stvo "Sviaz," 1977); and V. A. Omel'ko, *Nagradnye znaki obshchestvennykh organizatsii u muzeev*, vol. 1, *Nagrady za osvoenie kosmosa catalog* (Moscow: N.p., 2002).

36 For more detail on the history, distribution, and encouragement of collecting znachki, please see Cathleen S. Lewis, "From the Kitchen into Orbit: The Convergence of Human Spaceflight and Khrushchev's Nascent Consumerism," in *Into the Cosmos: Space Exploration and Soviet Culture*, ed. James T. Andrews and Asif F. Siddiqi (Pittsburgh: University of Pittsburgh Press, 2011), 213–39.

37 These were the official magazines of the Pioneer organization.

38 The voluntary Znanie Society promoted public scientific education and published this journal.

39 V. N. Il'inskii, *Znachki i ikh kollektsionirovanie (posobie dlia fileristov)* [Znachki and Collecting (A Guide to Collectors)], 1976, *Izdanie vtoroe pererabotannoe I dopolnenie* (Moscow: Izdatel'stvo "Sviaz," 1977) describes the history of the collection of the pins. Kruglov, *Chto takoe faleristika* [What Are Collectors of Pins], offers a history of the pins themselves, including useful information on their production and materials. V. N. Il'inskii, V. E. Kuzin, and M. B. Saukke, *Kosmonavtiki na znachkakh sssr* [Astronautics on Znachki in the USSR], *1957–1975: Katalog* (Moscow: Izdatel'stvo "Sviaz," 1977), is the most comprehensive catalog of space-related znachki.

40 Frederick C. Barghoorn, "Soviet Cultural Diplomacy Since Stalin," *Russian Review* 17, no. 1 (January 1958): 41–55.

41 Kruglov, *Chto takoe faleristika,* 10.

42 "Neobychnaia kollektsiia" [Unordinary Collection], *Ogonek* (Moscow), 12 February 1961, 26.

43 Il'inskii, *Znachki I ikh kollektsionirovanie (Posobie dlia fileristov),* 6.

44 These are the official terms that Russian falerists use in classifying their collections. Other historians use a more elaborate classification system that addresses all znachki, not focusing specifically on space-related ones; see Victor C. Seibert, "Falerists and Their Russian Znachki," *The Numismatist* (June 1979): 1, 198–202.

45 Il'inskii, Kuzin, and Saukke list at least twenty-three enterprises that manufactured space znachki as of 1975; see V. N. Il'inskii, V. E. Kuzin, and M. B. Saukke, *Kosmonavtika na znachkakh sssr, 1957–1975: Katalog* (Moscow: Izdatel'stvo "Sviaz," 1977), 143–44.

46 V. A. Omel'ko, *Nagradnye znaki obshchestvennykh organizatsii u muzeev* [Premium Znachki of Social Organizations and Museums], vol. 1, *Nagrady za osvoenie kosmosa Katalog* (Moscow, 2002), 6.

47 The "K" stands for kopek, the Russian penny.

48 Omel'ko, *Nagrady za osvoenie kosmosa katalog.* This book is a catalog of these awards. Although it does not list the recipient of each award, it does state the issuing organization, the purpose of the award, and the starting dates and cycles for each award.

49 Il'inskii, *Znachki i ikh kollektsionirovanie (Posobie dlia fileristov),* 9, 112, and 143. Tompak is a copper and zinc alloy that is an inexpensive alternative to copper and gold in costume jewelry. The Shcherbinsk Factory is an optical facility in the suburbs of Moscow.

50 Ibid., 9.

51 Childs, "Moscow Exhibits Stress Sputniks"; Waggoner, "Brussels Invites the World to Its Fair," 6; "U.S. and Soviet Agree to Exchange of Exhibits," *Washington Post and Times Herald,* 30 December 1958, A5; and "Text of Speeches by Nixon and Kozlov at Opening of Soviet Exhibition," *New York Times,* 30 June 1959, 16.

52 Sal'nikov, *Iurii Gagarin.*

53 Grant, "Socialist Construction of Philately in the Early Soviet Era," 493.

54 Il'inskii, Kuzin, and Saukke, *Kosmonavtiki na znachkakh sssr, 1957–1975: Katalog,* 6.

55 Il'inskii, *Znachki i ikh kollektsionirovanie (Posobie dlia fileristov),* 15.

56 Frank H. Winter, "The Silent Revolution or How R. H. Goddard May Have Helped Start the Space Age," paper presented at the Fifty-fifth International Astronautical Federation Congress, October 4–8, Vancouver, Canada, 2004. Winter goes into some detail how the image gained widespread international acceptance.

57 Il'inskii, Kuzin, and Saukke, *Kosmonavtiki na znachkakh sssr, 1957–1975: Katalog,* 143.

58 *Voskhod* launched on 12 October 1964, with commander Vladimir Komarov, en-

gineer Konstantin Feoktistov, and physician Boris Ëgorov. The 18 March 1965 launch of *Voskhod 2* carried Commander Pavel Belaev and pilot Aleksei Leonov.

59 Arthur Voyce, "Soviet Art and Architecture: Recent Developments," *Annals of the American Academy of Political and Social Science* 303, "Russia since Stalin: Old Trends and New Problems" (January 1956): 107.

60 Christina Kiaer, *Imagine No Possessions: The Socialist Objects of Russian Constructivism* (Cambridge, Mass.: The MIT Press, 2005).

61 The Union of Architects issued the 1955 Directive of the Party Design and Construction. Buchli, "Khrushchev, Modernism, and the Fight against 'Petit-Bourgeois' Consciousness in the Soviet Home," 161. "Ornamentalism" was the euphemism for Stalinist neoclassical, "wedding cake" architecture.

62 Ibid., 175.

63 The Znanie Society was an adult, public education voluntary organization founded in 1947. The society conducted public lectures on science and politics throughout the former Soviet Union. David C. Lee, "Public Organizations in Adult Education in the Soviet Union," *Comparative Education Review* 30, no. 3 (August 1986): 344–46.

64 Joanne M. Gernstein London, "A Modest Show of Arms: Exhibiting the Armed Forces and the Smithsonian Institution, 1945–1976" (Ph.D. diss., American Studies, George Washington University, 2000) is a complete recounting of the development of the desire to create a US military history museum into what became the National Air and Space Museum in 1965.

65 The signers were: I. B. Bardin, Vice-President of the Academy of Sciences; A. A. Blagonravov, Academic Secretary of the Department of Technical Sciences of the Academy of Sciences; L. I. Sedov, Member of the Academy of Sciences; A. N. Tupelov, Chief Designer of Tupelov aircraft; P. K. Oshchepkov, Radar Designer; L. A. Druzhkin, Corresponding Member of the Academy of Sciences; and I. I. Grai, member of the Academy of Sciences.

66 "Sozdat' muzei osvoeniia kosmosa" [To Create a Museum of the Conquest of Space], *Pis'mo v redaktsiiy, Literatura i zhizn'* (Moscow, USSR), 8 April 1959, No. 42 (158), 2.

67 Ibid., 2.

68 Brooks, *Thank You, Comrade Stalin!* 99, and "Sozdat' Muzei Osvoeniia Kosmosa," 2.

69 E. N. Kuzin, N. G. Belova, and T. V. Chugova, *Muzei kosmosa v kaluge* (Tula: Priok. kn. izd-va, 1986), 5.

70 A. Laskina, "Sozdat' muzei kosmosa" [To Create a Space Museum], *Pis'mo v redaktsiiu, Literatura i zhizn',* 24 April 1959, 3.

71 G. S. Vetrov, "S. P. Korolev i razvitie muzeev po kosmonavtike" [S. P. Korolev and the Development of the Space Museum], in *Trudy XXXII chtenii, posviashennikh razrabotke nauchnogo naslediia i razvitiiu idei k. e. Tsiolkovskogo* [The Proceedings of the 32nd Reading Related to the Scientific Research and ideas of K. E. Tsiolkovsky] (Kaluga, 15–18 sentiabria 1992 g.) (Moscow, USSR: IIET RAN, 1994), 195.

72   Gagarin, *Moi brat Iurii*, 331.

73   "Soviet Plans Space Museum," *New York Times*, 24 September 1967, 58.

74   Vetrov, "S. P. Korolev i razvitie muzeev po kosmonavtike," 195.

75   All these objects remain as part of the museum collection at the Energiia Museum in the city of Korolev. Energiia is the legacy organization of Korolev's original design bureau. Theo Pirard, "The Space Museum at RKK Energia," *Spaceflight* 42 (June 2000), 247–53.

76   Jamey Gambrell, "The Wonder of the Soviet World," *New York Review of Books* (New York), 22 December 1994, 31.

77   Ibid., 31.

78   K. Andrea Rusnock, "The Art of Soviet International Politics: Vera Mukhina's Worker and Collective Farm Woman in the 1937 Internationale Exposition," in *The Soviet Pavilion at the Exposition Internationale, Paris 1937*, chair David C. Fisher, AAASS National Convention, 17 November, New Orleans, 2007. Rusnock has researched the engineering and construction of the Mukhina sculpture that topped the Soviet pavilion in Paris in 1937.

79   McCannon, "Tabula Rasa in the North," 246.

80   Gambrell, "The Wonder of the Soviet World," 33.

81   Gambrell, "The Wonder of the Soviet World," 30; Vladimir Paperny, *Architecture in the Age of Stalin: Culture Two*, trans. John Hill and Roann Barris (Cambridge, UK: Cambridge University Press, 2002), 150–57.

82   Irina Cheredina, "Na meste Stalina sostavili raketu" [They Replaced the Statue of Stalin with a Rocket], *Ogonek* (Moscow), 12–18 March 2007, online.

83   Sonja D. Schmid, "Celebrating Tomorrow Today: The Peaceful Atom on Display in the Soviet Union," *Social Studies in Science* 36, no. 3 (30 June 2006), 340.

84   Gambrell dates the removal of the Stalin statue to the early 1950s, but that date is unlikely. VDNKh re-opened several times after World War II: in 1954, 1956, and finally in 1959. Khrushchev gave his "Secret Speech" at the 20th Party Congress, calling for de-Stalinization in February 1956 and the declaration for removing Stalin from the Lenin Mausoleum did not pass until the Twenty-Second Party Congress in October 1961. It is likely that the decision to remove the Stalin statue occurred during the renovations in the late 1950s. All evidence indicates that a Tupolev passenger aircraft first replaced the Stalin Statue in 1956. After renovations, the All-Union Agricultural Exhibition had become the Exhibition of Economic Achievements in 1959. Gambrell, 33. Taubman, *Khrushchev*, 513–15.

85   Childs, "Moscow Exhibits Stress Sputniks."

86   Ibid., A16.

87   Ibid., A16.

88   "Vostok Model is Shown to Public in Moscow," 10.

89   Konstantin Feoktistov, *Traektoriia zhizhi: mezhdu vchera i zavtra* [The Trajectory of Life: Between Yesterday and Tomorrow] (Moscow: Vagrius, 2000), 141–42.

90   "Vostok Revealed," *Spaceflight*, July 1965, 161.

91   Ibid., 161. Feoktistov, *Traektoriia zhizhi*, 143–44.

92   Ibid., 143–44.

93 "Vostok Revealed," 161.

94 The director of the pavilion claimed this label in 1971. Victor Bazykin, "Moscow's Permanent Space Exhibition," *Spaceflight* 3, no. 8 (August 1971): 295–97.

95 Ibid., 296.

96 "Soviet Space Monument Showing Rocket Unveiled," *New York Times*, 13 October 1964, 2.

97 Nina Tumarkin, *The Living and the Dead: The Rise and Fall of World War II in Russia* (New York: Basic Books, 1994), 1–3, and "Lev Kerbel and His Time (to the 85th Anniversary of the Sculptor's Birth)," Russian Culture Navigator, 09/03/2005, http://www.vor.ru/culture.cultarch238_eng.html.

98 The apartment blocks in the region were commonly known as the Korolev houses by the end of the 1960s. Chertok, *Rakety i Liudi: lunnaia gonka,* 13.

99 Vetrov, "S. P. Korolev i razvitie muzeev po kosmonavtike," 195.

100 Martovitskaia explains the long wait for the Memorial Museum of Cosmonautics. The architects of the original monument had not planned for an excavation underneath its foundation, and it was not until 1969 that the Moscow City Council approved the petition for the new museum. Anna Martovitskaia, "Muzei kosmonavtiki gotovitsia k rekonstruktsii" [Space Museum Prepared for Reconstruction], *Gazeta Kultura,* no. 14 (7524), 13–16 April 2006, Kultura Portal, 15 February 2006, http://www.kultura-portal.ru/tree_new/cultpaper/article.jsp?number=635&rubric_id=218.

101 My knowledge of the Zvezda Museum comes from a research visit that I took there to interview engineers about spacesuit development in 1996. Zvezda maintains a web page on its institutional history that includes photographs from its museum: OAO NPP "Zvezda," "Istoriia OAO NPP Zvezda," *OAO "NPP Zvezda,"* 2006, 29 October 2007, http://www.zvezda-npp.ru/histor.html.

102 Under Gorbachev during the mid- to late 1980s, all enterprises, even in the military and government, maintained an accounting of all funds and generated income out of their operating budgets. Under the rubric of *khozraschet [khoziaist-vennyi raschet,* self-financing], enterprises that had previously received generous government subsidies then turned to revenue-generating endeavors that included admission charges for tourists, especially foreign ones, and the sale of items in small shops. This type of revenue enhancement became an absolute necessity after the collapse of the Soviet Union.

103 After a speech on his flight on 24 June 1961, someone asked Yuri Gagarin about the center where the cosmonauts did all their training. He responded with the idea of creating a museum that would document their training and would be open for future generations to see. Russkii gosudarsvennyi nauchno-issledovatel'skii ispytatel'nyi tsentr podgotovki kosmonavtov imeni Iu. A. Gagarina (RGNIITsPK imeni Iu. A. Gagarina) [The Yuri Gagarin Russian State Scientific Research Center for the Testing and Training of Cosmonauts], 16 February 2007 http://www.gctc.ru/.

104 The Yuri Gagarin Russian State Scientific Research Center for the Testing and Training of Cosmonauts, http://www.gctc.ru/gagarin/default.htm.

105 Kamanin, *Skrytyi kosmos. Kniga tret'ia: 1967–1968,* 202.

106 Viacheslav Bakhtinov, "Interv'iu S Borisom Strugatskim," http://rusf.ru/abs/int0000.htm (accessed 28 September 2012).

107 Josephine Woll, *Real Images: Soviet Cinema and the Thaw,* ed. Richard Taylor, Kino: The Russian Cinema Series (London: I.B. Tauris Publishers, 2000), 83–99.

108 Personal Conversation with Klushantsev's daughter, 4 September 2009, in St. Petersburg, Russia.

109 Evgenii Kharitonov, "Kosmicheskaia odisseia Pavla Klushantseva" [The Space Odyssey of Pavel Klushantsev], in *Na ekrane—chudo,* ed. Evgenii Kharitonov and Andrei Shcherbak-Zhukov (Moscow: NII Kinoiskusstva, 2003), 19.

110 Kurt Maetzig, *Der schweigende stern* [The Silent Star], *(Bezmolvnaia Zvezda),* Yoko Tani, Oldrich Lukes, Ignacy Machowski, Julius Ongewe (Deutsche Film), 1959, 155 min.

111 Mikhail Kariukov and A. Kozyr', *Nebo zovet* [The Sky Calls], Pereverzev, Ivan; Shvorin, Aleksandr; Bartashevich, Konstantin; Borisenko, Larisa; Chernyak, V.; Dobrovolsky, Viktor (Gosudarstvenii komitet po kinematografii (Goskino), 1960), 77 min. American producer Roger Corman purchased the rights to the film and hired a young Francis Ford Coppola to rework the movie. *Battle Beyond the Stars* was an American interplanetary war movie with no reference to Cold War competition. Jimmy T. Murakami, *Battle Beyond the Stars,* Thomas, Richard; Vaughn, Robert, Saxon, John (New World Pictures, 1980), 104 min.

112 Pavel Klushantsev, *Planeta bur'* [Planet of Storms], Emel'ianov, V.; Sarantsev, Iu.; Zhzhenov, G.; Ignatova, K.: Vernov, G.; Teikh, G. (Lennauchfilm, Leningrad Popular Science Film Studio, 1962), 83 min. Like its immediate predecessor, this film, too, had a second cinematic life in American theaters, first as the 1965 *Voyage to the Prehistoric Planet* and then in 1968 as *Voyage to the Planet of Prehistoric Women.* Director Peter Bogdanovich created the second American version. Curtis Harrington, *Voyage to the Prehistoric Planet,* Rathbone, Basil; Domergue, Faith; Shannon, Marc (Roger Corman Productions, 1965), 78 min. and Peter Bogdanovich, *Voyage to the Planet of Prehistoric Women,* Van Doren, Mamie; Marr, Mary; Lee, Paige (The Filmgroup, 1968), 78 min.

113 Iosif Boyaksky and Ivan Ivanov-Vano, *Letaiushchii proletarii* [Flying Proletariat], Galich, Aleksandr (Soiuzmul'tfil'm, 1962), 16, 22 min.

114 Mikhail Karzhukov and Otar Koberidze, *Mechte navstrechu* [A Dream Come True], Gordeichik, Larisa; Borisenko, Boris; Koberidze, Otar; Kard, Peeter; Genesin, A.; (Odesskoi Kinostudii, 1963), 64 min.

115 Karzhukov and Kozyr', *Nebo zovet.*

Chapter 5. A Removal and Three Deaths

1 *Raketno-kosmicheskaia korporatsiia "Energiia" imeni S.P. Korolev* [S. P. Korolev

Energiia Rocket and Space Corporation], ed. Iurii Pavlovich Semenov (Moscow: Raketno-kosmicheskaia korporatsiia "energiia" imeni S.P. Korolev, 1996), 232–34. The document refers to orders of reorganization that left Korolev's design bureau at the head of the effort. Even though the Soviets were working on a Moon program the Politburo did not grant formal approval of the program until 1965. Siddiqi, *Challenge to Apollo,* 407, and *Raketno-kosmicheskaia korporatsiia "Energiia" imeni S.P. Korolev,* 232–34. Nevertheless, Soviet officials felt free to tease the West with references to their own Moon program. Joseph G. Whelan, "The Press and Khrushchev's 'Withdrawal' From the Moon Race," *The Public Opinion Quarterly* 32, no. 2 (Summer 1968): 234.

2 As discussed in Chapter 1, Khrushchev had been the first to recognize the political utility of spaceflight, and within ten days of *Sputnik,* he demanded a second launch to coincide with the fortieth anniversary of the Bolshevik Revolution. Chertok, *Rakety i Liudi: fili Podlipki Tiuratam,* 198.

3 Georgii Beregovoi's mission onboard *Soyuz 3* in late October 1968 was successful in that, unlike Vladimir Komarov, he did survive the mission. Celebratory press accounts notwithstanding, the *Soyuz 3* spacecraft failed in its assigned maneuvers and did not rendezvous and dock with a target vehicle.

4 Chertok, *Rakety i Liudi: lunnaia gonka,* 13–15.

5 Ibid., 13.

6 The Soviet Moon program remained an official secret at that time, but Chertok and other engineers understood that the failure to launch the lunar launch vehicle, the *N-1,* in July 1969, meant that the USSR would not get to the Moon. Vasili P. Mishin, Diaries, 1960–1974, Private Diaries and Notebooks, Dallas, Texas, Perot Foundation.

7 For a general discussion of the Thaw period, see Taubman, *Khrushchev,* 361–95; and Medvedev, *Khrushchev,* 83–103. For more detailed discussion of Khrushchev's impact on the arts, see Karl Edward Loewenstein, "The Thaw: Writers and the Public Sphere in the Soviet Union, 1951–1957" (Ph.D. diss., Duke University, 1999); Liudmila Alekseeva and Paul Goldberg, *The Thaw Generation: Coming of Age in the Post-Stalin Era* (Pittsburgh: University of Pittsburgh Press, 1993); Brooks, *Thank You, Comrade Stalin!* xvii–xviii; Susan Emily Reid, "Photography in the Thaw," *Art Journal* (Summer 1994): 33–39; and Woll, *Real Images.*

8 Catherine Merridale, *Night of Stone: Death and Memory in Twentieth Century Russia* (New York: Viking, 2001), 94–97. Merridale recounts the history of the creation of the mythology of the Red Martyrs.

9 Ibid., 94.

10 Ibid., 133–36. The new Soviet government initially proposed cremation as a solution to the unsanitary conditions of mass graves and as an alternative to religious services. However, the costs and cultural resistance to cremations put these ambitions on hold.

11 Korolev's immediate utility to the government was to supply spy satellites using the *Vostok* base vehicle. James J. Harford, *Korolev: How One Man Masterminded the Soviet Drive to Beat America to the Moon* (New York: Wiley, 1997), 190–99;

Siddiqi, *Challenge to Apollo,* 473–74; G. S. Vetrov and Boris Raushenbakh, eds., *S. P. Korolev i ego delo: svet i teni v istorii kosmonavtiki. Isbrannye trudy dokumenty* [S. P. Korolev and His Activities: Light and Shadows in the History of Spaceflight. A Collection of Documents], Komissiia po razrabotke nauchnogo naslediia pionerov osvoeniia kosmicheskogo prostranstva (Moscow: Nauka, 1998), 472.

12 Chertok, *Rakety i liudi: Fili podlipki tiuratam,* 199–201.

13 Laika was a passenger on board *Sputnik 2,* the first space ship just over one month after the first artificial satellite of the Earth.

14 Between the years 1961 and 1967, the USSR made four attempts to launch probes to Venus. The first two failed radio contact. *Venera 3* and *4* failed on entering the planet's atmosphere. The sole acknowledged attempt to send a probe to Mars in 1963 had failed communications, as well. *Luna 1* in 1958 missed the Moon. Of the lunar program, there were two complete successes: *Luna 2* landed, and *Luna 3* took photographs of the far side of the Moon in 1959. There was also a series of semi-successful lunar missions throughout the 1960s. *Spaceflight: A Smithsonian Guide,* eds. Valerie Neal, Cathleen S. Lewis, and Frank H. Winter (New York: Macmillan, 1995), 134, 161, and 166.

15 Iaroslav Kirilovich Golovanov, *Korolev: fakty i mify* [Korolev: Facts and Myths] (Moscow: Izd-vo "Nauka," 1994), 776; and B. V. Petrovskii, "Meditsinskoe zakliuchenie o bolezni i prichine smerti tovarishcha Korolev Sergeia Pavlovicha [Medical Conclusions about the Illness and Death of Sergei Pavlovich Korolev]," *Izvestiia* (Moscow), 16 January 1966, No. 13 (15101), 4.

16 Korolev, *Otets: kniga vtoraia* [Father: Volume 2] (Moscow: Nauka, 2002): 324–26. Iaroslav Kirilovich Golovanov, *Korolev: fakty i mify* [Korolev: Facts and Myths] (Moscow: Izd-vo "Nauka," 1994), 775–79. Each book provides detailed descriptions of Korolev's medical treatment including consultations with medical authorities and posthumous second-guessing.

17 Harford, *Korolev,* 49–63, and Golovanov, *Korolev,* 390–92.

18 The circumstances behind Korolev's anonymity remain unclear. It could be that he was as much a victim of bad timing as of Soviet secrecy. The identity of his contemporary, Igor Kurchatov, designer of the Soviet atomic bomb, was published in the Great Soviet Encyclopedia (BSE) in October/November 1954, see Schwartz, "Soviet Identifies Atomic Physicist," 28. It might be that Korolev was unfortunate that the timing of his contributions to spaceflight did not allow a timely disclosure.

19 Golovanov, *Korolev,* 779.

20 Brezhnev, "Akademik Sergei Pavlovich Korolev," *Pravda,* 4; and Leonid I. Brezhnev, "Akademik Sergei Pavlovich Korolev," *Izvestiia* (Moscow), 16 January 1966, No. 13 (15101), 4.

21 Brezhnev, "Akademik Sergei Pavlovich Korolev," *Pravda,* 4; and Brezhnev, "Akademik Sergei Pavlovich Korolev," *Izvestiia,* 4.

22 "Ot akademii nauk SSSR," *Pravda* (Moscow), 16 January 1966, No. 16 (17333), 4; and "Ot akademii nauk SSSR," *Izvestiia* (Moscow), 16 January 1966, No. 13 (15101), 4.

23 "Ot komissii po organizatsii pokhoron Sergeia Pavlovicha Koroleva," *Pravda* (Moscow), 16 January 1966, No. 16 (17333), 4.

24 Petrovskii, "Meditsinskoe zakliuchenie o bolezni i prichine smerti tovarishcha Koroleva Sergeia Pavlovicha," *Pravda* (Moscow), No. 13 (15101), 4; and B. V. Petrovskii, "Meditsinskoe zakliuchenie o bolezni i prichine smerti tovarishcha Korolev Sergeia Pavlovicha [Medical Conclusions about the Illness and Death of Sergei Pavlovich Korolev]," *Pravda* (Moscow), 16 January 1966, No. 16 (17333), 4.

25 See Boris E. Chertok, 1945, 1958–1988, Diaries and Notebooks, Washington, DC, Smithsonian National Air and Space Museum; Mishin, "Mishin Diaries," No. 13, December 1966 and April 1967; and Kamanin, *Skrytyi kosmos. Kniga tret'ia: 1967–1968*, 68–72.

26 Chertok, *Rakety i liudi: Goriachie dni kholodnoi voiny*, 443.

27 Ibid., 450. There are discrepancies in the stories recounting Gagarin's return to spaceflight. Contemporary stories about Komarov's death date Gagarin's return to flight status to early March 1967, although the only public official documents on the matter are dated after Komarov's death in December 1967 and did not mention Ustinov's approval. The quick turnaround of official approval for Gagarin's return to pilot training might be because those involved in the approval process assumed the issue was resolved nine months earlier.

28 Boris Chertok recounts an optimistic Gagarin throughout the flight until *Soyuz 1*'s landing. The official term in the press was "abnormal." Chertok, *Rakety i liudi: Goriachie dni kholodnoi voiny*, 450.

29 Chertok, *Rakety i liudi: Goriachie dni kholodnoi voiny*, 444.

30 Mishin, "Mishin Diaries," No. 13, December 1966 and April 1967; Chertok, *Rakety i liudi: Goriachie dni kholodnoi voiny*, 440; Kamanin, *Skrytyi kosmos. Kniga pervaia: 1960–1963*, 67.

31 Chertok, *Rakety i liudi: Goriachie dni kholodnoi voiny*, 445.

32 Mishin, "Mishin Diaries," No. 13, December 1966 and April 1967.

33 Chertok, *Rakety i liudi: Goriachie dni kholodnoi voiny*, 446; and Mishin, "Mishin Diaries," No. 13, December 1966 and April 1967.

34 *Raketno-kosmicheskaia korporatsiia "energiia" imeni S.P. Koroleva*, 183.

35 *Raketno-kosmicheskaia korporatsiia "energiia" imeni S.P. Koroleva (1946-1995)*, 182; and Kamanin, *Skrytyi kosmos. Kniga tret'ia: 1967–1968*, 73.

36 Mishin, "Mishin Diaries," No. 13, December 1966 and April 1967.

37 Ibid.

38 Harford, *Korolev*, 201–15.

39 Mishin, "Mishin Diaries," No. 13, December 1966 and April 1967.

40 "Vtoroi zvezdnyi reis kommunista Vladimira Komarova: uspeshnoe vpolnenie programmy namechennykh issledovanii" [The Second Celestial Flight of Cosmonaut Vladimir Komarov: Successful Completion of Planned Research], *Pravda* (Moscow), 24 April 1967, No. 14 (17796), 1.

41 TASS, "Polet novogo sovetskogo kosmicheskogo korablia" [The Flight of a New Soviet Spacecraft], *Pravda* (Moscow), 24 April 1967, No. 114 (17796), 1. The Soviet press was eager to publicize the breakthrough technology of the new space-

craft and published a two page spread on the *Soyuz* and its improvements on the *Voskhod*. Borzenko and Denisov, "'Voskhod'—'soiuz.'"

42 Mishin, "Mishin Diaries," No. 13, December 1966 and April 1967; Chertok, *Rakety i liudi: Goriachie dni kholodnoi voiny,* 440; Kamanin, *Skrytyi kosmos. Kniga tret'ia: 1967–1968,* 67.

43 *Pravda* (Moscow), 25 April 1967, No. 115 (17796), 1.

44 Ibid., 1.

45 Anna Gagarina died in the middle of June 1984. "Mother of Cosmonaut Gagarin," *Chicago Tribune,* 15 June 1984, D_A7.

46 Gagarina, *Slovo o syna,* 137.

47 Sal'nikov, *Iurii Gagarin*; Gagarin, *Moi brat Iurii,* 328.

48 Kamanin, *Skrytyi kosmos. Kniga pervaia: 1960–1963,* 57–58.

49 Ibid., 50–51.

50 Ponomareva, "Roman s kosmonavtikoi." Ponomareva was one of Valentina Tereshkova's backups for the flight of *Vostok 6.* Chertok, *Rakety i liudi: Goriachie dni kholodnoi voiny,* 441.

51 A copy of one letter was reprinted in the catalog to be sold at auction. Christie's East, *Space Exploration, Sale,* Auction, Wednesday, 9 May 2001 (New York: Christie's, 2001), 68. The Commandant of the Cosmonaut Corps Nikolai Kamanin published the content of the exchanges in his diaries. Kamanin, *Skrytyi kosmos. Kniga tret'ia: 1967–1968,* 151.

52 Mitroshenkov, *Zemliia pod nebom,* 442; and Kamanin, *Skrytyi kosmos. Kniga tret'ia: 1967–1968,* 195–96.

53 Mitroshenkov, *Zemliia pod nebom,* 442–47.

54 Sergei Mikhailovich Belotserkovskii, *Gibel' Gagarina: fakty i domysly* [The Death of Gagarin: Facts and Conjectures] (Moscow: Mashinostroenie, 1992), 26.

55 Belotserkovskii, *Gibel' Gagarina,* 56.

56 Ibid., 56.

57 Chertok, *Rakety i liudi: Goriachie dni kholodnoi voiny,* 509–512; and Kamanin, *Skrytyi kosmos. Kniga tret'ia: 1967–1968,* 225–26.

58 V. T. Kozyrev, *Eshche raz o gibeli Gagarina* [Once Again about the Death of Gagarin] (Moscow, 1998).

59 Of course, it is impossible to know the duration of classification, but a former KGB officer who had claimed involvement in the investigation made this assertion in an anonymous article, "Gagarina mogli ubit'" [They Could Have Killed Gagarin], *Literaturnaia gazeta* (Moscow), 25 March 1998, 3.

60 Merridale, *Night of Stone,* 26–29.

61 The list included the following members: D. F. Ustinov, L. V. Smirnov, P. V. Dement'ev, I. I. Yakubovskii, K. A. Vershinin, N. S. Zakharov, A. I. Mikoyan, I. I. Pstygo, I. I. Moroz, M. N. Mishuk, B. N. Ermin. The group included members of the Central Committee. Among them were the former Minister of the Defense Industries, First Deputy Minister of Defense, Marshal of the Air Force, a representative from the KGB, the chief designer of the Mikoyan-Gurevich Design Bu-

reau, the Deputy Commander of the Air Force, and many recipients of the Hero of the Soviet Union award.

62 Chertok, *Rakety i liudi: Goriachie dni kholodnoi voiny,* 504.

63 Kamanin, *Skrytyi kosmos. Kniga tret'ia: 1967–1968,* 202.

64 Ibid., 202.

65 There are extremes in the accuracy of the coverage of the investigation of Gagarin's death. Murasov interviewed participants in the space program and members of the public to assemble a complete listing of the supposed causes of his death. Boris Murasov, *Ubitstvo kosmonavta Iuriia Gagarina* [The Assassination of Cosmonaut Yuri Gagarin] (Moscow, 1995). Belotserkovskii, one of Gagarin's professors at the Zhukovskii Academy, attempts to reconstruct the investigation from the original procedures at the time of the cosmonaut's death. Sergei Belotserkovskii, *Gibel' Gagarina.*

66 Chertok, *Rakety i liudi: Goriachie dni kholodnoi voiny,* 505.

67 Kamanin, *Skrytyi kosmos. Kniga tret'ia: 1967–1968,* 332

68 Kozyrev, *Eshche raz o gibeli Gagarina.*

69 Kamanin, *Skrytyi kosmos. Kniga tret'ia: 1967–1968,* 202, and "Gagarina mogli ubit,'" 3.

70 Aleksandr Blokhnin, "I vse-taki: kak pogib Iurii Gagarin?" [And All the Same, How Did Yuri Gagarin Die?], *Izvestiia* (Moscow), 21 November 2006, 12.

71 Gagarina, *Slovo o syna,* 138.

72 Tsentral'nyi komitet KPSS, Presidium verkhnogo soveta SSSR, and Sovet Ministrov SSSR, "Ot tsentral'nogo komiteta kpss, presidium verkhnogo soveta SSSR i sovet ministrov SSSR" [From the Central Committee of the CPSU, the Presidium of the Supreme Soviet and the Council of Ministers of the USSR], *Pravda* (Moscow), 29 March 1968, No. 89 (18136), 1.

73 "Ot pravitel'stvennoi komissii po vyiasneniiu obstoiatel'stv gibeli letchika-kosmonavta sssr, geroia sovetskogo soiuza polkovnika Iu. A. Gagarin i geroia sovetskogo soiuza inzhener-polkovnika V. S. Seregina" [From the Leadership of the Commission for the Investigation into the Circumstances of the Deaths of the Pilot-Cosmonaut, Hero of the Soviet Union Yuri Gagarin and Hero of the Soviet Union Engineer-Colonel Seregin], *Pravda* (Moscow), 29 March 1968, No. 89 (18136), 1.

74 There have been numerous rumors that Gagarin was not the first in space, but only the first to survive. These rumors grew out of the overall and imperfect secrecy of the program. The unfortunate death of Valentin Bondarenko just weeks before Gagarin's flight gave credence to the idea that cosmonauts suffered a high mortality rate. Kamanin, *Skrytyi kosmos. Kniga pervaia: 1960–1963,* 33. Bondarenko's death coincided with the retrieval of the last test mission of the *Vostok* hardware before Gagarin's flight, which carried a prosthetic mannequin and broadcast phony telemetry back to Earth, Sotheby's, Russian Space History, Sale 6516, 12/11/93 auction catalogue (New York: Sotheby's, 1993), lot 10, n. p.

75 Scott W. Palmer, *Dictatorship of the Air: Aviation Culture and the Fate of Modern Russia,* Cambridge Centennial of Flight, eds. John Anderson and Von Hardesty

(Cambridge: Cambridge University Press, 2006), 233. Palmer reports that ". . . in keeping with the Soviet fashion of immortalizing heroes, both his hometown (Orenburg) as well as the island Udd were renamed in his honor." Chkalov died 5 December 1938 test flying a Polikarpov I-180. Palmer, *Dictatorship of the Air,* 233.

76 Brooks, *Thank You, Comrade Stalin!* (104) translates Voroshilov's eulogy dated 16 December 1938.

77 On 27 January 1967, *Apollo* astronauts Gus Grissom, Edward White, and Roger Chaffee died in a fire inside the *Apollo* spacecraft that was to carry them into Earth orbit later that year.

78 "Proshchania s geroiami" [Farewell to Heroes], *Izvestiia* (Moscow), 30 March 1968, No. 75 (15774), 1. The photographs in *Izvestiia* indicate that the attending crowd was at maximum.

79 Ibid., 94.

80 Ibid., 133.

81 *Ogonek,* April 1968, entire issue is dedicated to Gagarin's funeral.

82 Kamanin, *Skrytyi kosmos. Kniga tret'ia: 1967–1968,* 200.

83 "Proshchania s geroiami" [Farewell to Heroes], *Izvestiia* (Moscow), 30 March 1968, No. 75 (15774), 1.

84 Ibid., 1.

85 Kamanin, *Skrytyi kosmos. Kniga tret'ia: 1967–1968,* 202.

86 Russkii gosudarsvennyi nauchno-issledovatel'skii ispitatel'nyi tsentr podgotovki kosmonavtov imeni Iu. A. Gagarina (RGNIITsPK imeni Iu. A. Gagarina) [The Yuri A Gagarin Russian State Scientific-Research Institute for the Experimental Preparation of Cosmonauts], 16 February 2007 http://www.gctc.ru/.

87 Chkalov's birthplace, other cities, streets, and subway stations have borne his name since his death. Palmer, *Dictatorship of the Air,* 233.

88 Kamanin, *Skrytyi kosmos. Kniga tret'ia: 1967–1968,* 203.

89 Ibid., 203.

90 Ibid., 203.

91 Ibid., 204.

92 TASS, "Pervoprokhodtsu kosmicheskikh trass" [To the One Who First Stepped into Space], *Izvestiia* (Moscow), 5 July 1980, 2.

93 Ibid., 2.

94 Billington prefaces his classic work on Russian culture, *The Icon and the Axe,* explaining, "these two objects traditionally hung together on the wall of the peasant hut." James H. Billington, *The Icon and the Axe: An Interpretive History of Russian Culture,* 1966, paperback (New York: Vintage Books, 1970), vii.

95 Although the official report remains classified, unofficial partial versions have appeared over the years. Sergei Belotserkovskii published the first one. He was an engineer and professor at Moscow Aviation Institute. He published his volume on Gagarin's death in 1992. Belotserkovskii, *Gibel' Gagarina.*

96 Murasov, *Ubitstvo kosmonavta Iuriia Gagarina,* 9.

97 Ibid., 9.

98 Ibid., 9.

99 For example, Stalin used the assassination of Sergei Kirov in late 1934 as the pretense to start the Great Purge. Amy Knight, *Who Killed Kirov: The Kremlin's Greatest Mystery* (New York: Hill and Wang, 1999). Subsequent deaths became suspicious. The writer Maksim Gor'kii's death occurred under suspicious circumstances while he was under house arrest in 1936 and has always been the subject of rumor and conjecture in the absence of an official explanation. Stalin purportedly allowed Sergo Ordzhonikidze to commit suicide in 1937 to avoid trial as a concession to their long-standing friendship.

100 Murasov, *Ubitstvo kosmonavta Iuriia Gagarina,* 9.

101 Ibid., 15.

102 Siddiqi, *Challenge to Apollo,* 688–97.

103 US Central Intelligence Agency, Central Intelligence Agency (RG 263), Exposure #14, Corona/U-2 photographs, July 1969, Baikonur Cosmodrome After *N-1* Launch Failure, College Park, Maryland, US National Archives and Records Administration.

104 Chertok, *Rakety i liudi: lunnaia gonka,* 353–87; and *Raketno-kosmicheskaia korporatsiia "energiia" imeni S.P. Koroleva,* 248–52, and 258–63.

105 The fourth and final launch of the *N-1* took place on 23 November 1973. The program ended in 1974. *Raketno-kosmicheskaia korporatsiia "energiia" imeni S.P. Korolev,* 258.

106 Although many interviews and comments preceded it, this interview with Vasilli Mishin marked the first entirely attributed account of the lunar program. After its publication, many other engineers, including Mishin, began to publish their accounts of the lunar program. Vasili P. Mishin [interview], "Designer Mishin Speaks on Early Soviet Space Programmes and the Manned Lunar Project," *Spaceflight* 32 (1990): 104–06. Other, subsequent publications provided details on specific components of the program. V. M. Filin, *Vospominaniya o lunnon korable* [Memoirs about Lunar Spacecraft] (Moscow: Izdatel'stvo Kultura, 1992). I. B. Afanasyev, *Neizvestnie korabli* [Unknown Spacecraft], Kosmonavtika, astronomiya, znanie, vol. 12–91 (Moscow: Znanie, 1991).

107 Chertok, *Rakety i Liudi: lunnaia gonka,* 13. "The television coverage of the first steps on the Moon was rebroadcast to every country in the world except for the USSR and China. To see the broadcast from the United States, we had to go to NII-88, to which the images were broadcast via cable from the television center (Ostankino). Ostankino received the images from the Eurovision channel, but direct reception was forbidden. Once, a worker in Ostankino told me they had requested permission for a live broadcast, but Central Committee Secretary Mikhail Suslov had refused."

108 TASS, "Polet 'apollon-11'" [The Flight of Apollo 11], *Pravda* (Moscow), 18 July 1969, No. 199 (18612), 5.

109 TASS, "Sovetskaia avtomaticheskaia stantsiia 'luna-15' na okololynnoi orbite" [Soviet Automatic Station 'Luna 15' in Near Lunar Orbit], *Pravda* (Moscow), 18 July 1969, No. 199 (18612), 1.

110 V. Strel'nikov, "Zemliane na lune" [Landing on the Moon], *Pravda* (Moscow), 22 July 1969, No. 203 (18616), 5.

Chapter 6. Outpost in the Near Frontier

1 The sculptor was Pavel Bondarenko. He worked with the assistance of architects Yakov Belopolsky, F. M. Gazhevsky, and designer A. F. Sudakov.

2 David Remnick, *Lenin's Tomb: The Last Days of the Soviet Empire* (New York: Random House, 1993), 89, and Walter Laqueur, *Black Hundreds: The Rise of the Extreme Right in Russia* (New York: HarperCollins, 1993).

3 Remnick, *Lenin's Tomb,* 105; Martin Malia, *The Soviet Tragedy: A History of Socialism in Russia, 1917–1991* (New York: Free Press, 1994); and Orlando Figes, *A People's Tragedy: The Russian Revolution, 1891–1924* (New York: Penguin Books, 1998).

4 Emil Draitser, *Forbidden Laughter: Soviet Underground Jokes,* rev. ed. (Los Angeles: Almanac Pub. House: distributed by Maxim's Book Distributors, 1978).

5 This joke represents one distinct type of a political joke of the time—one that denigrates the leader's effectiveness. Another type, equally famous, criticized the leaders' intelligence. The following one was popular during the early 1970s:

> The Americans land on the moon, Brezhnev calls Soviet cosmonauts and gives an order:
> —By the end of this month Soviet spaceship must land on the sun!
> —Ok, Comrade General Secretary, but the problem is that we will burn alive, replied cosmonauts.
> —Do you think we are all stupid here in the Politburo! You are going to land there at night! replied Brezhnev

6 Polly Jones, "The Fire Burns On? The 'Fiery Revolutionaries' Biographical Series and the Rethinking of Propaganda in the Brezhnev Era," *Slavic Review* 74, no. 1 (Spring 2015): 32–56.

7 By the 1960s, *Novyi mir* gained the reputation as a dissenting literary and intellectual journal. The journal gained this fame from publishing Alexander Solzhenitsyn's novella *One Day in the Life of Ivan Denisovich* in 1962. *Novyi mir* was able to maintain its status as a publisher of controversial articles and in 1986 began publishing articles that overtly criticized the government of the USSR.

8 Chingiz Aitmatov, *The Day Lasts More Than a Hundred Years,* trans. John French (Bloomington: Indiana University Press, 1988).

9 Ibid., 343.

10 Ibid., 13.

11 Ibid., 29.

12 One such fictionalized account is the story of the Mankurt tribe that is known for the ability to take over the minds of their captives. One scholar interprets this as an allusion to Soviet mind control, James V. Wertsch. *Voices of Collective Remembering* (Cambridge: Cambridge University Press, 2002), 73.

13   Aitmatov, *The Day Lasts More Than a Hundred Years*, 342.

14   Ibid., 27.

15   Ibid., 28.

16   Ibid., 338.

17   Ibid., 338.

18   Ibid., 343.

19   Groys, *Ilya Kabakov*, 31–32.

20   Groys mistranslates the word "room" as "apartment" to avoid a prolonged explanation of the living situations of urban Soviet citizens.

21   Ibid., 5.

22   Ibid., 7.

23   Kamanin, *Skrytyi kosmos. Kniga tret'ia: 1967–1968*.

24   Sotheby's, in *Russian Space History, Sale 6516* (New York: Sotheby's, 1993).

25   For a more detailed history of the long history of the production of this film, see Elena Prokhorova, "White Sun of the Desert," in *The Russian Cinema Reader, Volume Two: The Thaw to the Present*, ed. Rimgaila Salys (Boston: Academic Studies Press, 2013), 126–33.

26   Rimgaila Salys provides a more detailed analysis of the broad cultural impact of the movie. Rimgaila Salys, "We have been Sitting Here for a Long Time," in *The Russian Cinema Reader, Volume Two: The Thaw to the Present*, ed. Rimgaila Salys (Boston: Academic Studies Press, 2013), 134–37.

27   S. Kovalev, "Sovereignty and the International Obligations of Socialist Countries," 26 September 1968, *Pravda*.

28   The Bulgarian mission with Georgi Ivanov on *Soyuz 33* failed to dock, but there was no interruption in the sequence to make up the mission.

29   Brian Harvey, *European-Russian Space Cooperation: From de Gaulle to ExoMars* (Chichester, UK: Springer Praxis, 2021), 1–39. Harvey describes the individual and organizational effort that culminated in de Gaulle's visit to Baikonur including its immortalization by a monument in Moscow.

30   Bettyann Holtzmann Kevles, *Almost Heaven: The Story of Women in Space* (New York: Basic Books, 2003).

31   Ponomareva, *Zhenskoe litso kosmosa*.

32   NASA Press Release for STS-41-G.

33   Bridger, "The Cold War and the Cosmos."

34   Kevles, *Almost Heaven*.

35   The full story of the rescue of the Salyut 7 space station did not come to the public until the 2017 release of the eponymous film. The director was American-educated Russian director Klim Shipenko. As much as the Soviet space establishment did not want to talk about the station and the mission to rescue it at the time, the Russian space establishment cooperated fully with the filmmaker, granting access to used training hardware for filming.

36   David Portree, *Mir Hardware Heritage* (Washington, DC: NASA, 2009).

37   During the lull of US launches of planetary probes that last for 15 years, the USSR focused on missions to the planet Venus. Over the years, Soviet Venus hardware improved to the point of attracting international attention to their missions that culminated in the *Vega 1* and *2* missions that combined sending experiments

through the atmosphere and to the surface of the planet followed by joining the multi-national armada to encounter Comet Halley in March 1986.

38 The story of *Soyuz T-13* crew of Dzhanibekov and Savinykh and their rescue of Salyut 7 became the subject of a Russian-made American-style special effects theatrical film by the same name. Klim Shipenko, dir., *Salyut 7*. CTB Films, Globusfilm, Lemon Films Studio, Telekanal Rossiya, 2017. Cosmonaut Viktor Savinykh wrote a companion memoir by the same name to fill out technical details glossed over in the film. Viktor Savinykh, *"Salyut 7" Zapiski s "mertvoi" stantsii* (Moscow: Eksmo, 2017).

39 Klim Shipenko, dir., *Salyut 7*, 2017.

40 In 2017, a multinational film created a fictional version of this episode from the perspective of a Cuban dissident. *Sergio y Sergei,* directed by Ernesto Daranas (Institudo Cubano del Arte e Industrias Cinematograficos), Mediapro, RTV Comercial, Wing and a Prayer Pictures, 2017).

41 The date of the first occurrence was in secret. Public knowledge of the blessings grew as the alliance between the Russian state and the Orthodox Church grew more public in other areas.

42 Khrushchev's attempts to secularize Soviet society through the substitution of secular ritualist for religious ones relied heavily on the space program at the time. One of the high points of this effort was the very public marriage between cosmonauts Valentina Tereshkova and Andriian Nikolayev. The marriage did not last, and the efforts faded with Khrushchev's removal from power. See Smolkin, *A Sacred Space is Never Empty,* 174–219.

43 Kristin Roth-Ey, "Finding a Home for Television in the USSR, 1950–1970," *Slavic Review* 66, no. 2 (Summer 2007): 279.

44 Anatoly Fedorovich Dobrynin, *In Confidence: Moscow's Ambassador to America's Six Cold War Presidents (1962–1986)* (New York: Times Books (Random House), 1995), 342.

45 Roth-Ey, "Finding a Home for Television in the USSR, 1950–1970," 298–301, publishes pictures of cartoons that illustrate the national boredom with Soviet television.

46 "If a story is about space and science, it is very likely to have visuals (81 percent). Here are some of the greatest achievements of the Soviet Union, and they are crucial to those who program the mass media. We should not think of the Soviet space program as in any way peripheral to their image of themselves or as problematic." Ellen Mickiewicz, *Split Signals: Television and Politics in the Soviet Union* (New York: Oxford University Press, 1988), 118.

47 Ibid., 119.

48 David Remnick, "Soviet TV Announces a Shake-Up: New Editor of Evening News to Bring It into Gorbachev Era," *Washington Post,* 6 January 1989, B4.

49 "Mikhail Gorbachev has seized on the medium of television to help persuade his people to work more energetically and, as he put it last fall, 'to feel at home in society.' Indeed, Mr. Gorbachev and his close adviser and propaganda chief Aleksandr N. Yakovlev believe that television, the first truly mass medium in Soviet

history, may be the instrument to break through the ossified deposits of bureaucratic power that this new Soviet leadership has found so resistant to change." Ellen Mickiewicz, "Soviet Viewers Are Seeing More, Including News of the U.S.," *New York Times*, 22 February 1987, H29.

50 "Soviet television announced a move today that would be the equivalent of the White House firing the heads of the three American network news divisions and replacing them all with a single figure whose background is a combination of 'Nightline' and 'Late Night with David Letterman.'" Remnick, "Soviet TV Announces a Shake-Up," B1.

51 "Poekhali!" had been Gagarin's word as the launch vehicle ignited to send him into orbit on 12 April 1961.

52 Much of what we know today about Soviet television in those later years comes from a United States Naval Academy intelligence contract that sought to record and analyze the changing landscape of Soviet television for a period from December 1988 to September 1991. United States Naval Academy, The US Naval Academy Collection of Soviet & Russian TV, September-February, 1988–1991, United States Library of Congress, Moving Image Research Center.

Chapter 7. Remembrance of Hopes Past

1 Alexei Yurchak, *Everything Was Forever until It Was No More: The Last Soviet Generation*, (Princeton: Princeton University Press, 2005), 282.

2 The "Z" article was a reference to George Kennan's "X" article (February 1946) also known as the long telegram that anonymously outlined the rationale of what would become the doctrine of containment. Z, "To the Stalin Mausoleum," *Daedalus* 119, no. 1 (1990): 295–344. http://www.jstor.org/stable/20025293.

3 Alexis de Tocqueville, *The Old Regime and the Revolution* (New York: Harper and Brothers), 1856.

4 Malia, *The Soviet Tragedy*.

5 Figes, *A People's Tragedy*.

6 Remnick, *Lenin's Tomb*.

7 These include the following scholarly surveys: Howard Schuman and Amy D. Corning, "Collective Knowledge of Public Events: The Soviet Era from the Great Purge to Glasnost," *American Journal of Sociology* 105, no. 4 (January 2000): 913–56, and *Russia's Sputnik Generation: Soviet Baby Boomers Talk About Their Lives*, ed. and trans. Donald J. Raleigh (Bloomington: Indiana University Press, 2006). Brief and informal surveys include Ekaterina Bershanskaia, "'On skazal "poekhali!" i makhnul rukoi" [He Said "Let's Go!" and Waved His Hand] *Surgutskaia Tribuna*, 12 April 2001, No. 069, 1; Anna Eroshova, Andrei Kabaninnikov, Chikin, and Nataliia Gracheva, "A chto dumaiut deti? Kto skazal "poekhali!" i makhnul rukoi?" [And What Do You Think? Who Said, "Let's Go!" and Waved His Hand?], *Komsomolskaia Pravda* (Moscow), 12 April 2001, 5.; and Tamara Gashimova, "Segodnia—den' kosmonavtiki. Kto vy, maior Gagarin?" [Today Is

Cosmonautics Day. Who Are You, Major Gagarin?], *Iakutiia,* 12 April 2002, No. 065, 1.

8 Exploration of the post-Soviet iconoclasm could quickly grow into a dissertation itself. Michael Wines, "Sculpting Soviet Giants, Watching Them Fall," *New York Times,* 14 June 2003, 11. Lev Kerbel was a prolific sculptor in the Soviet Union. His commissions included war memorials, but he is most famous for the busts of the Conquerors of Space that line the alley along the monument in Moscow. Kerbel created one of the last sculptures of Lenin erected in Moscow, the 1985 Lenin Monument in October Square. Nina Tumarkin begins her study of the cult of World War II in the Soviet Union with an interview with Lev Kerbel as well. Tumarkin, *The Living and the Dead,* 1–2.

9 With one exception, I have limited this discussion to late- and post-Soviet television, literature, and cinema. There are other areas of the creative arts that also address the Soviet space program of the 1960s, for example song and poetry. One area that scholars have explored has been the fine arts. For catalogues and reviews of two such examples, see Groys, *Ilya Kabakov,* and Alex Baker, Toby Kamps, and Svetlana Boym, *Space Is the Place,* eds. Alex Baker and Toby Kamps (New York: Independent Curators International, 2006).

10 Gagarin's orbit carried him a maximum of 315 kilometers above the Earth. The International Space Station orbits the Earth today at an altitude between 320 and 350 kilometers above the Earth.

11 In 1993, the United States and the Russian Federation signed an agreement to cooperate on the construction of the International Space Station, which had previously borne the designation *Alpha.* The first phase of cooperation took the form of a series of joint missions with the American space shuttle and the Russian space station Mir between 1994 and 1996. Lyndon B. Johnson Space Center. National Aeronautics and Space Administration, *A History of U.S. Space Stations,* tech. rept. no. IS-1997–06-ISS009JSC (Houston, Texas, 1997). The subsequent Russian decision to de-orbit the aging Mir space station in March 2001 left the Russians without an independent human spaceflight program for the first time in thirty years. "Proshchai, 'Mir!'" [Say Goodbye to Mir!], *Sankt-Peterburgskie vedomosti* (St. Petersburg), 24 March 2001, No. 054, 2.

12 Tumarkin, *The Living and the Dead.* Krylova focused tightly on the long-postponed monument to the memory of World War II. Anna Krylova, "Dancing on the Graves of the Dead: Building a World War II Memorial in Post-Soviet Russia," in *Memory and the Impact of Political Transformation in Public Space,* eds. Daniel J. Walkowitz and Lisa Maya Knauer (Durham, N.C.: Duke University Press, 2004), 83–102.

13 Tumarkin, *The Living and the Dead,* 175.

14 The location of the gravesite at Katyn forest, the timing of the German discovery, and their forthrightness in announcing the discovery have served as convincing evidence to the West that the Soviet Union was behind the executions; J. K. Zawwodny, *Death in the Forest: The Story of the Katyn Forest Massacre* (Notre Dame: University of Notre Dame Press, 1962). Historian Vojtech Mastny has ar-

gued that these events signified the beginning of the Cold War. Vojtech Mastny, *Russia's Road to the Cold War: Diplomacy, Warfare and the Politics of Communism, 1941–1945* (New York: Columbia University Press, 1979), 28. Gorbachev acknowledged Soviet guilt in the matter in April 1990. Tumarkin, *The Living and the Dead*, 181.

15 Taubman, *Khrushchev*, 272. Medvedev, *Khrushchev*, 87–88.

16 Krylova, "Dancing on the Graves of the Dead," 85.

17 Ibid., 93.

18 N. G. Andronnikov, "Les Monuments aux Héros de la Grande Guerre Nationale," *Revue International d'Histoire Militaire* 59 (1985): 294–95.

19 Krylova, "Dancing on the Graves of the Dead," 97.

20 "Yeltsin 'Categorically Against' Restoring Soviet Anthem," *Jamestown Foundation Monitor* 6, no. 228, 7 December 2000, http://www.jamestown.org, The Jamestown Foundation, 12/03/2008, http://www.jamestown.org/publications_details.php?volume_id=23&issue_id=1917&article_id=17856.

21 "The Rewriting of History," *Economist*, 8 November 2007, http://www.economist.com/world/europe/displaystory.cfm?story_id=10102921.

22 Vladimir Vladimirovich Putin, Russian Federation. President of Russia. Presidential Internet Resources, Speeches, 9 May 2000, File: Http://Www.Kremlin.Ru/Eng/Speeches/2000/05/09/0002_type127286_128408.Shtml, Speech at a Gala Reception Dedicated to the 55th Anniversary of Victory in the Great Patriotic War, The Kremlin, Moscow, Russia, Russian Federation. Office of the President. Vladimir Vladimirovich Putin, Russian Federation. President of Russia. Presidential Internet Resources, Speeches, 9 May 2007, File: Http://Www.Kremlin.Ru/Eng/Speeches/2007/05/09/1432_type82912type127286_127675.Shtml, Speech at the Military Parade Celebrating the 62nd Anniversary of Victory in the Great Patriotic War, Red Square, Moscow, Russia, Russian Federation. Office of the President.

23 Vladimir Vladimirovich Putin, Russian Federation. President of Russia. Presidential Internet Resources, Speeches, 9 May 2005, File: Http://Www.Kremlin.Ru/Eng/Speeches/2005/05/09/1506_type82912type127286_87850.Shtml, Speech at a Formal Reception Dedicated to the 60th Anniversary of Victory, The State Kremlin Palace, Moscow, Russia, Russian Federation. Office of the President.

24 The memorial was placed in the Aleksandrovskii Garden at the Kremlin Wall in December 1966, on the twenty-fifth anniversary of Gen. Georgii Zhukov's victory over the German Army outside Moscow.

25 "Soviet Space Monument Showing Rocket Unveiled," 2.

26 Michael Specter, "Moscow on the Make," *New York Times*, 1 June 1997, Sunday, Late Edition—Final, SM: 48–84.

27 Specter, "Moscow on the Make," 48–56.

28 Ibid., 75.

29 Schuman and Corning, "Collective Knowledge of Public Events," 913–56.

30 *Russia's Sputnik Generation.*

31 Raleigh lists seven of the twenty-nine by name. Ibid., xi.

32 Ibid., 5.

33 Schuman and Corning limited their research to six specific events, all but two of which took place in the late 1950s and early 1960s. The events included: the flight of Laika, the dog on *Sputnik 2*, in November 1957; the Cuban Missile Crisis in October 1962; the publication of Alexander Solzhenitsyn's *One Day in the Life of Ivan Denisovich* in November 1962; the Yezhovshchina of the 1930s; the Twentieth Congress of the CPSU in February 1956; and the Doctors' Plot in 1952–53. "Events like the dog Laika's orbit in *Sputnik 2* were favorable to the reputation of the Soviet regime and therefore publicized by the state-controlled media and schools. Knowledge of these events has a relatively low correlation with years of schooling, because most Russians were exposed to them early on regardless of their eventual educational attainment." Schuman and Corning, "Collective Knowledge of Public Events," 949.

34 Ibid., 931.

35 Ibid., 949.

36 Kommunisticheskaia partiia sovetskogo soiuza. *The Road to Communism: Documents of the 22nd Congress of the Communist Party of the Soviet Union. October 17–31, 1961* (Moscow, USSR: Foreign Languages Publishing House, 1961).

37 In his biography of Khrushchev, Richard Taubman cites Anastas Mikoyan's notes to demonstrate that the 1980 date was an arbitrary one, chosen for its distance: "Actually, Mikoyan later recalled, 'Khrushchev did not like statistics.' When Mikoyan resisted including twenty-year projections of steel output in the new program, Khrushchev replied, 'Nineteen-eighty won't arrive anytime soon,' which Mikoyan took to mean that Khrushchev 'didn't count on living until communism was fully constructed, so that it was not important to him whether the numbers were realistic or not.' What he wanted, Mikoyan said, was to 'impress the people. He did not understand that the people would demand an explanation if the promise were not fulfilled.'" Taubman, *Khrushchev,* 512.

38 *The Road to Communism,* 110–11.

39 *Russia's Sputnik Generation,* 167.

40 Ibid., 132.

41 These were the magazines for the Pioneer grade school youth groups, that combined national scouting with ideological indoctrination.

42 Bershanskaia, "'On skazal "poekhali!" i makhnul rukoi,'" 1.

43 Eroshova, Kabaninnikov, and Gracheva, "A chto dumaiut deti? Kto skazal "poekhali!" i makhnul rukoi?" 5.

44 Gashimova, "Segodnia—den' kosmonavtiki. Kto vy, maior Gagarin?" 1.

45 Eroshova, Kabaninnikov, and Gracheva, "A chto dumaiut deti? Kto skazal "poekhali!" i makhnul rukoi?" 5.

46 Bershanskaia, "'On skazal "poekhali!" i makhnul rukoi.'" 1.

47 Gashimova, "Segodnia—den' kosmonavtiki. Kto vy, maior Gagarin?" 1.

48 Ibid., 1.

49 Ibid., 1.

50 Eroshova, Kabaninnikov, and Gracheva, "A chto dumaiut deti? Kto skazal "poekh-ali!" i makhnul rukoi?" 5.

51 Svetlana Boym, "Kosmos: Remembrances of the Future," in *Kosmos: A Portrait of the Russian Space Age* (Princeton: Princeton Architectural Press, 2001), 82–99.

52 There are three published versions of this essay: Ibid., 82–99, and Svetlana Boym, *The Future of Nostalgia* (New York: Basic Books, 2001), 345. This excerpt is from the most recent and most extended version of the essay.

53 Titov, *700,000 Kilometres through Space,* (1961), 17.

54 Boym, *The Future of Nostalgia,* 345.

55 Ibid., 6.

56 Ibid., xvi.

57 Specter, "Moscow on the Make," 75.

58 Anatolii F. Britikov, *Russkii sovetskii nauchno-fantasticheskii roman* [The Russian Soviet Science Fantasy Novel] (Leningrad: Izdatel'stvo "nauka," 1970), 35–45. Pre-revolutionary writers, most notably V. Briusov, anticipated that spaceflight, as another stage of industrial development, would cause upheaval similar to nineteenth-century industrialization.

59 Clark, *The Soviet Novel,* 142–44.

60 Pelevin, *Omon Ra,* (Moscow: Tekst, 1993): 9. Omon's surname is a play on Dostoevsky's *The Brothers Karamazov. Karamazov* means "black stain." *Krivomazov* means "crooked stain." Simmons, "Fly Me to the Moon," 5.

61 OMON is the acronym for the Russian Police Special Weapons and Tactical Unit. The Russian phrase is Otriad militsii osobogo nazhacheniia. Pelevin, *Omon Ra,* 9.

62 Pelevin, *Omon Ra,* (Moscow: Tekst, 1993): 10. OVIR is the acronym for Otdel viz i registratsiia inostrannykh grazhdan [Department of Visas and Registrations for Foreign Citizens].

63 Ibid., 14.

64 Scott Palmer describes the official story of the life of Aleksei Mares'ev in his book, *Dictatorship of the Air:* "Still more inspiring was the extraordinary story of aviator Aleksei Mares'ev. Shot down in April 1942, Mares'ev survived the crash of his plane, only to suffer severe injuries to both of his feet. Unable to walk, he crawled between the German and Soviet front lines for eighteen days before being rescued by a band of partisans. Rushed to a Moscow military hospital, the fighter pilot's career appeared at an end when doctors were forced to amputate his gangrenous feet. Amazingly, Mares'ev overcame his disability. Only weeks after lifesaving surgery, he had learned to walk (and sometime later, dance) on his two prosthetic devices. Within a year, Mares'ev had become so adept that he was allowed to resume flying. He subsequently returned to active military service where he went on to record seven more kills. In August 1943, Mares'ev was awarded the title 'Hero of the Soviet Union.' A book-length account of his experiences titled *The Story of a Real Man* was published in 1946." Palmer, *Dictatorship of the Air,* 271.

65 Pelevin, *Omon Ra,* 39.

66 Lilya Kaganovsky, "How the Soviet Man Was (Un) Made," *Slavic Review* 63, no. 3 (Autumn 2004): 577–96.

67 Pelevin, *Omon Ra,* 66.

68 Ibid., 47.

69 Ibid., 93–98.

70 Ibid., 135.

71 "In addition, the NEP allowed the formation of the private joint-stock companies. Of these, the two most important were Rus and Mezhrabpom, which were later to form Mezhrabpom-Rus. Mezhrabpom was an abbreviation of International Workers' Aid, an organization established in Germany in 1921 by pro-Soviet and pro-Communist elements. Its original task was to help Soviet Russian fight famine." Peter Kenez, *Cinema and Soviet Society: From the Revolution to the Death of Stalin,* Kino: The Russian Cinema Series, ed. Richard Taylor (London: I.B. Tauris Publishers, 2006): 38. Yakov Protazanov's 1924 production of Alksey Tolstoy's book *Aelita* was clearly a priority for Sovkino, as the expense of the project revealed: "The production history of *Aelita* indicated that Protazanov prepared for his Soviet debut with great care and forethought, but without political foresight. Though schooled in the breakneck pace of pre-Revolutionary filmmaking, averaging more than ten films annually before the Revolution, he took over a year to complete *Aelita.* According to the handsome programs that were distributed at screenings of the picture, Protazanov shot 22,000 meters of film for the 2841-meter film (a 3:1 ratio was the norm) and employed a cast and crew of thousands." Youngblood, *Movies for the Masses: Popular Cinema and Soviet Society in the 1920s* (Cambridge: Cambridge University Press, 1992): 109. Advertising for the film, too, was unprecedented. Almost a year prior to its release, Soviet film newspapers and journals reported on the status of the production. In the weeks leading up to the opening in Moscow, *Pravda* advertised teasers for the prospective Moscow audiences. Aleksandr Ignatenko, *"Aelita": pervyi opyt sozdaniia blokbastera v Rossii* ["Aeilita": The First Experience with a Blockbuster in Russia] (St. Peterburg: Sankt-Peterburgskii gosudarstvennyi universitet kino i televideniia, 2007).

72 For a more complete recounting of this important interregnum in Russian film history, see Birgit Beumers, *A History of Russian Cinema* (London: Bloomsbury Academic, 2009): 217–40.

73 Aleksei Fedorchenko, *Pervye na lune* [First on the Moon], Vlasov, Boris; Slavnin, Aleksei; Osipov, Andrei; Otradnov, Anatolii; Ilinskaia, Viktoriia (Sverdlovsk Film Studio and Film Company Strana, 2005), 75 minutes. Ironically, this film won the "Best Documentary" award at the Venice Film Festival in 2005. The same year, it won "The Best Debut" prize at the Kinotaur festival, Sochi, Russia.

74 Aleksei Uchitel', *Kosmos kak predchuvstvie* [Space as Premonition], Mironov, Evgenii; Pegova, Irina; Tsyganov, Evgenii; Liadova, Elena (Rock Film Studio, 2005), 90 minutes. Uchitel' won the "Golden St. George" award at the Moscow International Film Festival in 2005 for this movie.

75 Aleksei German, *Bymazhnyi Soldat* (Lenfilm Studios, St. Petersburg, Russia, 2008), 118 minutes.

76 References to the female cosmonaut candidate as a girl and the nameless existence

of the dwarf could be poignant commentary on their marginality in Soviet society.

77 This scene was an homage to Sergei Eisenshtein's film version of the Nevskii story, completed in 1938 and withdrawn in 1939. In Fedorchenko's film, the invading dwarf Teutonic Knights bear the swastika-like crosses that Eisenshtein's attackers on Novgorod did. Sergei Eisenshtein, Aleksandr Nevskii, Cherkasov, Nikolai; Okhlopkov, Nikolai; Abrikosov, Andrei; Orlov, Dmitri; Novikov, Vasili (Mosfilm, 1938), 1:37 hour. This scene is one of many allusions to Soviet films in the movie. At one point, the cosmonauts go to see Zhuravlev's *Kosmicheskii reis* during their training.

78 Sergei Eisenshtein produced the 1938 film by the same name in which the 13th century Russian hero fought Teutonic knights who wore regalia that closely resembled swastikas.

79 Katerina Clark, "Aleksei Uchitel', Dreaming of Space [Kosmos kak predchuvstvie] (2005)," *KinoKultura*, October 2005, 10 September 2005, 08/11/2006, http://www.kinokultura.com/reviews/R10-05kosmos.html.

80 Bulat Shalvovich Okudzhava (Okudjava) was born on 9 May 1924, in Moscow, Soviet Union. His father, named Shalva Okudjava, was a ranking member of the Communist Party in Tbilisi, Georgia; he came to Moscow for a career but was arrested and executed in 1937, during the "Great Terror" under the dictatorship of Joseph Stalin. His mother was imprisoned in the Gulag camp for 18 years. Young Okudzhava was raised by his grandmother in Tbilisi. He volunteered for the Red Army during the Second World War and served in the frontlines as infantry; he was severely wounded and decommissioned in 1944. In 1950 he graduated from the Philological Department of the Tbilisi State University. Okudzhava was not allowed to return to live in Moscow until after the death of Joseph Stalin. From 1950 to 1956 he worked as a schoolteacher in the village of Shamorodino, and then in the town of Kaluga, Russia. There he published his first collection of poetry in 1956. Okudzhava returned to Moscow in 1956, and gradually developed a reputation as an independent poet, a free thinker. In 1959 he published his second collection, *Ostrova* [Islands]. Initially, Okudzhava was not as widely recognized as his contemporaries Yevgeni Yevtushenko, Bella Akhmadulina, Andrei Voznesensky, Robert Rozhdestvensky, Vasili Aksyonov, and others who would be later called the '60s generation. Okudzhava picked up a guitar and earned popularity as a singer-songwriter during the cultural Thaw which was initiated by Nikita Khrushchev. He wrote beautiful lyrics for the favorite film *Beloe solntse pustyni* [The White Sun of the Desert], 1970.

81 Translation from the Russian by Michael Slager. Lyrics Translate. https;//lyricstranslate.com/en/bumazhniy-soldat-paper-soldier.html.   30   September   2018.

82 Ekaterina Danilova, "Na meste Stalina postavili raketu" [A Rocket Replaced Stalin], *Ogonek* (Moscow), 12 March 2007, 24.

83 State funding stopped in 1991, and in 1992 Yeltsin changed the name to the Vserossisskii vystavochnyi tsentr (All-Russian Exhibition Center, VVTs) Benjamin

Forest and Juliet Johnson, "Unraveling the Threads of History: Soviet-Era Monuments and Post-Soviet National Identity in Moscow," *Annals of the Association of American Geographers* 92, no. 3 (2002), 535.

84 Gambrell, "The Wonder of the Soviet World," 33. Sergei Leskov and Iurii Snegirev, "Sud nad VDNKh zakonchilsia" [The Jury Is in on VDNKh], *Izvestiia* (Moscow), 13 March 1997, No. 047, 1.

85 Leskov and Snegirev, "Sud nad VDNKh zakonchilsia" [The Jury Is in on VDNKh], 1.

86 The repossession of objects and exhibits from the Kosmos pavilion was not an orderly process. Within a few years, commentators were uncertain as to which objects returned to their origin of manufacture, which were sold at auction and which remained on the premises at VDNKh/VVTs. Boris Ustiugov, "Na VVTs nashli *Sputnik* i partbilet Gagarin" [They Found *Sputnik* and Gagarin's Party Card at the All-Russian Exhibition Center], *Izvestiia* (Moscow), 16 June 2004, No. 105, 11 and 12.

87 Energiia now publicly advertises its museum for paid admission. The museum features Gagarin's original *Vostok* spacecraft and Leonov's *Voskhod,* spacecraft among other hardware that previously only specialists could see. Also, among the displays is the mockup of the *Apollo-Soyuz* test project that had resided at the Kosmos Pavilion of VDNKh for over twenty years. Access for a foreigner is difficult, requiring advance clearance through Energiia's security office. S. P. Korolev Rocket and Space Corporation Energia. "The Museum," 26 October 2007, http://www.energia.ru/english/energia/history/museum/museum.html; Pirard, "The Space Museum at RKK Energia," 247–53.

88 Cathleen S. Lewis, "The Rise, Fall and Unlikely Rebirth of VDNKh and the Buran Shuttle, 1988–2014" in *Scientific Heritage at World Exhibitions and Beyond. The Long XX Century,* Marco Beretta, Elena Canadelli, Laura Ronzon, eds. SISP, 2019: 182–202.

89 "Vladimir Putin Visiting Cosmos Pavilion at VDNKh exhibition on Cosmonautics Day," Vladimir Putin visited the renovated Cosmos pavilion at the VDNKh exhibition. 12 April 2018, http://en.kremlin.ru/events/president/news/57245.

90 Specter, "Moscow on the Make," 48–56.

91 Boym, "Kosmos: Remembrances of the Future," 49.

92 Boym, *The Future of Nostalgia,* 346.

Chapter 8. Epilogue

1 Yuri's Night, official webpage. https://yurisnight.net. Santa Rosa, California, 95403, Legal Designed on Earth. Made with star stuff. Rock the planet. "Yuri's Night is a celebration of the power of space to bring the world together." 18 August 2016.

2 The 1961 musical piece has music and lyrics by composer and science-fiction poet Oleg Aleksandrovich Sokolov-Tobolsky. Popular baritone Eduard Labkovskii sings the song with the backing of the civilian Soviet Song orchestra. The title is a

misnomer as it is a musical break from the traditional celebratory songs that retell heroic battles of previous generations. The march differs from previous patriotic songs dedicated to aviation. It has neither a civil defense nor a military theme. "Gagarin's March" does not speak of wartime but paints a picture of a bright and enthusiastic trek into the Soviet future with Yuri Gagarin at the lead. In this case, Gagarin is leading the homeland to a new optimistic world.

3 Both films focus exclusively on Gagarin and Leonov's respective times in space. Pavel Parkhomenko, dir., *Gagarin: First in Space* [*Gagarin. Pervyi v kosmose*]. Kremlin Films, 2013, and Dmitryi Kiselev, *Spacewalk* [*Vremya pervykh*] Bazelevs Production, 2017.

Bibliography

Primary Sources

Archival Sources

Institut mediko-biologicheskikh problem [Institute for Biomedical Problems], (Khimki, Russia). Museum archives.

National Aeronautics and Space Administration (Washington, DC) NASA History Office (RG16580), U.S. Space Park, New York World's Fair, 1964–65.

National Archives and Records Administration (College Park, Maryland) Central Intelligence Agency (RG 263) Project Corona Images.

National Archives and Records Administration—Southwest Region (Ft. Worth, Texas) National Aeronautics and Space Administration. Lyndon B. Johnson Space Center (RG 25516580) Project Mercury Files.

Raketno-kosmicheskaia korporatsiia imeni S. P. Koroleva [S. P. Korolev Energiia Rocket and Space Corporation (legacy corporation of Korolev's design bureau)], (Korolev, Russia).

Rossiiskii gosudarstvennyi arkhiv nauchno-tekhnicheskoi dokumentatsii [Russian State Archive for Scientific and Technical Documentation], (Moscow, Russia). Kosmos Pavilion files.

Rossiskii gosudarstvennyi nauchno-issledovatel'skii ispitael'nyi tsentr podgotovki kosmonavtov [Russian State Scientific-Research Test Spaceflight Training Center], (Star City, Russia) Museum and museum archives.

Smithsonian Institution. Smithsonian Institution Archives (Washington, DC) Office of the Secretary, Record Group 50, 190.

University of Chicago Library (Chicago, Illinois). Department of Special Collections. John A. Simpson Papers.

University of Washington Libraries. Special Collections Division (Seattle, Washington) John Glenn's *Friendship 7* in Seattle. Photographs.

Zvezda, otkrytoe aktsionernoe obshchestvo nauchno-proizvodstvennoe predpriiatie [Open Stock Society (LLP), Research and Development Production Enterprise "Zvezda"], (Tomilino, Russia) Museum.

Diaries, Memoirs, Notebooks, and Oral Histories

Alpert, Yakov. Interview by David H. DeVorkin, November 6, 1988, National Air and Space Museum, Washington, DC.

Chertok, Boris E. 1945, 1958–1988. Diaries and Notebooks. Washington, DC, Smithsonian National Air and Space Museum.

———. *Rakety i liudi* [Rockets and People]. Moscow: Mashinostroenie, 1995.

———. *Rakety i liudi: fili Podlipki Tiuratam* [Rockets and People: Beginnings at Podlipki and Tiuratam]. Moscow: Mashinostroenie, 1996.

———. *Rakety i liudi: goriachie dni kholodnoi voiny* [Rockets and People: The Hot Days of the Cold War]. Moscow: Mashinostroenie, 1997.

———. *Rakety i liudi: lunnaia gonka* [Rockets and People: The Moon Race]. Moscow: Mashinostroenie, 1999.

Feoktistov, Konstantin. 1958–1959. Notebooks. Denver, Colorado, Stephenson Family.

Filin, V. M. *Vospominaniya o lunnon korable* [Memoirs about the Lunar Spacecraft]. Moscow: Izdatel'stvo Kultura, 1992.

Gagarin, Iurii Alekseevich. *Doroga v kosmos* [The Road to Space]. Moscow: Detskaia Literatura, 1978.

———. *Est' Plamia! Stat'i, rechi, pis'ma, interv'iu* [Here Is the Flame! Essays, Speeches, Letters, Interviews]. Moscow: Molodaia Gvardiia, 1968.

———. *Psikhologiia i kosmos* [Psychology and Space]. English translation by Boris Belitsky. Moscow: Mir Publishers, 1970.

Gazenko, Oleg. "Soviet Space Medicine, Session Two, Museum Gallery Tour." Oral history interview, ed. Cathleen Lewis. 9551. Institute for Biomedical Problems, vol. transcript. Moscow, USSR, 28 November 1989. Smithsonian Videohistory Program.

Gazenko, Oleg, Abraham Genin, and Evgenii Shepelev. "Soviet Space Medicine, Session One, Group Interview." Oral history interview, ed. Cathleen S. Lewis. 9551. Institute for Biomedical Problems, vol. transcript. Moscow, USSR, 28 November 1989. Smithsonian Videohistory Program.

Genin, Abraham. "Soviet Space Medicine, Session Three, Museum Gallery Tour." Oral history interview. 9551. Institute for Biomedical Problems, vol. transcript. Moscow, USSR, 29 November 1989. Smithsonian Videohistory Program.

———. *Traektoriia zhizhi: Mezhdu vchera i zavtra* [The Trajectory of Life: Between Yesterday and Tomorrow]. Moscow: Vagrius, 2000.

Kamanin, Nikolai Petrovich. *Moia biografiia tol'ko nachinaetsia* [My Biography Has Only Begun], 1935.

———. *Skrytyi kosmos. Kniga chetvertaia* [Secret Space. Book Four]. Moscow: Infotekst, 1997.

———. *Skrytyi kosmos. Kniga pervaia: 1960–1963* [Secret Space. Book One: 1960–1963]. Moscow: Infortekst—IF, 1995.

———. *Skrytyi kosmos. Kniga tret'ia: 1967–1968 gg* [Secret Space. Book Three: 1967–1968]. Moscow: Novosti Kosmonavtiki, 1999.

———. *Skrytyi kosmos. Kniga vtoraia: 1964–1966 gg* [Secret Space. Book Two: 1964–1966]. Moscow: Infortekst, 1997.

Leonov, Aleksei A., Andrei Sokolov, and I. A. Golovanov. *Chelovek i vselennaia* [Man and the Universe]. Moscow: Izobrazitel'noe iskusstvo, 1976.

Mishin, Vasili P. [Interview]. "Designer Mishin Speaks on Early Soviet Space Programmes and the Manned Lunar Project." *Spaceflight* 32 (1990): 104–06.

———. Diaries. 1960–1974. Private Diaries and Notebooks. Dallas, Texas, Perot Foundation.

Mishin, Vasili P., and N. I. Panitskiy. *Osnovy aviatsionnoy i rakety* [The Foundations of Aviation and Rocketry]. Moscow: MAI, 1998.

Nikolaev, Adriian Grigor'evich. *Vstretimsia na orbite* [We Will Meet in Orbit]. Moscow: Voennoe Isdatel'stvo ministerstva oborony SSSR, 1966.

Ponomareva, Valentina Leonidovna, Moscow. "Roman s kosmonavtikoi" [A Novel with Spaceflight] 1995 (unpublished memoir).

———. *Zhenskoe litso kosmosa* [The Women's Face of Space]. Edited by L. V. Golovanov. Moscow: Gelios, 2002.

Raketno-kosmicheskaia korporatsiia "Energiia" imeni S. P. Koroleva [S. P. Korolev Energiia Rocket and Space Corporation]. Edited by Iurii Pavlovich Semenov. Moscow: Raketno-kosmicheskaia korporatsiia "Energiia" imeni S.P. Koroleva, 1996.

Raketno-kosmicheskaia korporatsiia "Energiia" imeni S.P. Koroleva na rubezhe dvukh vekov, 1996–2001 [S. P. Korolev Energiia Rocket and Space Corporation on the Edge of Two Centuries, 1996–2001]. Edited by Iurii Pavlovich Semenov. Korolev: Raketno-kosmicheskaia korporatsiia Energiia im. S.P. Koroleva, 2001.

Scott, David, and Alexei Leonov. *The Other Side of the Moon: Our Story of the Cold War Space Race*. New York: Thomas Dunn Books, 2004.

Shepelev, Evgenii. "Soviet Space Medicine, Audio Interview with Evgenii Shepelev." Oral history interview. 9551. Institute for Biomedical Problems, vol. transcript. Moscow, USSR, 30 November 1989. Smithsonian Videohistory Program.

"Soviet Space Medicine, Visual Tour." Oral history interview. 9551. Institute for Biomedical Problems, vol. transcript. Moscow, USSR, 30 November 1989. Smithsonian Videohistory Program.

Titov, German. *Golubaia moia planeta* [My Blue Planet]. Moscow: Voenizdat, 1973.

———. *Na zvezdnykh i zemnykh orbitakh* [On Starry and Earth Orbits]. Moscow: Detskaia literatura, 1987.

Titov, Gherman. *Gherman Titov: First Man to Spend a Day in Space*. Documents of Current History, vol. 21. New York: Crosscurrents Press, 1962.

Titov, Gherman, and Martin Caidin. *I Am Eagle!* Indianapolis: Bobbs-Merrill Co., 1962.

Titov, Herman. *700,000 Kilometres through Space: Notes by Soviet Cosmonaut No. 2*. Translated by R. Daglish. Edited by Nikolai Kamanin. Moscow: Foreign Languages Publishing House.

Cosmonaut Biographies

Adzhubei, Aleksei, and Azizian Ateik Kegamovich. *Utro kosmicheskoi ery* [The Dawn of Spaceflight]. Moscow: Gos Polit, 1961.

Avdeev, Aleksei. *Na zemli on takoi* [On Earth He Is]. Moscow: Detskaia literatura, 1983.

Beregovoi, Georgii Timofeevich. *Nebo nachinaetsia na zemle* [The Sky Begins on Earth]. Moscow: Voenizdat, 1976.

———. *Tri vysoty* [Three Altitudes]. Moscow: Voenizdat, 1986.

Berezovoi, A. N., et al. *S dumoi o zemle* [With the Thought of the Earth]. Moscow: Politizdat, 1987.

Burchett, Wilfred G. *Cosmonaut Yuri Gagarin, First Man in Space*. London: A. Gibbs & Phillips, 1961.

Burchett, Wilfred G., and Anthony Purdy. *Gherman Titov's Flight into Space*. London: Hamilton, 1962.

Dikhtiar', A. *Prezhde chem prozvuchalo: "Poekhali"* [Before Anyone Said, "Let's Go"]. Moscow: Politizdat, 1987.

Gagarin, Valentin Alekseevich. *Moi Brat Iurii: Povest* [My Brother Yuri: A Story]. Moscow: Moskovskii Rabochii, 1979.

———. *My Brother Yuri: Pages from the Life of the First Cosmonaut*. Moscow: Progress, 1973.

Gagarina, Anna Timofeeva. *Slovo o Syna* [A Word About My Son]. Moscow, USSR: Molodiia gvardiia, 1983.

Golovanov, Iaroslav Kirilovich. *Our Gagarin*. Moscow: Progress Publishers, 1978.

Gol'tsev, V., comp, and Dmitrii Fedorovich Mamleev, comp. *Sem'sot tysiach kilometrov v kosmose* [Six Hundred Thousand Kilometers in Space]. 1961.

Gorshkov, Valentin Sergeevich. *My—deti zemli* [We Are the Children of Earth]. Leningrad: Lenizdat, 1986.

Grib, V. F. *Pervye v mire* [The First in the World]. Moscow: Planeta, 1987.

Gubarev, Vladimir S. *Vek kosmosa: stranitsy letopisi* [The Era of Space: Pages of a Chronicle]. Moscow: Sovetskii pisatel', 1985.

Ilina, L., V. B. Grekhov, and IA. A. Zuperman. *Ulybka Gagarina literaturnyi sbornik* [Gagarin's Smile, A Literary Collection]. Samara: Izd-vo "Samarskii dom pechati," 2001.

Ishlinskii, A. Iu., ed. "A. G. Nikolaev," 482–85. Moscow: Nauka, 1987.

———. *Akademik S.P. Korolev: Uchenyi, inzhener, chelovek. Tvorcheskii portret po vospominanaiiam sovremennikov* [Academician S. P. Korolev: Scientist, Engineer, Man: A Creative Portrait based on the memories of contemporaries]. Moscow: Nauka, 1987.

Kamanin, Nikolai Petrovich. *Letchiki i kosmonavty* [Pilots and Cosmonauts]. Moscow, USSR: Politizdat, 1971.

———. *Pervyi grazhdanin vselennoi* [The First Citizen of the Universe]. Moscow: n.p., 1962.

———. *Semero na orbite* [Seven in Orbit]. Moscow: n.p., 1969.

———. *Sotvori sebia* [Conduct of Life]. Moscow: Molodaia gvardiia, 1982.

———. *Starty v nebo* [Launch into the Sky]. Moscow: n.p., 1976.

Kokhov, A. *Nasha "chaika"* [Our "Seagull"]. Moscow: Izogiz, 1963.

Komolov, Vadim, and Iurii Dokuchev. *Kosmonavt i ego rodina* [The Cosmonaut and his Homeland]. Moscow: Izdatel'stvo agenstva pechati "Novosti," 1967.

Koptev, IU. I., and S. A. Nikitin. *Kosmos: sbornik* [Space: A Collection]. Leningrad: Detskaia literatura, 1987.

Kotysh, Nikolai Timofeevich. *Zhdite nas, zvezdy* [Wait for Us, Stars]. Moscow: n.p., 1962.

Kramarov, Grigorii Moiseevich. *Pervoe v mire obshchestvo kosmonavtiki* [The First Society of Cosmonauts]. Moscow: n.p., 1962.

Kuznetskii, M. I. *Gagarin na kosmodrome Baikonur* [Gagarin at the Baikonur Cosmodrome]. Krasnoznamensk: Reklamno-izdatel'skii dom "VLADI," 2001.

———. *Titov: vtoroi kosmonavt planety* [Titov: The Second Cosmonaut]. Krasnoznamensk: Vladi, 2005.

Markelova, Larisa Pavlovna. *Vsegda v polete* [Forever in Flight]. Moscow: n.p., 1977.

Mitroshenkov, Viktor. *Zemliia pod nebom: khronika zhizni Iuriia Gagarina* [The Earth Under the Skies: A Chronicle of the Life of Yuri Gagarin]. 2nd edition. Moscow: Sovetskaia Rossiia, 1987.

Murav'ev, A. I., and R. P. Smirnova, comp. *Novyi polet v kosmos* [The New Flight into Space]. Moscow: n.p., 1961.

Nechaiuk, P. L. *Den gagarina* [The Day of Gagarin]. Moscow: "Sovremennik," 1986.

Nestorova, Valentina Fedorovna, Nikolai Aleksandrovich Kuz'michev, Oleg Aleksandrovich Mikhailov, and Ivan Grigorevich Borisenko. Gosudarstvennyi arkhivnyi fond SSSR. *Iurii Gagarin fotodokumenty gosudarstvennogo arkhivnogo fonda SSSR* [Yuri Gagarin Archival Photographic Documents from the State Archive Collections of the USSR]. Moscow: Izd-vo "Planeta," 1986.

Obukhova, Lidiia Alekseevna. *Liubimets veka* [The Favorite of the Century], 1977.

———. *Zvezdnyi syn zemli* [The Starry Son of Earth], 1974.

Petrov, E. A. *Kosmonavty* [Cosmonauts], 1962.

Pochta kosmonavtov [Cosmonaut Post Office]. Moscow: Sovetskaia rossiia, 1970.

Popov, Leonid Ivanovich, I. I. Kas'ian, and N. A. Kuz'michev. *Chetvero iz kosmicheskoi sem'i* [Four of the Space Family]. Moscow: Mashinostroenie, 1991.

Popovich, Pavel Romanovich. *Beskonechnye dorogi vselennoi* [The Never-Ending Road to the Universe]. Moscow: Sovetskii pisatel', 1985.

Prishchepa, Vladimir Iosifovich. *20 Let Poletu Gagarina Sbornik Statei* [Twenty Years Since the Flight of Gagarin, a Collection of Essays]. Moscow: Izd-vo "Znanie," 1981.

Rebrov, Mikhail Fedorovich. *Kosmonavty* [Cosmonauts], 1977.

Sharpe, Mitchell R. *Yuri Gagarin; First Man in Space*. [Huntsville, Ala.]: Strode Publishers, [1969].

———. *"It is I, Sea Gull," Valentina Tereshkova, First Woman in Space*. New York: Crowell, 1975.

Sokolov, Vasilii Dmitrievich, and R. P. Smirnova. *Snova k zvezdam!* [Once Again to the Stars!]. 1964.

Stepanov, Viktor. *Iurii Gagarin. Zhizn' zamechatel'nykh liudei* [Yuri Gagarin. The Lives of Notable People]. Moscow: Molodaia gvardiia, 1987.

Suvorov, Vladimir. *The First Manned Spaceflight: Russia's Quest for Space*. Edited by Alexander Sabelnikov. Commack, NY: Nova Science Publishers, 1997.

Umanskii, S. P. *Chelovek na kosmicheskoi orbite* [Man in Space Orbit]. Moscow: Mashinostroenie, 1974.

Usachev, Iurii Vladimirovich. *Dnevnik kosmonavta* [Diary of a Cosmonaut]. Moscow: Geleos, 2004.

Ustinov, Iurii S. *Bessmertie Gagarina* [Eternal Gagarin]. Moscow: Geroi otechestva, 2004.

———. *Kosmonavt No. 3 Andriian Nikolaev* [Cosmonaut No. 3 Adriian Nikolaev]. Moscow: "Geroi otechestva," 2004.

Yuri Gagarin: The First Cosmonaut. Moscow: Novosti Press Agency Pub. House, 1977.

Zaliubovskaia, Mariia. *Syn zemli i zvezd: liricheskaia povest' o Gagarin* [Son of the Earth and the Stars: A Lyrical Story about Gagarin]. Second edition. Kiev: Izdatel'stvo TsK LKSMU, 1984.

Newspapers

Chicago Tribune
Christian Science Monitor
Izvestiia
Komsomolskaia pravda
Literatura i zhizn'
Literaturnaia gazeta
Moscow Times
New York Times
Pravda
Washington Post
Washington Post and Times Herald

Periodicals

Aviatsiia i kosmonavtika
Filateliia sssr
Krylia rodiny
Life magazine
The Numismatist
Ogonek
Pioner
Priroda
Smena
Sovetskii kollektsionner
Spaceflight
Time

Science and Spaceflight Fiction

Aitmatov, Chingiz. *The Day Lasts More Than a Hundred Years.* Translated by John French. Bloomington: Indiana University Press, 1988.

Bogdanov, Aleksandr (Aleksandr Aleksandrovich Malinovsky). *Inzhener menni* [The Engineer Menni]. Moscow: Self-published, 1913.

———. *Krasnaia Zvezda. Roman—utopiia* [Red Star. A Utopian Novel]. St. Petersburg: Self-published, 1908.

——. *Red Star: The First Bolshevik Utopia.* Translated by Charles Rougle. Edited by Loren Graham and Richard Stites. Soviet History, Politics, Society, and Thought. Bloomington: Indiana University Press, 1984.

Efremov, Ivan Antonovich. *Andromeda Nebula.* Moscow, USSR: Foreign Languages Publishers House, 1957.

Pelevin, Viktor. *Omon Ra.* Moscow, Russia: Vagrius, 2001.

Tolstoy, Aleksey Nikolayevich. *Aelita.* Translated by Antonnia W. Bouis. Edited by Theodore Sturgeon. Macmillan's Best of Science Fiction. New York: Macmillan, 1981.

Tsiolkovskii, Konstantin Eduardovich. *Put' k zvezdam. Sbornik nauchno-fantasticheskikh proivedenii* [The Road to Space. A Collection of Fantasy Works]. Moscow: Izdatel'stvo Akademii Nauka SSSR, 1960.

Films

Boyarsky, Iosif, and Ivan Ivanov-Vano. *Letaiushchii Proletarii [The Flying Proletarian].* Aminmated. Soyuzmultfilm, USSR, 1962. 16 min. 22 sec.

Columbia Broadcasting System. *Meeting of the Astronauts* [Motion Picture]. [n.p.]: CBS News, 1962.

Erin, A. *Polet k Tysiacham Solnts [Flight to a Thousand Suns].* Birtsev, V; Ess, Misha. Lennauchfilm, 1963. 120 min.

Fedorchenko, Aleksei. *Pervye na lune* [First on the Moon] Vlasov, Boris; Slavnin, Aleksei; Osipov, Andrei; Otradnov, Anatolii; Ilinskaia, Viktoriia. Sverdlovsk Film Studio and Film Company Strana, 2005. 75 min.

German, Aleksei, dir. *Bymazhnyi Soldat* [Paper Soldier]. Lenfilm Studios, St. Petersburg, Russia, 2008. 118 min.

Kara, Yuri. *Korolev.* Sergei Astakhov, Natalya Fateeva, Viktoriia Toltogavnova. L.S. D., Russia, 2008. 130 minutes.

Kariukov, Mikhail, and A. Kozyr'. *Nebo zovet* [The Sky Calls]. Pereverzev, Ivan; Shvorin, Aleksandr; Bartashevich, Konstantin; Borisenko, Larisa; Chernyak, V.; Dobrovolsky, Viktor. Gosudarstvenii komitet po kinematografii (Goskino), 1960. 77 min.

Kiselyov, Dmitry. *Vremia Pervykh* [Spacewalk]. Yevgeny Mironov, Konstantin Khabensky, Vladimir Ilyin. Bazelevs, 2017. 140 min.

Klushantsev, Pavel. *Doroga k zvezdam* [Road to the Stars]. Lennauchfilm, 1957. 52 min.

——. *Luna.* Lennauchfilm, 1965.

——. *Mars.* Lennauchfilm, 1965–1968.

——. *Planeta bur'* [Planet of Storms]. Emel'ianov, V.; Sarantsev, Iu.; Zhzhenov, G.; Ignatova, K.: Vernov, G.; Teikh, G. Leningrad Popular Science Film Studio, 1962. 83 min.

Klushantsev, Pavel, and N. Leshchenko. *Vselennaia* [The Universe]. Lennauchfilm, 1951.

Maetzig, Kurt. *Der schweigende stern* [The Silent Star], [Bezmolvnaia zvezda]. Tani, Yoko; Lukes, Oldrich; Machowski, Igancy; Ongewe, Julius. Deutsche Film (DEFA), 1959. 155 min.

Motyl, Vladimir. *Beloe Solntse Pustyni* [White Sun of the Desert]. Valentin Ezhov, Pavel Luspekaev, Spartak Mishulin. Lenfilm, 1970. 84 min.

Parkhomenko, Pavel. *Gagarin. Pervyy v Kosmose* [Gagarin: First in Space]. Yaroslave
Zhalnin, Mikhail Filippov, Vladimir Steklov. Kremlin Films, 2013. 108 min.

Sal'nikov, Iurii, Director. *Iurii Gagarin*. Film. Moscow: Ekran, 1969.

Shestobitov, E. *Tumannost' andromedy (Chast' i. pleniki zheleznoi zvezdy)* [The Andromeda Nebula]. K/st. im. A Dovzhenko, 1967. 97 min.

Shipenko, Klim, dir., *Salyut 7*. CTB Films, Globus-film, Lemon Films Studio, Telekanal
Rossiya, 2017.

Uchitel', Aleksei. *Kosmos kak predchuvstvie* [Space as Premonition]. Mironov, Evgenii;
Pegova, Irina; Tsyganov, Evgenii; Liadova, Elena. Rock Film Studio, 2005. 90 min.

Secondary Sources

A Note on Secondary Sources and Their Role in This Book

The opening of archives and increased availability of sources and materials have not been the only changes in the field of Soviet and Russian space history. In the twenty-first century, the history of science and technology broadened general-ly, which for space historians led to a shift from the internal workings of a space program to examinations that include the methods of intellectual, cultural, and social history. One of the first scholars to do this was James Andrews, whose bi-ography of Soviet pre-spaceflight legend Konstantin Tsiolkovsky challenged the iconography of a legend. Asif Siddiqi in *The Rockets' Red Glare* has gone further to dismantle the Soviet legends about Tsiolkovsky that persist in modern Rus-sia. Others, such as Slava Gerovitch, have applied the methodology of the social construction of technology to improve the understanding of the functioning of organizations that executed the first human spaceflights. Andrew Jenks applied his understanding of Russian and Soviet aesthetic iconography to dissect the life and legend of the first man in space, Yuri Gagarin. These new histories of the space program utilized the existing methodologies within the Soviet stud-ies field in ways that others had not done before. One notable use was Sheila Fitzpatrick's study of the changed rules of everyday life in revolutionary Rus-sia, which Gerovitch has applied to the space community to understand the behavior of those involved. He has produced not one, but two volumes and the subject. The first was collected interviews from the space program and the second an analysis of the contrast between the public and private lives of those individuals. Similarly, Asif Siddiqi has produced a social history of the nascent space program in *The Rockets' Red Glare*. He has examined the interactions be-tween the cultural and institutional origins of human spaceflight in Russia and the Soviet Union, seeking insight to influences that literature and the arts had on the early planners who produced *Sputnik* and Gagarin's flight.

This new scholarship has not only focused solely on the Soviet space program. European and American scholars have recently taken the opportunity to reexamine the global experience of the early space age through a series of conferences. These conferences have generated collected essays on a wide range of topics on European and American programs, providing new perspectives on the early years of the space age that transcend the Cold War paradigm. Scholars have published collected articles, often from international conferences that looked at US and Soviet experiences as well as the impact of the Space Race on Europe by incorporating the history of art, popular culture, and gender studies into the field. These new studies have broadened our understanding of the impact of the space age on society and culture, touching the everyday life of global populations and spurred new monographs. The most recent addition is an expansion of an article in *Soviet Space Culture*, Iina Kohonen's study of the visual imagery of the early Soviet space program. I rely heavily on works such as these to identify the creators and audiences of these works and aspire to build upon those changes that have taken place in the field, and seek to apply and expand them to cover more than a century of history of Russian and Soviet cosmonauts.

One striking feature of these new histories has been the documentation of the existence of fatally flawed founding myths of the heroic Soviet space program. In his book, *Voices of the Soviet Space Program*, Gerovitch demonstrated the extent to which lies about basic facts became ingrained in the culture of the space program, from the earliest times. On the one hand lying served the purpose of maintaining state secrets; on the other hand, it fostered cynicism that would ultimately encourage those who would degrade the image of the Russian cosmonaut. This culture of lying is but one flaw that was immanent in the founding of the Red Stuff that has contributed to its reassessment in recent years. Each of these flaws developed into a crack in the history and memory of the space program. This book examines the changes over decades in three areas from which first cracks appeared.

This book builds on existing scholarship of the history of the Soviet human spaceflight program by incorporating the emerging research in the field of late- and post-Soviet history. For example, Stephen Wagg and David Andrews's edited volume on sports and the Cold War examines the role of sports in late twentieth and early twenty-first century Soviet and Russian culture. Another field is the study of the culture of death, mourning, and World War II that intertwine in Soviet history, notably Catherine's Merridale's treatise on death and memory in the USSR. Of similar significance is Nina Tumarkin's dissection of the Soviet experience during the Great Patriotic War and the repeated reexaminations of the World War II experience. Others include Donald Raleigh's oral interview research on Soviet baby boomers, the generation most closely touched by the peak of Soviet space achievements, and Kathleen Smith's study of the creation of memory in late Soviet politics parallels the work that Gerovitch has done about the spaceflight community.

Cosmonaut also draws from the recent scholarship on the material culture, architecture, and film of the USSR and Russia and their changing roles in Russian society over a century. The Soviet state franchised the expression of the Red Stuff in three ways—through collectibles, the erection of monuments and museums, and film. Each of these fields was significant in three ways. First, they all began as state-sanctioned popularization of a vision of the Soviet cosmonaut. Each started with an official message within its medium of an official vision. Collectibles such as stamps and pins relied on official sanction to begin distribution and became a state-sanctioned portrayal of events. Architecture and design were closely monitored as politically charged aesthetic choices. Film in the USSR was one of the earliest arts that Lenin declared to be the "most important." The second common characteristic was the fact that each of these sectors began disseminating visions of the Red Stuff outside the traditional space industry. Thirdly, each distinguished itself in how it began to generate autonomous approaches to the image of the Russian cosmonaut.

Recent growth in the field of studying the material culture of the USSR has been of great utility to this book. Among them are Graham Robert's edited volume, *Material Culture in Russian and the USSR: Things, Values, Identities,* which includes chapters on the government's attempt to manipulate Soviet cultural life through things, and the successes and failures of that effort. Natalya Chernyshova's study of consumerism during the Brezhnev years examines the overlooked success and stability fostered in the guided consumerism of that time. Another perspective on material culture is found in Alexey Golubev's insightful work, *The Things of Life: Materiality in Late Soviet Russia,* which provocatively proposes that Soviet society manipulated itself through interaction with material culture and how the life cycle of artifacts of the era progressed in response. The Soviet and later Russian film industry emerged as an independent entity after generations of official guidance on subject matter and technique. Film historians such as Birgit Beumers have analyzed the divergence and continual links that the film industry has nurtured as Soviet power waned and Russian nationalism has risen in its place. All of the above-mentioned and many more monographs and articles have contributed to the broad methodological foundation of this book.

Some of the timeliest examinations of the processes of reevaluation of the post-Soviet experience have been journalistic efforts that explain the rapidly shifting reality that overtook the USSR in the 1980s and transformed the former Soviet Union afterward. First among these works is David Remnick's *Len-*

in's Tomb. His book recounts the rapid deterioration and sluggish responses from the USSR in its final years when Remnick was a foreign correspondent assigned to Moscow. During the period of the most intense national introspection, philosopher, and literary scholar, Svetlana Boym has used her personal life to guide readers through the collective and individual desire for fashioning comforting memories of the past through places. Recently, the *Guardian's* foreign correspondent Shaun Walker has written *The Long Hangover* to explain Vladimir Putin's effort to restore national confidence to Russia through encouraging the selective memory of international and domestic history. Alexei Yurchak's anthropological study of the last Soviet generation synthesizes the turbulent and contradictory influences that besieged the post-Stalin generation and their idiosyncratic responses to the paradoxes of everyday life. No study of the current situation in Russia would be complete without mentioning the prolific work of former Soviet journalist Masha Gessen. She has dedicated her career to the unmasking of Vladimir Putin. Her first book, *The Man without a Face*, is a portrait of the man before his rise to national power. Her insightful scrutiny of Putin in his earliest days provides clues as to his motivations and expectations of post-Soviet Russia. Gessen's recent book *The Future Is History: How Totalitarianism Reclaimed Russia* expounds on the current public culture of Russia and how Russians coped with the disappointments of unfulfilled expectations of the Soviet era.

The following bibliography of secondary sources combines the old with the new, connecting the traditional, technical analyses of the Soviet space program from times when the only available measures were from orbital tracking and public announcements to contemporary scholarship that approaches the twentieth- and twenty-first century Soviet history from more oblique angles.

Abramov, I. P., G. I. Severin, A. IU. Stoklitskii, R. K. Sharipov, et al. *Skafandry i sistemy dlia raboty v otkrytom kosmose* [Spacesuits and Systems for Work in Open Space]. Moscow: Mashinostroenie, 1984.

Adams, Mark B. "Red Star: Another Look at Aleksandr Bogdanov." *Slavic Review* 48, no. 1 (Spring 1989): 1–15.

Afanasyev, I. B. *Neizvestnie korabli* [Unknown Spacecraft]. Kosmonavtika, Astronomiya, Znanie, vol. 12–91. Moscow: Znanie, 1991.

Akademiia Nauka SSSR. Komissiia po pazpabotke nauchnogo naslediia pionerov osvoeniia kosmicheskogo prostranstva. *Materialy po istorii kosmicheskogo korabliu "Vostok" (K 30-Letniiu pervogo poleta cheloveka v kosmicheskoe prostranstvo)* [Materials on the History of the Spacecraft "Vostok" (Contributions to the 30th Anniversary of the First Human Spaceflight)]. Edited by Boris V. Raushenbakh. Moscow: Nauka, 1991. 214.

Alekseev, S.M., and S. P. Umanskii. *Vysotnye i kosmicheskie skafandry* [High Altitude Pressure and Spacesuits]. Moscow: Mashinostroenie, 1973.

Alekseev, V.A., et al. *Kosmicheskoe sotrudnichestvo* [Space Cooperation]. Moscow: Mashinostroenie, 1987.

Alekseeva, Liudmila, and Paul Goldberg. *The Thaw Generation: Coming of Age in the Post-Stalin Era*. Pittsburgh: University of Pittsburgh Press, 1993.

Alexander, Catherine. "The Factory: Fabricating the State." *Journal of Material Culture* 5, no. 2 (2000): 177–95.

Allen, George V. "Are the Soviets Winning the Propaganda War?" *Annals of the American Academy of Political and Social Science* 336 (July 1961): 1–11.

Allyn, Bruce J., James G. Blight, and David A. Welsh. *Back to the Brink: Proceedings of the Moscow Conference on the Cuban Missile Back to the Brink: Proceedings of the Moscow Conference on the Cuban Missile Crisis, January 27–28, 1989*. Edited by Georgy Shakhnazarov. CSIA Occasional Paper No. 9. Lanham, Md.: University Press of America, 27–28 January 1992.

———. "Essence of Revision." *International Security* 14, no. 3 (Winter 1989/90): 138–42.

Amalrik, A. *Will the Soviet Union Survive Until 1984*. New York: Harper and Row, 1970.

Anderson, Benedict. *Imagined Communities: Reflections on the Origin and Spread of Nationalism*. 1983. Revised edition. London: Verso, 1993.

Andrew, Christopher, and Vasili Mitokhin. *The World Was Going Our Way: The KGB and the Battle for the Third World*. New York: Basic Books, 2005.

Andrews, James T. "K. E. Tsiolkovskii, Ascribed Identity, and the Politics of Constructing Soviet Space Mythology, 1917–1957." In *Ascribed Identity, Soviet Mythology, and the Politics of Space Culture, 1917–Present*, chair Harley D. Balzer. American Association for the Advancement of Slavic Studies National Convention. Washington, DC, 16 November 2006.

———. *Science for the Masses: The Bolshevik State, Public Science, and the Popular Imagination in Soviet Russia, 1917–1934*. College Station, TX: Texas A&M University Press, 2003.

Andronnikov, N. G. "Les monuments aux heros de la grande guerre nationale" [Monuments to the Heroes of the Great Patriotic War]. *Revue International d'Histoire Militaire* 59 (1985): 292–304.

Androunas, Elena. "The Struggle for Control over Soviet Television." *Journal of Communications* 41, no. 2 (Spring 1991): 185–200.

Archeo-Biblio Base: Archives in Russia. February 1998. 26 June 1998 http://www.iisg.nl/~abb/index.html.

Arlazorov, Mikhail Saulovich. *Protazanov*. Moscow, USSR, 1973.

Atwill, William D. *Fire and Power: The American Space Program as Postmodern Narrative*. Athens: University of Georgia Press, 1994.

Aucoin, Amanda Wood. "Deconstructing the American Way of Life: Soviet Responses to Cultural Exchange and American Information Activity During the Khrushchev Years." PhD diss., History, University of Arkansas, 2001.

Baevskii, R. M. *Fiziologicheskie metody v kosmonavtike* [Physiological Methods in Space]. Moscow: Nauka, 1965.

Bagrov, Peter. "Ermler, Stalin, and Animation: On the Film 'The Peasants' (1934)." *KinoKultura*, 15 January 2007. http://www.kinokultura.com/2007/15-bagrov.shtml.

Baidukov, Georgii. *Chkalov.* Moscow: Molodaia gvardiia, 1991.

―――. *Over the North Pole.* Translated by Jessica Smith. New York: Harcourt, Brace and Company, 1938.

―――. *Russian Lindbergh: The Life of Valery Chkalov.* Translated by Peter Belov. Washington, DC: Smithsonian Institution, 1991.

Bailes, Kendall E. "Science, Philosophy and Politics in Soviet History: The Case of Vladimir Vernadskii." *Russian Review* 40, no. 3 (1981): 278–99.

―――. "Technology and Legitimacy: Soviet Aviation and Stalinism in the 1930s." *Technology and Culture* 17, no. 1 (January 1976): 55–81.

―――. *Technology and Society under Lenin and Stalin: Origins of the Soviet Technical Intelligentsia, 1917–1941.* Studies of the Russian Institute, Columbia University. Princeton, NJ: Princeton University Press, 1978.

Baker, Alex, Toby Kamps, and Svetlana Boym. *Space Is the Place.* Edited by Alex Baker and Toby Kamps. New York: Independent Curators International, 2006.

Baker, Mary T., and Ed McManus. "History, Care and Handling of America's Spacesuits: Problems in Modern Materials." *Journal of the American Institute of Conservation* 31 (1992): 77–85.

Baker, Norman L. *Soviet Space Log, 1957–1967.* Washington, DC: Space Publications, 1967. 59.

Balzer, Harley. "Education, Science and Technology." In *The Soviet Union Today,* ed. James Cracraft. Chicago: University of Chicago Press, 1988.

Baranova, M. P. *Rags, Borya and the Rocket; A Tale of Home-Less Dogs and How They Became Famous.* Translated by Anne Hansen. Moscow: Progress Publishers, 1964.

Barber, John, and Mark Harrison, eds. *The Soviet Defence-Industry Complex from Stalin to Khrushchev.* New York: St. Martin's Press, 2000.

Barghoorn, Frederick C. "Soviet Cultural Diplomacy since Stalin." *Russian Review* 17, no. 1 (January 1958): 41–55.

Barthes, Roland. *Mythologies.* Translated by Annette Lavers. New York: Hill and Wang, 1972.

Bartos, Adam. *Kosmos: A Portrait of the Russian Space Age.* Princeton: Princeton Architectural Press, 2001.

Basile, Giovanni Maniscalco. "The Utopia of Rebirth: Aleksandr Bogdanov's *Krasnaia zvezda.*" *Canadian-American Slavic Studies* 18, no. 1–2 (Spring-Summer 1981): 54–62.

Baudry, Patrick. *Le rêve spatial inachevé de Youri Gagarine au voyageur universel* [The Spaceflight Dream that Gagarin Did Not Achieve for Universal Space Travel]. Raconter l'Histoire. Paris: Tallandier, 2001.

Beliakov, Aleksandr Vasil'evich. *Dva pereleta* [Two Flights]. Moscow: n.p., 1939.

Belk, Russell. *Collecting in a Consumer Society.* London: Routledge, 1995.

Belotserkovskii, Sergei Mikhailovich. *Diplom Gagarina* [The Gagarin Diploma]. Moscow: Molodaia gvardiia, 1986.

―――. *Gibel' Gagarina: Fakty i domysly* [The Death of Gagarin: Facts and Conjectures]. Moscow: Mashinostroenie, 1992.

———. *Pervoprokhodtsy vselennoi: zemlia—kosmos—zemlia* [The First Travelers into Space: Earth—Space—Earth]. Moscow: Mashinostroenie, 1997.

Benson, Morton. "Russianisms in the American Press." *American Speech* 37, no. 1 (February 1962): 41–47.

Bentley, Eric. *Century of Hero-Worship.* Gloucester, MA: P. Smith, 1969.

Benvenuto, Francesco. *Stakhanovism and Stalinism.* CREES Discussion Papers. Soviet Industrialisation Project Series. Birmingham, England: Centre for Russian and East European Studies, University of Birmingham, 1989.

Berg, Raisa. *Acquired Traits. Memoirs of a Geneticist from the Soviet Union.* New York: Penguin Books, 1990.

Bergaust, Erik. *The Russians in Space.* New York: Putnam, 1969.

Bergman, Jay. "Valerii Chkalov: Soviet Pilot as New Soviet Man." *Journal of Contemporary History* 33, no. 1–2 (January 1998): 135–52.

Biatkin, Andrei. "Cherez ternii—k zruteliu" [From Thorns to Spectators]. *Mir fantastiki* 2, no. 2 (October 2003). http://www.mirf.ru/articles/art290.htm.

Biriukov, IU. V. *Materialy po istorii kosmicheskogo korablia "Vostok"* [Material on the History of the *Vostok* Spacecraft]. Edited by Boris V. Raushenbakh. Komissiia po razrabotke nauchnogo naslediia pionerov osvoeniia kosmicheskogo prostranstva. Moscow: Nauka, 1991.

Bittner, Stephen V. "Review of *Ideologicheskie komissii TsK KPSS, 1958–1964: Dokumenty.*" *Kritika: Explorations in Russian and Eurasian History* 3, no. 2 (2002): 356–361.

Blagonravov, A.A., ed. *U.S.S.R. Achievements in Space Research (First Decade in Space, 1957–1967).* Translated by Joint Publications Research Service. JPRS (Series): 47, 311. Washington, DC: Joint Publications Research Service, 1969. 615.

Blaine, James Cyril Dickson. *The End of an Era in Space Exploration: From International Rivalry to International Cooperation.* San Diego, CA: American Astronautical Society: distributed by Univelt, 1976.

Blom, Philipp. *To Have and to Hold: An Intimate History of Collectors and Collecting.* Woodstock: The Overlook Press, 2003.

Blum, Martin. "Remaking the East German Past: *Ostalgie,* Identity, and Material Culture." *Journal of Popular Culture* 34, no. 3 (Winter 2000): 229–53.

Bogdanov, Aleksandr (Aleksandr Aleksandrovich Malinovsky). *The Struggle for Viability: Collectivism Through Blood Exchange.* Edited and translated by Douglas W. Heustis. Philadelphia, PA: Xlibris Corp., 2001.

Bogdanov, Nikolai. "Literary Characters Influence Life of Soviet Children." *Journal of Educational Sociology* 35, no. 4 (December 1961): 162–64.

Bogomolov, Yuri. "The Revitalization of the Soviet Film Industry." *Journal of Communication* 41, no. 2 (Spring 1991): 39–45.

Bollinger, Martin J. *Stalin's Slave Ships: Kolyma, The Gulag Fleet and the Role of the West.* New York: Praeger, 2003.

Boltyanskii, G. M., ed. *Lenin i kino* [Lenin and film]. Moscow/Leningrad, 1925.

Bonnell, Victoria E. *Iconography of Power: Soviet Political Posters under Lenin and Stalin.* Berkeley: University of California, 1997.

Boorstin, Daniel J. *The Image: A Guide to Pseudo-Events in America.* New York: Vintage Books, 1992.

Borodina, Evgeniia. "Interv'iu (Interview)." *Metro,* 20 July 1999, 1.

Borsody, Stephen. *The Tragedy of Central Europe: Nazi and Soviet Conquest and Aftermath.* Revised edition. Yale Russian and East European Publications. New Haven: Yale Concilium on International and Area Studies, 1980.

Boss, V. *Newton and Russia: The Early Influence, 1698–1796.* Cambridge, MA: Harvard University Press, 1972.

Boym, Svetlana. *Common Places: Mythologies of Everyday Life in Russia* (Library of African Adventure; 3) (Cambridge, Harvard University Press, 1995).

———. *The Future of Nostalgia.* New York: Basic Books, 2001.

———. "Kosmos: Remembrances of the Future." In *Kosmos: A Portrait of the Russian Space Age,* 82–99. Princeton: Princeton Architectural Press, 2001.

———. "Nostalgia, Moscow Style: Authoritarian Postmodernism and Its Discontents." *Harvard Design Magazine,* no. 13 (Winter/Spring 2001): 1–8.

Bridger, Susan. "The Cold War and the Cosmos: Valentina Tereshkova and the First Woman's Space Flight." In *Women in the Khrushchev Era,* ed. Melanie Ilic, Susan E. Reid, and Lynne Attwood, 222–37. New York: Palgrave Macmillan, 2004.

Britikov, Anatolii F. *Russkii sovetskii nauchno-fantasticheskii roman* [The Russian Soviet Science Fantasy Novel]. Leningrad: Izdatel'stvo "nauka," 1970.

Brontman, Lazar. *The Heroic Flight of the Rodina.* Moscow: Foreign Languages Publishing House, 1938.

———. *Vladimir Kokkinaki.* Moscow: Voenizdat, 1939.

Brooks, Jeffrey. *Thank You, Comrade Stalin! Soviet Public Culture from Revolution to Cold War.* Princeton: Princeton University Press, 2000.

———. *When Russia Learned to Read. Literacy and Popular Literature, 1861–1917.* Princeton: Princeton University Press, 1985.

Buchli, Victor. "Khrushchev, Modernism, and the Fight against 'Petit-Bourgeois' Consciousness in the Soviet Home." *Journal of Design* 10, no. 2 (1997): 161–76.

Bulkeley, Rip. "Harbingers of Sputnik: The Amateur Radio Preparations in the Soviet Union." Paper presented at the Sputnik: Fortieth Anniversary Conference, October 1997.

———. *The Sputniks Crisis and Early United States Space Policy: A Critique of the Historiography of Space.* Bloomington: Indiana University Press, 1991.

Burdzhalov, E.N. *Russia's Second Revolution. The February 1917 Uprising in Petrograd.* Translated by D. J. Raleigh. Bloomington: Indiana University Press, 1987.

Bureau International des Expositions. 16 February 2007. http://www.bie-paris.org/main/index.php?lang=1.

Bushnell, John. "Urban Leisure Culture in Post-Stalin Russia: Stability as a Social Problem." In *Soviet Society and Culture, Essays in Honor of Vera S. Dunham,* ed. Terry L. Thompson and Richard Sheldon. Boulder, CO: Westview Press, 1988.

Bykov, L. T., M. S. Yegorov, and P. V. Tarasov. *High Altitude Aircraft Equipment.* Translated by W. J. Fiedler. Edited by O. J. Marstrand. New York: Pergamon Press, 1961.

Caidin, Martin. *Red Star in Space.* New York: Crowell-Collier Press, 1963.

Campbell, Joseph. *The Hero with a Thousand Faces*. Princeton, NJ: Princeton University, 1949.

Chaput, Patricia R. "Culture in Grammar." *Slavic and East European Journal* 41, no. 3/16/96 auction catalog (Autumn 1997): 403–14.

Chatterjee, Choi, Lisa A. Kirschenbaum, and Deborah A. Field. *Russia's Long Twentieth Century: Voices, Memories, Contested Perspectives*. London: Routledge, 2014.

Cheredina, Irina. "Na meste Stalina postavili raketu" [In Place of Stalin, They Placed a Rocket]. *Ogonek* (Moscow), 12–18 March 2007, online.

Chernigovskogo, V. N., ed. *Problemy kosmicheskoi biologii, tom xxv* [The Problems of Space Biology, Volume XXV]. Moscow: Nauka, 1974.

Chernyshova, Natalya. *Soviet Consumer Culture in the Brezhnev Era*. London: Routledge, 2013.

Chertovskoi, E.E. *Stratosfernyi skafandry* [Stratospheric Pressure Suits]. Moscow: Voenizdat, 1940.

———. *Stratostaty* [Stratospheric Ballons]. Moscow: ONTI, 1936.

Chkalov, Valerii Pavlovich. *Nash transpoliarnyi reis: Moskva-Severnyi Polius-Severnaia Amerika* [Our Transpolar Course: Moscow—The North Pole—North America]. Moscow: OGIZ, gosudarstvennoe izdatel'stvo politicheskoi literatury, 1938.

Christensen, Peter G. "Women as Princesses or Comrades: Ambivalence in Yakov Protazanov's 'Aelita' (1924)." *New Zealand Slavonic Journal* (2000): 107–22.

Christie, Ian. "Protazanov: A Timely Case for Treatment." *KinoKultura*, 9 July 2005. http://www.kinokultura.com/articles/jul05-christie.html.

Christie's East. *Space Exploration, Sale*. Auction, Wednesday, 9 May 2001. New York: Christie's, 2001.

Chu, Julian J. "Building the Future on the Past: The Expansion of Soviet Space." *Harvard International Review* 13, no. 1 (Fall 1990): 55–58.

Chugunova, Nina. "Cosmonauts Number Zero." *Bulletin of Atomic Scientists* 50, no. 3 (May/June 1994): 16.

Clark, Katerina. "Aleksei Uchitel', Dreaming of Space [Kosmos kak predchuvstvie] (2005)." *KinoKultura*, 10 September 2005. http://www.kinokultura.com/reviews/R10–05kosmos.html.

———. "Changing Image of Science and Technology in Soviet Literature." In *Science and the Soviet Social Order*, ed. Loren Graham. Program in Science, Technology, and Society, 259–98. Cambridge, MA: Harvard University Press, 1990.

———. "The Mutability of the Canon: Socialist Realism and Chingiz Aitmatov's I Dol'she Veka Dlitsia Den'." *Slavic Review* 43, no. 4 (Winter 1984): 573–87.

———. "Socialist Realism and the Sacralizing of Space" In *The Landscape of Stalinism. The Art and Ideology of Soviet Space*, ed. Evgeny Dobrenko and Eric Naiman, 3–18. Seattle: University of Washington Press, 2003.

———. *The Soviet Novel: History as Ritual*. Chicago: University of Chicago Press, 1981.

Collins, Martin J., and the Division of Space History. *Space Race: The U.S.-U.S.S.R. Competition to Reach the Moon*. San Francisco: Pomegranate, 1999.

Colomina, Beatriz, AnnMarie Brennan, and Jeannie Kim, eds. *Cold War Hothouses: In-*

venting Postwar Culture, from Cockpit to Playboy. New York: Princeton University Press, 2004.

Compton, W. David, and Charles D. Benson. *Living and Working in Space: A History of Skylab.* NASA SP-409. Washington, DC: NASA, 1983.

Condee, Nancy, and Vladimir Padunov. "A 'Reading Room' for Research on Russian Cinema." Review of *Chital'nyi zal: kritiko-bibliograficheskii zhurnal o kino* [The reading room: a critical bibliographical journal about film], ed. Aleksandr Troshin (Moscow: Eizenshteinovskii tsentr issledovanii kinokul'tury, 1995–). *The Russian Review* 57 (January 1998): 104–106.

Coopersmith, Jonathan. *The Electrification of Russia, 1880–1926.* Ithaca, N. Y.: Cornell University Press, 1992.

Corn, Joseph J. *The Winged Gospel: America's Romance with Aviation, 1900–1950.* New York: Oxford University Press, 1983.

Cox, Donald W. *The Space Race; From Sputnik to Apollo, and Beyond.* Philadelphia: Chilton Books, 1962.

Crouch, Tom D. *Aiming for the Stars: The Dreamers and Doers of the Space Age.* Washington, DC: Smithsonian Institution Press, 1999.

———. *A Dream of Wings: Americans and the Airplane, 1875–1905.* New York: Norton, 1981.

Daniloff, Nicholas. *The Kremlin and the Cosmos.* New York: Knopf, 1972.

Davydov, I. V. *Triumf i tragediia: sovetskoi kosmonavtiki* [Triumph and Tragedy: Soviet Spaceflight]. Moscow: Globus, 2000.

Day, Dwayne A. "Cover Stories and Hidden Agenda: Early American Space and National Security Policy." Paper presented at Reconsidering Sputnik: Forty Years since the Soviet Satellite. Washington, DC, 30 September–1 October 1997.

Derevianko, Iu. M. "Resultaty kliniko-fiziologicheskogo issledovania kosmonavta No. 1, Iu A. Gagarina v TsNIAGe" [Results of the Clinical and Physiological Investigations of Cosmonaut No. 1, Yuri A. Gagarin at TsNIAG]. Moscow, Russia, 1999.

DeVorkin, David H. *Science with a Vengeance: How the Military Created the US Space Sciences after World War II.* New York: Springer-Verlag, 1992.

Dobbs, Michael. "Can't Verify, Can't Trust." Review of *Khrushchev's Cold War: The Inside Story of an American Adversary,* by Aleksandr Fursenko and Timothy Naftali (New York: Norton, 2006). *The Washington Post* 1 February 2007: C3.

Dobrenko, Evgeny, and Eric Naiman, eds. *The Landscape of Stalinism: The Art and Ideology of Soviet Space.* Seattle: University of Washington Press, 2003.

Dobrynin, Anatoly Fedorovich. *In Confidence: Moscow's Ambassador to America's Six Cold War Presidents (1962–1986).* New York: Times Books (Random House), 1995.

Dorogi v Kosmos. Vospominaniia Veteranov Raketno-Kosmicheskoi Tekhniki i Kosmonavtiki, tom 1 [The Road to Space: Memoirs of the Veterans of Rocket Space Technology and Spaceflight, Vol. 1]. Moscow: Izdatel'stvo MAI, 1992.

Dorogi v Kosmos. Vospominaniia Veteranov Raketno-Kosmicheskoi Tekhniki i Kosmonavtiki, tom 2 [The Road to Space: Nemoirs of the Veterans of Rocket Space Technology and Spaceflight, vol. 2]. Moscow: Izdatel'stvo MAI, 1992.

Du Quenoy, Paul. "The Role of Foreign Affairs in the Fall of Nikita Khrushchev in October 1964." *International History Review* (Canada) 25, no. 2 (June 2003): 253–504.

Dunham, Vera. *In Stalin's Time: Middleclass Values in Soviet Fiction*. Cambridge, UK: Cambridge University Press, 1976.

Dunlop, John B. *The Faces of Contemporary Russian Nationalism*. Princeton: Princeton University Press, 1983.

Dyson, Freeman. *Imagined Worlds*. Cambridge, MA: Harvard University, 1997.

Dziarzhauny muzei belaruskai ssr, and I. M. Pakhomenko. *Znachki, zhetony dobrovol nykh obshchestv, 1920–1941 gg. Katalog* [Pins, Medals of Voluntary Societies, 1920–1941. Catalog]. Minsk: "Polymiia," 1987.

Edelman, Robert. *Serious Fun: A History of Spectator Sport in the USSR*. New York: Oxford University, 1993.

Eksler, Isaak Borisovich. *Geroicheskii perelet* [Heroic Flight]. Moscow: Gosudarstvennoe voennoe izdatel'stvo narkomata oborony soiuza, 1939.

Emerson, Caryl. "Soviet Civilization: Its Discontents, Disasters, Residual Fascinations." *The Hudson Review* 44, no. 4 (Winter 1992): 574–84.

Engel, Barbara Alpern. *Women in Russia, 1700–2000*. Cambridge: Cambridge University Press, 2004.

Ezell, Edward Clinton, and Linda Neuman Ezell. *The Partnership: A History of the Apollo-Soyuz Test Project*. NASA SP-409. Washington, DC: NASA, 1978.

Feoktistov, Konstantin. *Kosmicheskaia tekhnika: Perspektivy razvitiia* [Space Technology: Perspectives on its Development]. Moscow: Izdatel'stvo MGTU im. N. E. Baumana, 1997.

The Film Factory: Russian and Soviet Cinema in Documents, 1896–1939. 1988. Translated by Richard Taylor. Edited by Richard Taylor and Ian Christie. Paperback. London: Routledge, 1994.

Findley, John M. *Magic Lands: Western Cityscapes and American Culture after 1940*. Berkeley: University of California, 1992.

Finney, Ben, Vladimir Lytkin, and Liudmila Alepko. "Tsiolkovsky's 'Album of Space Voyages': Visions of a Space Theorist Turned Film Consultant." 1997. In *Proceedings of the Thirty-First History Symposium of the International Academy of Astronautics, Turin, Italy, 1997*, ed. Donald C. Elder and George S. James. Vol. 26, *History of Rocketry and Astronautics*. AAS History Series, 3–16. San Diego, CA: Univelt, 2005.

Fitzpatrick, Sheila, ed. *Cultural Revolution in Russia, 1928–31*. Bloomington: Indiana University Press, 1984.

———. *Tear off the Masks! Identity and Imposture in Twentieth-Century Russia*. Princeton: Princeton University Press, 2005.

Forest, Benjamin, and Juliet Johnson. "Unraveling the Threads of History: Soviet-Era Monuments and Post-Soviet National Identity in Moscow." *Annals of the Association of American Geographers* 92, no. 3 (2002): 524–47.

Forest, Benjamin, Juliet Johnson, and Karen Till. "Post-Totalitarian National Identity: Public Memory in Germany and Russia." *Journal of Cultural Geography* 50, no. 3 (September 2004): 357–80.

Fursenko, Aleksandr, and Timothy Naftali. *Khrushchev's Cold War: The Inside Story of an American Adversary.* New York: W. W. Norton, 2006.

Gak, A. M., comp. *Samoe vazhnoe iz vsekh iskusstv. Lenin o kino.* Sbornik dokumentov i materialov [The Most Important Art. Lenin on Film. A Collection of Documents and Materials]. Edited by I. S. Smirnov. 2d ed. Moscow: Izdatel'stvo "Iskusstvo," 1973.

Galin, Boris Abramovich. *Stalinskie sokoly* [Stalinist Falcons], 1937.

Gambrell, Jamey. "The Wonder of the Soviet World." *The New York Review of Books* (New York), 22 December 1994, 31–33.

Gardner, James H., and Peter S. LaPaglia, eds. *Public History: Essays from the Field.* Malabar, Florida: Kreiger Publishing Company, 1999.

Gatland, Kenneth William. *Spacecraft and Boosters; The First Comprehensive Analysis of More Than Seventy U.S. and Soviet Space Launchings.* London: Iliffe Books, 1964.

Gauthier, Yves. *Gagarine, Ou, le Reve Russe de l'Espace* [Gagarin, or the Russian Dream of Space]. Paris: Flammarion, 1998.

Gavriushin, Nikolai Konstantinovich. "The Cosmic Route to 'Eternal Bliss' (K. E. Tsiolkovskii and the Mythology of Technocracy)." *Russian Studies in Philosophy* (Moscow) Summer 1995: 36–47.

Gazenko, O. G., and A. A. Gyurdzhian. "Physiological Effects of Gravitation." In *Life Sciences and Space Reserach (LSSR) IV,* ed. A. H. and M. Florkin Brown, 3–17. Washington, DC: Spartan Books, 1966.

Geertz, Clifford. "Centers, Kings, and Charisma: Reflections on the Symbolics of Power." In *Local Knowledge.* New York: Basic, 1983.

Gelber, Steven M. "Free Market Metaphor: The Historical Dynamics of Stamp Collecting." *Comparative Studies in Society and History* 34, no. 4 (October 1992): 742–69.

Geppert, Alexander C. T., ed. *Imagining Outer Space: European Astroculture in the Twentieth Century.* Palgrave Studies in the History of Science and Technology. New York: Palgrave Macmillan, 2012.

Gerasimova, M. I., and A. G. Ivanov. *Zvezdnyi put* [The Starry Path]. Moscow: Izd-vo polit. lit-ry, 1986.

Gerchuk, Iurii. "The Aesthetics of Everyday Life in the Khrushchev Thaw in the USSR (1954–64)." In *Style and Socialism: Modernity and Material Culture in Post-War Eastern Europe,* ed. Susan E. Reid and David Crowley, 81–100. Oxford: Berg, 2000.

Gerovitch, Slava. "'New Soviet Man' Inside Machine: Human Engineering, Spacecraft Design, and the Construction of Communism." In *The Self as Project: Politics and the Human Sciences,* ed. Greg Eghigian, Andreas Killan and Christine Leuenberger. OSIRIS, vol. 22, 135–57. Chicago: University of Chicago Press, 2007.

———. *Soviet Space Mythologies: Public Images, Private Memories, and the Making of a Cultural Identity* (Russian and East European Studies). Pittsburgh: University of Pittsburgh Press, 2015.

———. *Voices of the Soviet Space Program: Cosmonauts, Soldiers, and Engineers Who Took the USSR into Space.* Palgrave Studies in the History of Science and Technology. New York: Palgrave Macmillan, 2014.

Gessen, Masha. *The Future Is History: How Totalitarianism Reclaimed Russia.* New York: Riverhead Books, 2017.

———. *The Man without a Face: The Unlikely Rise of Vladimir Putin.* New York: Riverhead Books, 2012.

Gies, Joseph. "Shows That Make Dreams Come True." *Los Angeles Times* (Los Angeles), 12 March 1961, E20–22.

Gillespie, David. *Early Soviet Cinema: Innovation, Ideology and Propaganda.* Short Cuts: Introductions to Film Studies. London; New York: Wallflower, 2000.

Gil'zin, Karl Aleksandrovich. *Sputniks and After.* Translated by Pauline Rose. London: Macdonald, 1959.

Ginzbursky-Blum, Bella. "Review of Yuri Norstein and 'Tale of Tales': An Animator's Journey, by Clare Kitson." *Russian Review* 66, no. 1 (2007): 134.

Glad, John. *Extrapolations from Dystopia: A Critical Study of Soviet Science Fiction.* Princeton: Kingston Press, 1982.

Glazkov, Iu. N. *Zemlia nad namy* [The Earth is Above Us]. Moscow: Mashinostroenie, 1986.

Gleason, Abbott, Peter Kenez, and Richard Stites, eds. *Bolshevik Culture: Experiment and Order in the Russian Revolution.* Bloomington: Indiana University Press, 1985.

Glushko, V. P., ed. *Kosmonavtika: entsiklopediia* [Cosmonautics: An Encyclopedia]. Moscow: Sovetskaia Entsiklopediia, 1985.

Golovanov, Iaroslav Kirilovich. *Korolev: fakty i mify* [Korolev: Facts and Myths]. Moscow: Izd-vo "Nauka," 1994.

Golubev, Alexey. *The Things of Life: Materiality in Late Soviet Russia.* Ithaca: Cornell University Press, 2020.

Gorbatov, Boris. *Vladimir Konstantinovich Kokkinaki.* Moscow: Gosudarstvennyi Izdatel'stvo politicheskoi literatury, 1938.

Gough, Maria. *The Artist as Producer: Russian Constructivism in Revolution.* Berkeley: University of California Press, 2005.

Graham, Loren. "Bogdanov's Inner Message," translated by Charles Rougle. In *Red Star: The First Bolshevik Utopia,* ed. Loren R. Graham and Richard Stites, 241–53. Bloomington: Indiana University Press, 1984.

———. "Bogdanov's Red Star: An Early Bolshevik Science Utopia." In *Nineteen Eighty-Four: Science Between Utopia and Dystopia,* ed. Everett Mendelsohn and Helga Nowotny, 111–24. Dordrecht: Reidel, 1984.

———. *The Ghost of the Executed Engineer: Technology and the Fall of the Soviet Union.* Cambridge, MA: Harvard University Press, 1993.

———. "The Place of the Academy of Sciences in the Overall Organization of Soviet Science." In *Soviet Science and Technology,* ed. J. R. Thomas and U. M. Kruse-Vaucienne. Washington, DC: Published for the National Science Foundation by George Washington University, 1977.

———. *Science in Russia and the Soviet Union: A Short History.* New York: Cambridge University Press, 1993.

Grant, Jonathan. "The Socialist Construction of Philately in the Early Soviet Era." *Comparative Studies in Society and History* 37, no. 3 (July 1995): 476–93.

Grigor'ev, A. I., R. M. Baevskii, and N. Iu. Galeev. *Vasilii Vasil'evich Parin i ego rol' v razvitii kosmicheskoi meditsiny i fiziologii* [Vasilli Vasil'evich Parin and his Role in the Development of Space Medicine and Physiology]. Moscow: Firma "Slovo," 2004.

Grigor'ev, Grigorii Karlovich. *Sledy v Nebe* [Traces in the Sky]. Moscow: Izdatel'stvo DOSAAF, 1960.

Grishanov, N. G., and S. P. Umanskii. "Skafandr letchika i kosmonavta" [Spacesuits of Pilots and Cosmonauts]. *Aviatsiia i Kosmonavtika*, no. 7 (1965).

Gritsai, Olga, and Herman van der Wusten. "Moscow and St. Petersburg, a Sequence of Capitals, a Tale of Two Cities." *GeoJournal* 51 (2000): 33–45.

Groys, Boris. *Ilya Kabakov: The Man Who Flew into Space from His Apartment*. Installation Review. London: Afterall Books, 2006.

Guggenheim Museum. "Russia! Opening New Spaces: 1980 to the Present." In *Guggenheim Museum Arts Curriculum Online*. Guggenheim Museum. 24 September 2007, http://www.guggenheim.org/artscurriculum/lessons/russian_L9.php.

Gumbert, Heather L. "Soviet Cosmonauts Visit the Berlin Wall: The Spatial Contradictions of the Cold War." In *The Cultural Impact of the Cold War Cosmonaut*, chair Diane P. Koenker. American Association for the Advancement of Slavic Studies National Convention. Washington, DC, 17 November 2006.

Gurney, Gene. *Cosmonauts in Orbit; The Story of the Soviet Manned Space Program*. New York: F. Watts, 1972.

Haddow, Robert H. *Pavilions of Plenty: Exhibiting American Culture Abroad in the 1950s*. Washington, DC: Smithsonian Institution, 1997.

Haeseler, Dietrich. "Alexei Leonov's Way to Space: Airlock of Voskhod-2." *Spaceflight* 36 (August 1994): 280–82.

Hall, Rex, and David J. Shayler. *The Rocket Men: Vostok and Voskhod, the First Soviet Manned Spaceflights*. Chichester: Springer-Verlag, 2001.

Harford, James J. *Korolev: How One Man Masterminded the Soviet Drive to Beat America to the Moon*. New York: Wiley, 1997.

Harvey, Brian. *European-Russian Space Cooperation: From de Gaulle to ExoMars*. Chichester, UK: Springer Praxis, 2021.

———. *The New Russian Space Programme: From Competition to Collaboration*. Wiley-Praxis Series in Space Science and Technology. Chichester, UK: John Wiley, 1996.

Hazan, Barukh. *Olympic Sports and Propaganda Games: Moscow 1980*. New Brunswick, NJ: Transaction Books, 1982.

Hilton, Alison. "Humanizing Utopia: Paradoxes of Soviet Folk Art." *Kritika: Explorations in Russian and Eurasian History* 3, no. 3 (Summer 2002): 459–71.

Hobsbawm, Eric. "Introduction: Inventing Tradition." In *The Invention of Tradition*, ed. Eric Hobsbawm and Terence Ranger, 1–14. Cambridge: Cambridge University Press, 1984.

Holloway, David. *Stalin and the Bomb: The Soviet Union and Atomic Energy, 1939–1956*. New Haven: Yale University Press, 1994.

Horton, Andrew J. "Science Fiction of the Domestic: Iakov Protazanov's *Aelita*." In *Russian Science Fiction Literature and Cinema: A Critical Reader*, edited by Anindita Banerjee, 166–177. Boston: Academic Studies Press, 2018.

Hosking, Geoffrey Alan. *Empire and Nation in Russian History*. The Fourteenth Charles Edmondson Historical Lectures. Waco, TX: Baylor University, 1992.

Howe, Barbara J, and Emory L. Kemp, eds. *Public History: An Introduction.* Malabar, FL: Robert E. Krieger Publishing Company, 1986.

Ignatenko, Aleksandr. *"Aelita": pervyi opyt sozdaniia blokbastera v rossii* [Aelita: The First Experience with the Creation of a Blockbuster in Russia]. St. Petersburg: Sankt-Peterburgskii gosudarstvennyi universitet kino i televideniia, 2007.

Il'in S. B. *Vsenarodnaia akademiia* [National Academy]. Moscow, USSR: Politizdat, 1986.

Il'inskii, V. N. *Znachki i ikh kollektsionirovanie (posobie dlia fileristov)* [Pins and Their Collecting (A Guide for Collectors)]. 1976. Izdanie vtoroe pererabotannoe i dopolnenie. Moscow: Izdatel'stvo "Sviaz'," 1977.

Il'inskii, V. N., V. E. Kuzin, and M. B. Saukke. *Kosmonavtiki na znachkakh SSSR, 1957–1975: Katalog* [Spaceflight in Pins of the USSR, 1957–1975]. Moscow: Izdatel'stvo "Sviaz'," 1977.

Intelligence and Security Committee. *The Mitrokhin Inquiry Report.* Presented to Parliament by the Prime Minister by Command of Her Majesty no. Cm 4764. June 2000.

International A. Bogdanov Institute. *Bibliografia osnovykh rabot A. A. Bogdanova* [Bibliography of the Works of A. A. Bogdanov]. 01/30/2007 http://www.bogdinst.ru/bogdanov/bibliogr.htm.

Ivanov, D. I., and A. I. Khromushkin. *Sistemy zheneobespecheniia cheloveka pri vysotnykh i kosmicheskikh poletakh* [Systems of Life Support During High-Altitude and Space Flights]. Moscow: Mashinostroenie, 1968.

James, D. Clayton. "American and Japanese Strategies in the Pacific War." In *Makers of Modern Strategy from Machiavelli to the Nuclear Age,* ed. Peter Paret. Princeton, NJ: Princeton University Press, 1986.

James, Peter N. *Soviet Conquest from Space.* New Rochelle, NY: Arlington House, 1974.

Jenks, Andrew L. *Collaboration in Space, and the Search for Peace on Earth.* New York: Anthem Press, 2022.

———. *The Cosmonaut Who Couldn't Stop Smiling: The Life and Legend of Yuri Gagarin.* Dekalb: Northern Illinois University Press, 2012.

Johnson, Nicholas L. *Handbook of Soviet Manned Space Flight.* AAS Science and Technology Series; v. 48. San Diego: Univelt, 1980.

———. *The Soviet Year in Space 1988.* Colorado Springs, CO: Teledyne Brown Engineering, 1989.

Joravsky, David. "Political Authorities and the Learned Estate in the U.S.S.R." In *Soviet Science and Technology,* ed. J. R. Thomas and U. M. Kruse-Vaucienne. Washington, DC: Published for the National Science Foundation by George Washington University, 1977.

Josephson, Paul R. "Atomic-Powered Communism: Nuclear Culture in the Postwar USSR." *Slavic Review* 55, no. 2 (Summer 1996): 297–324.

———. "'Projects of the Century' in Soviet History: Large-Scale Technologies from Lenin to Gorbachev." *Technology and Culture* 36, no. 3 (July 1995): 519–59.

———. *Red Atom: Russia's Nuclear Power Program from Stalin to Today.* Pittsburgh: University of Pittsburgh Press, 2005.

———. "Rockets, Reactors, and Soviet Culture." In *Science and the Soviet Social Order*, ed. Loren R. Graham, 168–91. Cambridge, MA: Harvard University Press, 1990.

JSC Zvezda. *Kazbek-U Anti-G Seat*. Technical Data Sheet. JSC Zvezda.

———. *Pingvin (Penguin)-3 Muscle and Bone System Loading Suit*. Technical Data Sheet. JSC Zvezda.

Kaganovsky, Lilya. "How the Soviet Man Was (Un)Made." *Slavic Review* 63, no. 3 (Autumn 2004): 577–96.

Kaiser, Daniel, ed. *The Workers Revolution in Russia 1917, The View from Below*. New York: Cambridge University, 1987.

Kamanin, Nikolai Petrovich. *Eksperimentalnaia kosmicheskaia stantsiia* [Experimental Space Stations]. Moscow, USSR, 1969.

Kapitsa, Petr Leonidovich. *Peter Kapitsa on Life and Science*. Edited by Albert Parry. New York: Macmillan, 1968.

Kelly, Catriona. *Comrade Pavlik: The Rise and Fall of a Soviet Boy Hero*. London: Granta, 2005.

———. "'Thank You for the Wonderful Book' Soviet Child Readers and the Management of Children's Reading, 1950–75." *Kritika: Explorations in Russian and Eurasian History* 6, no. 4 (Fall 2005): 717–53.

Kenez, Peter. *The Birth of the Propaganda State: Soviet Methods of Mass Mobilization, 1917–1929*. Cambridge [Cambridgeshire]; New York: Cambridge University Press, 1985.

———. *Cinema and Soviet Society: From the Revolution to the Death of Stalin*. Kino: The Russian Cinema Series, ed. Richard Taylor. London: I.B. Tauris Publishers, 2006.

Kerber, Lev L. *Stalin's Aviation Gulag: Memoir of Andrei Tupolev and the Purge Era*. Translated by Paul Mitchell. Smithsonian History of Aviation & Spaceflight, ed. Von Hardesty. Washington, DC: Smithsonian Institution Press, 1996.

Kevles, Bettyann Holtzmann. *Almost Heaven: The Story of Women in Space*. New York: Basic Books, 2003.

Khachatur'iats, L. C., and E. V. Khrunov. *Pobezhdnaia nevesomost* [The Conquest of Space]. Moscow: Znanie, 1985.

Kharitonov, Evgenii. "Kosmicheskaia odisseia Pavla Klushantseva" [The Space Odyssey of Pavel Klushantsev]. In *Na ekrane—chudo* [On the Screen are Marvels], ed. Evgenii Kharitonov and Andrei Shcherbak-Zhukov, 17–22. Moscow: NII Kinoiskusstva, 2003.

Kharitonov, Evgenii, and Andrei Shcherbak-Zhukov. *Na ekrane—chudo: otechestvennaia kinofantastika i kinoskazka (1909–2002): Materialy k populiarnoi entsiklopedii* [On the Screen are Marvels: National Space Fantasies and Tales 1902–2002)]. Moscow: NII Kinoiskusstva, 2003.

Khromushkin, A. I. *Skfandry i kislorodno-spasatel'naia apparatura dlia vysotnykh poletov* [Spacesuits and Oxygen Preservation Devices for High-Altitude Flights]. Moscow: Oborongiz, 1949.

Khrushchev, Nikita Sergeevich, ed. *Khrushchev Remembers: The Glasnost Tapes*. Edited and translated by Jerrold L. Schecter and Vyacheslav V. Luchkov. Boston: Little, 1990.

———, ed. *Memoirs of Nikita Khrushchev*. University Park, PA: Pennsylvania State University, 2004.

Khvat, Lev Borisovich. *Besprimernyi perelet* [Unparalleled Flight]. Moscow: n.p., 1936.

Kiaer, Christina. *Imagine No Possessions: The Socialist Objects of Russian Constructivism*. Cambridge, MA: The MIT Press, 2005.

Kit, Boris V. *U.S.S.R. Space Program; Manpower, Training, and Research Developments*. College Park, Maryland: University of Maryland, Department of Physics and Astronomy, 1964.

Knight, David B. "Review of *Politics and Place-Names: Changing Names in the Late Soviet Period*, by John Murray." *Slavic Review* 62, no. 1 (2003): 195–96. https://doi.org/10.2307/3090518.

Kohonen, Iina. *Picturing the Cosmos: A Visual History of Early Soviet Space Endeavor*. Bristol, England: Intellect Ltd., 2017.

Kojevnikov, Alexei. *Stalin's Great Science: The Times and Adventures of Soviet Physicists*. London: Imperial College Press, 2004.

Kolesnikov, Iu. V. *Kosmos—zemle: nauchno-khudoshestvennaia literature* [Space on the Earth: Scientific-Artistic Literature]. Moscow: Detskaia Literatura, 1987.

Kolm, Suzanne Lee. "Women's Labor Aloft: A Cultural History of Airline Flight Attendants in the United States, 1930–1978." PhD diss., History, Brown University, 1995.

Kommunisticheskaia partiia sovetskogo soiuza. *The Road to Communism: Documents of the 22nd Congress of the Communist Party of the Soviet Union. October 17–31, 1961*. Moscow, USSR: Foreign Languages Publishing House, 1961.

Korol, Alexander G. *Soviet Education for Science and Technology*. Westport, CT: Greenwood Press, 1974.

Koroleva, Nataliia Sergeevna. *Otets* [My Father]. Moscow, Russia: Nauka Leningradskoe otd-nie, 2001.

Kosloski, Lillian D. "Conservation-Conscious Manikins for Space Suit Displays—A Curatorial Viewpoint." *Ars Textrina*, no. 11 (1989): 77–106.

———. *U.S. Space Gear: Outfitting the Astronaut*. Washington, DC: Smithsonian, 1994.

Kosloski, Lillian, and Maura J. Mackowski. "The Wrong Stuff." *Final Frontier*, May/June 1990.

Kosmodem'ianskii, A. A. *Konstantin Eduardovich Tsiolkovskii (1957–1935)*. Edited by A. S. Fedorov. 2d ed. Moscow: Nauka, 1988.

Kotek, Joel. *Students and the Cold War*. New York: St. Martin's Press, Inc., 1996.

Kovalov, Oleg. "Aleksei Fedorchenko, 'First on the Moon (Pervye Na Lune).'" In *KinoKultura* 11 January 2006, http://www.kinokultura.com/2006/11r-firstmoon1.shtml.

Kozyrev, V. T. *Eshche raz o gibeli Gagarina* [Once Again about the Death of Gagarin]. Moscow, 1998.

Kreiger, F. J. *Behind the Sputniks: A Survey of Soviet Space Science*. Washington, DC: Public Affairs Press, 1958.

Krylova, Anna. "Dancing on the Graves of the Dead: Building a World War II Memorial in Post-Soviet Russia." In *Memory and the Impact of Political Transformation in Public Space*, ed. Daniel J. Walkowitz and Lisa Maya Knauer, 83–102. Durham: Duke University Press, 2004.

Kryshtanovskaya, Olga, and Stephen White. "From Soviet Nomenklatura to Russian Elite." *Europe-Asia Studies* 48, no. 5 (July 1996): 711–33.

Kubasov, V. N., V. A. Taran, and S. N. Maksimov. *Professional'naia podgotovka kosmonavtov* [Cosmonauts' Professional Preparation]. Moscow: Mashinostroenie, 1985.

Kuzin, E. N. "Dvadtsat' piat' let gosudarstvennomu muzeiu istorii kosmonavtiki imeni k. e. Tsiolkovskogo" [Twenty-Five Years of the Tsiolkovsky State Museum of the History of Spaceflight]. In *Trudy XXXII chtenii, posviashennikh razrabotke nauchnogo naslediiq i razvitiiu idei k. e.tsiolkovskogo (Kaluga, 15–18 sentiabria 1992 g.),* 103–11. Moscow, USSR: IIET RAN, 06/11/27, 1994.

Kuzin, E. N., N. G. Belova, and T. V. Chugova. *Muzei kosmosa v kaluge* [The Space Museum in Kaluga]. Tula: Priok. kn. izd-va, 1986.

Lane, Christel. *The Rites of Rulers: Ritual in Industrial Society: The Soviet Case.* Cambridge: Cambridge University Press, 1981.

Laqueur, Walter Z., and George Lichtheim, eds. *The Soviet Cultural Scene, 1956-1957.* New York: Atlantic Books, 1958.

Lavitsov, A. A. *Osnovy aviatsionnoi i kosmicheskoi meditsiny* [The Fundamentals of Space Medicine]. Moscow: Voenizdat, 1975.

Lee, David C. "Public Organizations in Adult Education in the Soviet Union." *Comparative Education Review* 30, no. 3 (August 1986): 344–58.

"Lev Kerbel and His Time (to the 85th Anniversary of the Sculptor's Birth)." In *Russian Culture Navigator.* 09/03/2005, http://www.vor.ru/culture.cultarch238_eng.html.

Levant, Alex. "The Soviet Union in Ruins." M. A. thesis, Methodologies for the Study of Western History and Culture M. A. Program, Trent University, 1999.

Levine, Alan J. *The Missile and Space Race.* Westport, CT: Praeger, 1994.

Levine, Lawrence W. "Folklore of Industrial Society: Popular Culture and Its Audience." *American Historical Review,* December 1992: 1369–448.

Lewin, Moshe. *The Gorbachev Phenomenon: A Historical Interpretation.* Berkeley: University of California Press, 1988.

Lewis, Cathleen S. "The Birth of Soviet Space Museums: Creating the Earthbound Experience during the Golden Years of the Soviet Space Programme, 1957-68." In *Showcasing Space,* eds. Martin Collins and Douglas Millard. Artefacts Series. Studies in the History of Science and Technology, 142–58. East Lansing, Michigan: Michigan State University Press, 2005.

Leyda, Jay. *Kino: A History of the Russian and Soviet Film.* 1960. Third. Princeton: Princeton University Press, 1983.

Lindquist, Galina. "Spirits and Souls of Business: New Russians, Magic and the Esthetics of Kitsch." *Journal of Material Culture* 7, no. 3 (2002): 329–43.

Linenthal, Edward Tabor. *Changing Images of the Warrior Hero in America: A History of Popular Symbolism.* Studies in American Religion, Vol. 6. New York: E. Mellen, 1982.

Linz, Susan J. "World War II and Soviet Economic Growth, 1940-1953." In *The Impact of World War II on the Soviet Union,* ed. Susan J. Linz. Totowa, NJ: Rowman and Allenheld, 1985.

Lobanov, Vitalii Nikolaevich, and Dmitrii Aleksandrovich Chernikov. *V Kosmose—"Vostok 2"* [In Space—Vostok 2]. Moscow: n.p., 1961.

Lodder, Christina. *Russian Constructivism*. New Haven, CT: Yale University Press, 1983.

Loewenstein, Karl Edward. "The Thaw: Writers and the Public Sphere in the Soviet Union, 1951–1957." PhD diss., Duke University, 1999.

London, Joanne M. Gernstein. "A Modest Show of Arms: Exhibiting the Armed Forces and the Smithsonian Institution, 1945–1976." PhD diss., American Studies, George Washington University, 2000.

Lubrano, Linda, and Susan Gross Solomon, eds. *The Social Context of Soviet Science*. Boulder, CO: Westview Press, 1980.

Luehrmann, Sonja. "Recycling Cultural Construction: Desecularisation in Postsoviet Mari El." *Religion, State & Society* 33, no. 1 (March 2005): 35–55.

Lule, Jack. "Roots of the Space Race: Sputnik and the Language of U.S. News in 1957." *Journalism Quarterly* 68, no. 1–2 (Spring-Summer 1991): 76–86.

Lyndon B. Johnson Space Center. National Aeronautics and Space Administration. *A History of U.S. Space Stations*. Tech. Rept. no. IS-1997–06-ISS009JSC. Houston, Texas, 1997.

Makarenkov, P. E. *Raketa Gagarina* [Gagarin's Rocket]. Smolenskoe oblastnoe: Smiadyn,' 2003.

Makoveeva, Irina. "Soviet Sports as a Cultural Phenomenon: Body and/or Intellect." *Studies in Slavic Cultures,* no. III (July 2002): 9–32.

Makushin, Ruslan. *Lev Kerbel: "Ia liubil sovetskuiu vlast"* [Lev Kerbel: I Loved Soviet Power]. Interview with Lev Kerbel. 2003. 09/03/2005, http://www.peoples.ru/art/sculpture/kerbel.

Malia, Martin. *The Soviet Tragedy: A History of Socialism in Russia, 1917–1991*. New York: Free Press, 1994.

Malina, Frank J. "On the Visual Fine Arts in the Space Age." *Leonardo* 3, no. 3 (July 1970): 323–25.

Mallan, Lloyd. *Suiting up for Space*. New York: John Day, 1971.

Mally, Lynn. "Utopias Lost and Found: In Search of Soviet Culture." *Radical History Review* 59 (Spring 1994): 181–89.

Mamleev, Dmitrii Fedorovich, comp. *Sem'sot tysiach kilometrov v kosmose* [700,000 Kilometers in Space]. Moscow: n.p., 1961.

Marinin, I. A., S. Kh. Shamsutdinov, and A. V. Glushko. *Sovetskie i Rossiiskie kosmonavty: 1960–2000*. Spravochik [Soviet and Russian Cosmonauts: 1960–2000, A handbook]. Moscow: OOO Informatsionno-izdatel'skii dom "Novosti kosmonavtiki," 2001.

Martovitskaia, Anna. "Muzei kosmonavtiki gotovitsia k rekonstruktsii" [The Museum of Spaceflight Begins Reconstruction]. *Gazeta Kultura*, no. 14 (7524, April 2006): 13-16. URL: http://www.kultura-portal.ru/tree_new/cultpaper/article.jsp?number=635&rubric_id=218.

Mastny, Vojtech. *Russia's Road to the Cold War: Diplomacy, Warfare and the Politics of Communism, 1941–1945*. New York: Columbia University Press, 1979.

McCannon, John. "Positive Heroes at the Pole: Celebrity Status, Socialist-Realist Ideals and the Soviet Myth of the Arctic, 1923–39." *Russian Review* 56 (July 1997): 346–65.

———. *Red Arctic: Polar Exploration and the Myth of the North in the Soviet Union, 1932–1939*. New York: Oxford University Press, 1998.

———. "Tabula Rasa in the North: The Soviet Arctic and Mythic Landscapes in Stalinist Popular Culture." In *The Lands of Stalinist: The Art and Ideology of Soviet Space,* eds. Evgeny Dobrenko and Eric Naiman, 241–60. Seattle: University of Washington Press, 2003.

———. "Technological and Scientific Utopias in Soviet Children's Literature, 1921–1932." *Journal of Popular Culture* 34, no. 4 (Spring 2001): 153–69.

McClelland, James C. "The Utopian and the Heroic: Divergent Paths to the Communist Education Ideal." In *Bolshevik Culture: Experiment and Order in the Russian Revolution,* eds. Abbott Gleason, Peter Kenez and Richard Stites. Bloomington, Ind.: Indiana University Press, 1985.

———. "Utopianism versus Revolutionary Heroism in Bolshevik Policy: The Proletarian Culture Debate." *Slavic Review* 39, no. 3 (September 1980): 403–42.

McDougall, Walter A. . . . *the Heavens and the Earth: A Political History of the Space Age.* New York: Basic Books, 1985.

McDowell, Jennifer. "Soviet Civil Ceremonies." *Journal for the Scientific Study of Religion* 13, no. 3 (September 1974): 265–79.

Medvedev, Roy. *Khrushchev.* Translated by Brian Pearce. Garden City, NY: Anchor Press/Doubleday, 1984.

Merridale, Catherine. *Night of Stone: Death and Memory in Twentieth Century Russia.* New York: Viking, 2001.

Mevedev, A. K. *Osvoenie aerokosmicheskogo prostranstva: Proshloe, nastoiashchee, budyshchee* [Mastery of Aerospace: Past, Present and Future]. Moscow: IIET RAN, 1997. 202.

Mickiewicz, Ellen. "How Soviet TV Focuses Its Cameras on Eastern Europe." *The New York Times,* 31 December 1989, 29.

———. "Soviet Viewers Are Seeing More, Including News of the U.S." *The New York Times,* 22 February 1987, H29, 34.

———. *Split Signals: Television and Politics in the Soviet Union.* New York: Oxford University Press, 1988.

Miller, Daniel. *Material Culture and Mass Consumption.* Oxford: Basil Blackwell, Inc., 1987.

Miller, Frank J. *Folklore for Stalin: Russian Folklore and Pseudofolklore of the Stalin Era.* Armonk, NY: M. E. Sharpe, 1990.

Miller, Jamie. "The Purges of Soviet Cinema, 1929–38." *Studies in Russian and Soviet Cinema* 1, no. 1 (2007): 5–26.

Miller, Jay. *Soviet Space: An Exhibition in the City of Ft. Worth.* Ft. Worth, TX: Ft. Worth Museum of Science and History Association, 1991.

Minde, George F.I.I. "Reform of the Soviet Military under Khrushchev and the Role of America's Strategic Modernization." In *Reform in Russia and the U.S.S.R.: Past and Prospects,* ed. Robert O. Crummey, 186–206. Urbana: University of Illinois Press, 1989.

Moltz, James Clay. "Managing International Rivalry on High Technology Frontiers: U.S.-Soviet Competition and Cooperation in Space." PhD diss., University of California at Berkeley, 1989.

Mozhayev, Alexander. "Exhibition Soviet Style." *Russian Life* (Moscow), March/April 2005, 48–53.

Murasov, Boris. *Ubiitstvo kosmonavta Iuriia Gagarina* [The Assassination of Cosmonaut Yuri Gagarin]. Moscow, 1995.

Muzei kosmosa v kaluge [Space Museum in Kaluga]. Edited by S. D. Oshevskii. Tula: Priokskoe knizhnoe izdatel'stvo, 1986.

Neary, Rebecca Balmas. "Mothering Socialist Society: The Wife-Activists' Movement and the Soviet Culture of Daily Life, 1934–41." *Russian Review* 58, no. 3 (July 1999): 396–412.

Newkirk, Dennis. *Almanac of Soviet Manned Space Flight.* Houston: Gulf Publishing Company, 1990.

Norlander, David. "Khrushchev's Image in the Light of Glasnost and Perestroika." *The Russian Review* 52 (April 1993): 248–164.

Oberg, James E. *Red Star in Orbit.* New York: Random House, 1981.

Odom, William E. *The Soviet Volunteers: Modernization and Bureaucracy in a Public Mass Organization.* Princeton: Princeton University Press, 1973.

Official Site of the Brussels World's Fair, 1958. 1/22/2004, https://atomium.be/expo58.

Oinas, Felix J. "Folklore and Politics in the Soviet Union." *Slavic Review* 32 (1973): 45–56.

Okey, Robin. *Eastern Europe 1740–1985: Feudalism to Communism.* 2d ed. Minneapolis: University of Minnesota Press, 1986.

Omel'ko, V. A. *Nagradnye znaki obshchestvennykh organizatsii u muzeev.* Vol. tom 1, *Nagrady za osvoenie kosmosa catalog* [Award Pins of Social Organizations and Museums, Volume 1, Awards for the Conquest of Space, Catalog]. Moscow: n.p., 2002.

O'Neill, Mark A. "Air Combat on the Periphery: The Soviet Air Force in Action During the Cold War, 1945–89." In *Russian Aviation and Air Power in the Twentieth Century,* eds. Robin Higham, John T. Greenwood and Von Hardesty, 208–35. London: Frank Cass, 1998.

———. "The Other Side of the Yalu: Soviet Politics in the Korean War, Phase One, 1 November 1950–13 October 1951." PhD diss., History, Florida State University, 1996.

O'Toole, Thomas. "The Man Who Didn't Walk on the Moon." *New York Times Magazine* 17 July 1994: 26–29.

Palmer, Scott W. *Dictatorship of the Air: Aviation Culture and the Fate of Modern Russia.* Cambridge Centennial of Flight, eds. John Anderson and Von Hardesty. Cambridge: Cambridge University Press, 2006.

———. "Soviet Air-Mindedness as an Ideology of Dominance." *Technology and Culture* 41, no. 1 (January 2000): 1–26.

Paperny, Vladimir. *Architecture in the Age of Stalin: Culture Two.* Translated by John Hill and Roann Barris. Cambridge, UK: Cambridge University Press, 2002.

Parin, V. V., and V. I. Iazdovskii. "Put' sovetskoi kosmicheckoi fiziologii" [The Path of Soviet Space Physiology]. *Fiziologicheskii zhurnal sssr* XLVII, no. 10 (1961): 1217–26.

Parin, V. V., F. P. Kosmolinskii, and B. A. Dumkov. *Kosmicheskaia bologiia I meditsina* [Space Biology and Medicine]. Moscow: Prosveshchenie, 1970.

Parin, V. V., Yu. M. Volykhin, and P. V. Vassilyev. "Manned Space Flight—Some Scientific Results." In *Life Sciences and Space Research III. A Session of the 5th International*

Space Symposium, Florence, 12–16 May 1964, ed. M. Florkin, 3–17. Amsterdam: North Holland, 1965.

Parry, Albert. *The Russian Scientist.* New York: Macmillan, 1973.

Pennington, Reina. "From Chaos to the Eve of the Great Patriotic War, 1921–41." In *Russian Aviation and Air Power in the Twentieth Century,* eds. Robin Higham, John T. Greenwood and Von Hardesty. Studies in Air Power. London: Frank Cass, 1998.

———. *Wings, Women, and War: Soviet Airwomen in World War II Combat.* Lawrence: University Press of Kansas, 2001.

Pereskokov, A., P. Egorov, and M. Raevskii. *Potolok letchika* [The Limits of a Pilot]. Moscow: Voenizdat, 1931.

Petrone, Karen. *Life Has Become More Joyous, Comrades: Celebrations in the Time of Stalin.* Bloomington, Indiana: Indiana University Press, 2000.

Petrov, G. I., ed. *Osvoenie kosmicheskogo prostranstva v SSSR* [Mastery of Space in the USSR]. Translated by Stefania Dhingra. New Delhi: Published for the National Aeronautics and Space Administration and the National Science Foundation by Amerind Publishing Company, 1973.

Pipes, Richard. *The Formation of the Soviet Union.* Cambridge, Massachusetts: Harvard University Press, 1964.

———, ed. *The Russian Intelligentsia.* New York: Columbia University, 1961.

Pirard, Theo. "The Space Museum at RKK Energia." *Spaceflight* 42 (June 2000): 247–53.

Pomerantsev, Peter. *Nothing Is True and Everything Is Possible. The Surreal Heart of the New Russia.* New York: Public Affairs Press, 2014.

Popescu, Julian. *Russian Space Exploration: The First 21 Years.* Henley-on-Thames, England: Gothard House Publications, 1979.

Popovich, Pavel Romanovich. "Assotsiatsia muzeev kosmonavtiki" [The Association of Space Museums]. *Aviatsiia i Kosmonavtika* (Moscow), March 1990, 42–43.

Prokhorov, Alexander. "The Redemption of Lunar Reality: Aleksei Fedorchenko's First on the Moon (Pervye Na Lune), 2005." In *KinoKultura: New Russian Cinema.* Film Review. 2006. 2 November 2006, http://www.kinokultura.com/2006/11r-firstmoon2 .shtml.

Raleigh, Donald J. *Russia's Sputnik Generation: Soviet Baby Boomers Talk about their Lives.* Bloomington: Indiana University Press, 2006.

Raushenbakh, Boris V., and G. A. Skuridin, eds. *S. P. Korolev: Soviet Spacecraft Designer.* Washington, DC: National Aeronautics and Space Administration, 1977.

Reid, Susan E. "All Stalin's Women: Gender and Power in Soviet Art of the 1930s." *Slavic Review* 57, no. 1 (Spring 1998): 133–73.

———. "Cold War in the Kitchen: Gender and the De-Stalinization of Consumer Taste in the Soviet Union under Khrushchev." *Slavic Review* 61, no. 2 (Summer 2002): 211–52.

———. "Destalinization and Taste, 1953–1963." *Journal of Design History* 10, no. 2 (1997): 177–201.

———. "Socialist Realism in the Stalinist Terror: The *Industry of Socialism* Art Exhibition, 1935–41." *The Russian Review* 60 (April 2001): 153–84.

Reid, Susan E., and Lynne Attwood, eds. *Women in the Khrushchev Era.* Studies in Russian and East European History and Society. Houndmills, Basingstoke, England; New York: Palgrave Macmillan, 2004.

Reid, Susan E., and David Crowley, eds. *Style and Socialism: Modernity and Material Culture in Post-War Europe.* Oxford: Berg, 2000.

Reid, Susan Emily. "De-Stalinization and the Remoderization of Soviet Art: The Search for a Contemporary Realism, 1953–1963," PhD diss., University of Pennsylvania, 1996.

———. "Photography in the Thaw." *Art Journal* (Summer 1994): 33–39.

Remnick, David. "Soviet TV Announces a Shake-Up: New Editor of Evening News to Bring It into Gorbachev Era." *The Washington Post,* 06 January 1989, B1, B4.

Riabchikov, Evgenii Ivanovich. *Russians in Space.* Translated by Guy Daniels. Garden City, NY: Doubleday, 1971.

Richers, Julia, Eva Mauer, Monica Ruthers, and Carmen Scheide, eds., *Soviet Space Culture: Cosmic Enthusiasm in Socialist Societies.* Basingstoke, England: Palgrave Macmillan, 2011.

Riordan, James. *Soviet Sport Background to the Olympics.* New York: Washington Mews Books, 1980.

Riskin, Andrei. "Shkola dlia glukhikh stal 'kosmicheskoi' [A School for the Blind Has Become Space-Like]. V muzee internata sobrano 13 Tysiach Eksponatov." *NG Kollektsiia,* 12 June 2000, 6.

Roberts, Graham H., ed. *Material Culture in Russian and the USSR: Things, Values, Identities.* London: Bloomsbury Academic Press, 2017.

Rollberg, Peter. "Soviet Literary Culture in the 1970s: The Politics of Irony." *World Literature Today* 69, no. 1 (1 January 1995): 176–77.

Romanov, Aleksandr. *Konstructor kosmicheskikh korablei* [The Designer of Spacecraft]. English edition. Moscow: Novosti Press Agency Publishing House, 1976: 118.

Roth-Ey, Kristin. "Finding a Home for Television in the USSR, 1950–1970." *Slavic Review* 66, no. 2 (Summer 2007): 278–306.

Ruble, Blaire A. "The Expansion of Soviet Science." *Knowledge* 2, no. 4 (1981): 529–53.

Rusnock, K. Andrea. "The Art of Soviet International Politics: Vera Mukhina's Worker and Collective Farm Woman in the 1937 Internationale Exposition." In *The Soviet Pavilion at the Exposition Internationale, Paris 1937,* chair David C. Fisher. AAASS National Convention. New Orleans, 17 November 2007.

Russell, I. Willis. "Among the New Words." *American Speech* 37, no. 2 (May 1962): 145–47.

Russia's Sputnik Generation: Soviet Baby Boomers Talk about Their Lives. Edited and translated by Donald J. Raleigh. Bloomington: Indiana University Press, 2006.

Russkii gosudarsvennyi nauchno-issledovatel'skii ispitatel'nyi tsentr podgotovki kosmonavtov imeni Iu. A. Gagarina (RGNIITsPK imeni Iu. A. Gagarina) [The Yuri A Gagarin Russian State Scientific-Research Institute for the Experimental Preparation of Cosmonauts]. 16 February 2007. http://www.gctc.ru/.

Sagdeev, Roald Z. *The Making of a Soviet Scientist: My Adventures in Nuclear Fusion and Space from Stalin to Star Wars.* Edited by Carl Sagan and Susan Eisenhower. New York: John Wiley & Sons, 1994.

Santy, Patricia A. *Choosing the Right Stuff: The Psychological Selection of Astronauts and Cosmonauts.* Westport: Praeger, 1994.

Saukke, M. "Chelovek v kosmose" [Man in Space]. *Filateliia Sssr* (Moscow) 1968, 12–15.

Savinykh, V. *Zemlia zhdet i nadeetsia* [The Earth Waits and Hopes]. Perm: Permskoe knizhnoe izdatel'stvo, 1983.

Schmid, Sonja D. "Celebrating Tomorrow Today: The Peaceful Atom on Display in the Soviet Union." *Social Studies in Science* 36, no. 3 (June 30, 2006): 331–65.

Schmidt, Paul. "Constructivism and Film." *Soviet Union* 3, no. 2 (1976): 283–93.

Schuman, Howard, and Amy D. Corning. "Collective Knowledge of Public Events: The Soviet Era from the Great Purge to Glasnost." *American Journal of Sociology* 105, no. 4 (January 2000): 913–56.

Schwan, Fred. "Remembering World War II." *The Numismatist* 108, no. 9 (September 1995): 1107–13.

Seibert, Victor C. "Falerists and Their Russian Znachki." *The Numismatist* June 1979: 1198–202.

Seredina, E. V. *Planetarii—ochag nauchno-ateisticheskoi propaganda (v pomoshch' uchiteliu i kul'tprosvetrabotniku)* [Planetarium—Centers of Scientific Atheistic Propaganda (A Guide for Teachers and Cultural Workers)]. Kursk: Oblastnoe upravlenie kultury kurskii planetarii, 1960.

Shatnoff, Judith. "Expo 67: A Multiple Vision." *Film Quarterly* 21, no. 1 (Autumn 1967): 2–13.

Sheldon, Charles S. *United States and Soviet Rivalry in Space: Who Is Ahead, and How Do the Contenders Compare.* Washington, DC: Library of Congress, Legislative Reference Service, 1969.

Shelton, William Roy. *Soviet Space Exploration; The First Decade.* New York: Washington Square Press, 1968.

Shklovsky, Iosif. *Five Billion Vodka Bottles to the Moon: Tales of a Soviet Scientist.* Translated by Mary Fleming Zirin and Harold Zirin. 1st. New York: Norton, 1991.

Shlapentokh, Dmitry, and Vladimir Shlapentokh. *Soviet Cinematography: Ideological Conflict and Social Reality, 1918–1991.* New York: A De Gruyter, 1993.

Siddiqi, Asif A. "Before Sputnik: Early Satellite Studies in the Soviet Union, 1947–1957, Part 1." *Spaceflight* 39 (October 1997): 334–37.

———. "Before Sputnik: Early Satellite Studies in the Soviet Union, 1947–1957, Part 2." *Spaceflight* 39 (November 1997): 389–92.

———. *Challenge to Apollo.* Washington, DC: NASA, 2000.

———. *The Rockets' Red Glare: Spaceflight and the Russian Imagination, 1857–1957.* Cambridge: University of Cambridge Press, 2014.

———. "Soviet Space Programme: Organisational Structure, 1940s–1950s." *Spaceflight* 36 (August 1994): 283–86.

———. "Soviet Space Programme: Organisational Structure in the 1960s." *Spaceflight* 36 (September 1994): 314–20.

Siegelbaum, Lewis H. *Stakhanovism and the Politics of Productivity in the USSR.* Cambridge: Cambridge University Press, 1988.

Simmons, Cynthia. "Fly Me to the Moon: Modernism and the Soviet Space Program in Viktor Pelevin's 'Omon Ra.'" *Harriman Review* 12, no. 4 (November 2000): 4–9.

Simmons, Ernest J., ed. *Continuity and Change in Russian and Soviet Thought.* New York: Russell and Russell, 1967.

Siniavskii. "Bez skidok" [Without Allowances]. *Voprosy literatury* I (1960): 47–91.

Smart, Christopher. *The Imagery of Soviet Foreign Policy and the Collapse of the Russian Empire.* Westport, CT: Praeger, 1995.

Smith, Kathleen Elizabeth. "Coming to Terms with Previous State-Sponsored Repression: Civic Activism and Regime Responses in the Soviet Union." PhD diss., University of California, Berkeley, 1994.

———. *Mythmaking in the New Russia: Politics and Memory During the Yeltsin Era.* Ithaca, NY: Cornell University Press, 2002.

Smolders, P. L. L. *The Russians in Space.* Lutterworth Press, 1973.

Smolkin, Victoria. *A Sacred Space Is Never Empty. A History of Soviet Atheism.* Princeton: Princeton University Press, 2018.

Sochor, Zenovia A. "On Intellectuals and the New Class." *The Russian Review* 49 (1990): 283–92.

———. *Revolution and Culture: The Bogdanov-Lenin Controversy.* Studies of the Harriman Institute, eds. Joseph F. Berliner, Seweryn Bialer and Sheila Fitzpatrick. Ithaca, NY: Cornell University Press, 1988.

Sotheby's. *Russian Space History, Sale 6516.* 12/11/93 auction catalog. New York: Sotheby's, 1993.

———. *Russian Space History, Sale 6753.* 3/16/96 auction catalog. New York: Sotheby's, 1996.

Spaceflight: A Smithsonian Guide. Edited by Valerie Neal, Cathleen S. Lewis, and Frank H. Winter. New York: Macmillan, 1995.

Spasskii, V. A. *Fizziologo-gigienncheskoe obespeshenie poletov v skafandre* [Physio-Hygienic Safeguards for Spaceflight]. Moscow: Medgiz, 1940.

Specter, Michael. "Moscow on the Make." *The New York Times* (New York), 1 June 1997, Sunday, Late Edition—Final, SM: 48–84.

Stites, Richard, ed. *Culture and Entertainment in Wartime Russia.* Bloomington: Indiana University Press, 1995.

———. *Revolutionary Dreams: Utopian Vision and Experimental Life in the Russian Revolution.* New York: Oxford University, 1989.

———. *Russian Popular Culture: Entertainment and Society Since 1900.* Cambridge, UK: Cambridge University Press, 1992.

———. "Stalin: Utopian or Antiutopian? An Indirect Look at the Cult of Personality." In *The Cult of Power: Dictators in the Twentieth Century,* ed. Joseph Held. New York: Columbia University Press, 1983.

———. "World Outlook and Inner Fears in Soviet Science Fiction." In *Science and the Soviet Social Order,* ed. Loren Graham. Program in Science, Technology, and Society, 299–324. Cambridge, MA: Harvard University Press, 1990.

Stoiko, Michael. *Soviet Rocketry: Past, Present, and Future.* New York: Holt, Rinehart and Winston, 1970.

Striedter, Jurij. "Three Postrevolutionary Russian Utopian Novels." In *The Russian Novel from Pushkin to Pasternak*, ed. John Garrard, 177–201. New Haven: Yale University Press, 1983.

Strukov, Vlad, and Helena Goscilo, eds. *Russian Aviation, Space Flight, and Visual Culture*. London: Routledge, 2017.

Suny, Ronald Grigor. *The Revenge of the Past: Nationalism, Revolution and the Collapse of the Soviet Union*. Stanford, CA: Stanford University Press, 1993.

Suvin, Darko. *Metamorphoses of Science Fiction*. New Haven: Yale University Press, 1979.

———. "The Utopian Tradition in Russian Science Fiction." *Modern Language Review* 66, no. 1 (1971): 139–59.

Sweeting, C. G. *Combat Flying Equipment: U.S. Army Aviators' Personal Equipment, 1917–1945*. Shrewsbury, UK: Airlife, 1989.

Swenson, Loyd S., Jr., James M. Grimwood, and Charles C. Alexander. *This New Ocean: A History of Project Mercury*. Washington, DC: National Aeronautics and Space Administration, 1966.

Swift, Anthony. "The Soviet World of Tomorrow at the New York World's Fair, 1939." *The Russian Review* 57 (July 1998): 364–79.

Talenskii, N. A. Editorial. *Voennaia mysl'* No. 9 (September 1953). Reprinted in *The Soviet Art of War: Doctrine, Strategy, and Tactics*. Edited and translated by Harriet Fast Scott and William F. Scott, 123–136, Boulder, CO: Westview Press.

Taubman, William. "The Correspondence: Khrushchev's Motives and His Views of Kennedy." *Problems of Communism*. Spring 1992: 14–18.

———. *Khrushchev: The Man and His Era*. New York: W. W. Norton, 2003.

Terins, John. "Political Changes to Change U.N. Philatelic Market." *Stamps* 230, no. 8 (24 February 1990): 278.

Thompson, Kristin. "Government Policies and Practical Necessities in the Soviet Cinema of the 1920s." In *The Red Screen: Politics, Society and Art in Soviet Cinema*, ed. Anna Lawton, 19–41. New York: Routledge, 1992.

Tower, Samuel A. "Looking into New Soviet Issues." *The New York Times* (New York), 22 June 1975, 143.

Treskin, A., and Valerii Shteinbakh. *Istoriia Olimpiiskikh igr medali, znachki, plakaty* [The History of the Olympic Games in Medals, Pins and Posters]. Moscow: TERRA-Sport Olimpiia Press, 2001.

Tsymbal, N. A. *First Man in Space the Life and Achievement of Yuri Gagarin: A Collection*. Moscow: Progress Publishers, 1984.

Tumarkin, Nina. *Lenin Lives! The Lenin Cult in Soviet Russia*. Cambridge, Massachusetts: Harvard University Press, 1983.

———. *The Living and the Dead: The Rise and Fall of World War II in Russia*. New York: Basic Books, 1994.

Umanskii, S. P. *Snariazhenie letchika i kosmonavtika* [Equipment for Pilots and Cosmonauts]. Moscow: Voenizdat, 1967.

"Under New Management." *Economist*, 8 October 1994, 21–23.

United States. Central Intelligence Agency. *Civil Defense in the USSR*. National Intelligence Estimate no. NIE-60. 14 April 1952.

———. *Probable Warning of Soviet Attack on the U.S. Through Mid-1957.* Special National Intelligence Estimate no. 11-8-54. 14 September 1954.

United States. Department of the Army. Office, Chief of Research and Development. *USSR: Missiles, Rockets and Space Effort; A Bibliographic Record, 1956–1960.* Department of the Army Pamphlet 70-5-8. Washington, DC: Department of the Army, 1960.

United States. House of Representatives, Eight-sixth Congress, first session. Committee on Science and Astronautics. *Soviet Space Technology. Hearings Before the Committee on Science and Astronautics and Special Subcommittee on Lunik Probe.* Washington, DC: United States Government Printing Office, 1959.

United States. Library of Congress. Legislative Reference Service. *Soviet Space Programs: Organization, Plans, Goals, and International Implications: Staff Report Prepared for the Use of the Committee on Aeronautical and Space Sciences, United States Senate.* Washington, DC: United States Government Printing Office, 1962.

United States. Library of Congress. Legislative Research Service. *Soviet Space Programs, 1962-65: Goals and Purposes, Achievements, Plans, and International Implications; Staff Report Prepared for the Use of the Committee on Aeronautical and Space Sciences, United States Senate.* Washington, DC: United States Government Printing Office, 1966.

United States. Library of Congress. Science Policy Research Division. *Review of the Soviet Space Program, with Comparative United States Data.* Washington, DC: United States. Government Printing Office, 1967.

United States. National Aeronautics and Space Administration. *Astronaut Selection and Training.* NASA Facts. KSC 35-81, 1985. 4.

van der Grijp, Paul. "Passion and Profit: The World of Amateur Traders in Philately." *Journal of Material Culture* 7, no. 1: 23–47.

van Ree, Erik. "Heroes and Merchants: Stalin's Understanding of National Character." *Kritika: Explorations in Russian and Eurasian History* 8, no. 1 (Winter 2007): 41–65.

Vereshchetin, V. S., et al. *Orbity sotrudnichestva* [Orbits of Cooperation]. Moscow: Mashinostroenie, 1983.

Vetrov, G. S. "S. P. Korolev i razvitie muzeev po kosmonavtike" [S. P. Korolev and the Development of the Space Museums]. In *Trudy XXXII Chtenii, Posviashennikh razrabotke nauchnogo nasledii i razvitiiu idei k. e. tsiolkovskogo (Kaluga, 15–18 Sentiabria 1992 g.),* 195–200. Moscow, USSR: IIET RAN, 06/11/27, 1994.

———, ed. *S. P. Korolev i ego delo: svet i teni v istorii kosmonavtiki. Isbrannye Trudy Dokumenty* [S. P. Korolev and His Activities: Light and Shadows in the History of Spaceflight. A Collection of Documents]. Edited by Boris Raushenbakh. Komissiia Po Razrabotke Nauchnogo Naslediia Pionerov Osvoeniia Kosmicheskogo Prostranstva. Moscow: Nauka, 1998.

Vilenskii, Ezra Samoilovich, and K. Taradankin. *Vzlet* [Flight]. Moscow: Izdatel'stvo Detskoi Literatury, 1939.

Viola, Lynne. *Best Sons of the Fatherland.* New York: Oxford University Press, 1987.

Vladimirov, Leonid. *Rossiia bez prikaz i umolchanii* [Russia Without Orders and Suppression]. Frankfurt: Posev, 1973.

———. *The Russian Space Bluff.* Translated by David Floyd. London: Tom Stacey, Ltd., 1971. 192.

Vlasov, Viktor Gureevich, and Evgenii Mikhailovich Gortinskii. *Esli ty sobiraesh marki, znachki* [If You Collect Stamps and Pins]. Moscow: n.p., 1975.

Volgyes, Ivan. "Personality Structure and Change in Communist Systems: Dictatorship and Society in Eastern Europe." In *The Cult of Power: Dictators in the Twentieth Century,* ed. Joseph Held. New York: Columbia University, 1983.

Volkov, Aleksandr Ivanovich, and Nikolai Ivanovich Shtan'ko. *Vetv' sibirskogo kedra* [A Branch of the Siberian Cedar]. Moscow: n.p., 1962.

Voyce, Arthur. "Soviet Art and Architecture: Recent Developments." *Annals of the American Academy of Political and Social Science* 303, Russia since Stalin: Old Trends and New Problems (January 1956): 104–15.

Vtoroi Sovetskii iskutvennyi sputnik zemli [The Second Soviet Earth Satellite]. Moscow: VDNKh SSSR, 1984. NASM Technical Reference Files.

Vujosevic, Tijana. "The Flying Proletarian: Soviet Citizens at the Threshold of Utopia." *Grey Room,* 59, Spring 2015, 78–101.

Walker, Shaun. *The Long Hangover: Putin's New Russia and the Ghosts of the Past.* New York: Oxford University Press, 2018.

Walkowitz, Daniel J., and Lisa Maya Knauer, eds. *Memory and the Impact of Political Transformation in Public Space.* Durham: Duke University Press, 2004.

Weitekamp, Margaret A. *Right Stuff, Wrong Sex: America's First Women in Space Program.* Baltimore: Johns Hopkins University Press, 2004.

Wertsch, James V. *Voices of Collective Remembering.* Cambridge: Cambridge University Press, 2002.

Wheelon, Albert D., and Sidney N. Graybeal. "Intelligence for the Space Race (Prospects and Methodology for Supporting a Symbolic Olympian Technological Duel)." *Studies in Intelligence* 5 (Fall 1961).

Whelan, Joseph G. "The Press and Khrushchev's 'Withdrawal' from the Moon Race." *The Public Opinion Quarterly* 32, no. 2 (Summer 1968): 233–50.

Widdis, Emma. "To Explore or Conquer? Mobile Perspectives on the Soviet Cultural Revolution." In *The Landscape of Stalinism: The Art and Ideology of Soviet Space,* ed. Evgeny Dobrenko and Eric Naiman, eds., 219–40. Seattle: University of Washington Press, 2003.

Winter, Frank H. *Prelude to the Space Age: The Rocket Societies: 1924–1945.* Washington, DC: Smithsonian, 1983.

Wolfe, Tom. *The Right Stuff.* New York: Bantam Books, 1980.

Woll, Josephine. *Real Images: Soviet Cinema and the Thaw.* Kino: The Russian Cinema Series, ed. Richard Taylor. London: I.B. Tauris Publishers, 2000.

Youngblood, Denise J. "The Fate of Soviet Popular Cinema during the Stalin Revolution." *Russian Review* 50, no. 2 (April 1991): 148–62.

———. *Movies for the Masses: Popular Cinema and Soviet Society in the 1920s.* Cambridge: Cambridge University Press, 1992.

Yurchak, Alexei. *Everything Was Forever, Until It Was No More: The Last Soviet Generation.* Princeton: Princeton University Press, 2005.

Zaehringer, Alfred J. *Soviet Space Technology.* New York: Harper, 1961.

Zaloga, Steven J. "Most Secret Weapon: The Origins of Soviet Strategic Cruise Missiles, 1945–60." *Journal of Soviet Military Studies* 6, no. 1 (June 1993): 262–73.

———. *Target America: The Soviet Union and the Strategic Arms Race, 1945–1964.* Novato, CA: Presidio, 1993.

Zubkova, Elena. *Russia After the War: Hopes, Illusions, and Disappointments, 1945–1957.* Translated by Hugh Ragsdale. Edited by Donald J. Raleigh. The New Russian History Series. Armonk, New York: M. E. Sharpe, 1998.

Zubok, Vladislav. *A Failed Empire: The Soviet Union in the Cold War from Stalin to Gorbachev.* The New Cold War History, ed. John Lewis Gaddis. Chapel Hill: University of North Carolina Press, 2007.

Zubok, Vladislav, and Hope M. Harrison. "The Nuclear Education of Nikita Khrushchev." In *Cold War Statesmen Confront the Bomb: Nuclear Diplomacy since 1945,* ed. John Lewis Gaddis, 141–70. Oxford: Oxford University Press, 1999.

Index

CATHLEEN S. LEWIS is curator of International space programs and spacesuits at the Smithsonian Institution's National Air and Space Museum, specializing in Soviet and Russian history. Her role as curator has been to reinterpret the collections and research on the history of spaceflight for new generations and diverse audiences through scholarly research, artifacts, exhibitions, public presentations, and all forms of media. She also studies and curates the histories of astrobiology and Blacks in aviation and spaceflight in the United States and abroad. Lewis completed both bachelor's and master's degrees in Russian and East European studies at Yale University and her PhD in history at George Washington University.

9 781683 403708